WHAT GOOD ARE BUGS?

GILBERT WALDBAUER

WHAT GOOD ARE BUGS?

INSECTS IN THE WEB OF LIFE

HARVARD UNIVERSITY PRESS

Cambridge, Massachusetts

London, England

2003

Illustrations by Meredith Waterstraat

Library of Congress Cataloging-in-Publication Data
Waldbauer, Gilbert.
What good are bugs? : insects in the web of life / Gilbert Waldbauer.
p. cm.
Includes bibliographical references (p.).
ISBN 0-674-01027-2 (hardcover : alk. paper}
1. Insects—Ecology.   I. Title.

QL496.4 .W35 2003
595.717—dc21   2002027335

To the naturalists who came before me, especially my teachers
Aretas A. Saunders, Charles P. Alexander, William R. Horsfall,
Gotftried S. Fraenkel, and Leigh E. Chadwick

CONTENTS

WHAT GOOD ARE BUGS?

MACROCOSM

If all the insects, or even just some critically important ones, were to disappear from the earth—if there were none to pollinate plants, serve as food for other animals, dispose of dead organisms, and do other ecologically essential tasks—virtually all of the terrestrial ecosystems on earth, those webs of life consisting of communities of interdependent organisms interacting with their physical environment, would unravel. There is no way to predict what would replace them. But there is no doubt that without insects the world would be radically different and far less friendly to us humans, assuming that we could survive at all. The ecosystems of earth are our only home and the source of our food. Even city dwellers survive only because there are greener ecosystems beyond their habitats of concrete and brick. Among these green webs of life are wild forests and grasslands that absorb carbon dioxide from the atmosphere and are havens for millions of species of our fellow living creatures. A few insects, about 1.5 percent of the over 900,000 currently known species and an infinitesimal percent of the 9,000,000 or more insects that have yet to be discovered and named, are inimical to our interests. They transmit diseases and damage our crops, but the good services of the great multitude of insects, those that even the pushiest insecticide salesman would not call pests, immensely exceeds the harm done by the few destructive ones.

Yet most people are not aware of our dependence upon insects and know little or nothing about them. We tend to fear the unknown. Thus many people are suspicious of almost all insects and, to the great detriment of our collective ecological conscience, look upon them, with only a few exceptions, as our natural enemies. When I took a group of campers on a nature walk in New Hampshire, one of them, a boy of about twelve, revealed just how sus-

picious he was of creatures that were strange to him. At the sight of an unfamiliar animal, perhaps a large beetle or a walkingstick, he would ask, "What does it do to you?" The irrational aversion that some uninformed people display when faced with insects and their multilegged relatives was appallingly expressed—with what I hope is rare vehemence—by a woman walking along a Connecticut beach with her husband. When they came upon a harmless horseshoe crab crawling on the sand, I heard her shriek, "Look! It's alive. Kill it!" Unfortunately, people seldom consider the inescapable fact that insects are offspring of the same creation that produced us and such benignly viewed creatures as sea turtles, robins, dogs, and cats. Thus we tend not to see insects—as well as their relatives, such as horseshoe crabs, spiders, and centipedes—as integral and necessary constituents of the web of life.

Ecosystems are formidably complex, and ecologists are only beginning to understand them. The community of life in an ecosystem is a complex, all-pervasive, and often marvelously intricate web of interdependent and interacting plants and animals, each one of which has its own "occupation," or function, in the system. Green plants are the indispensable foundation of virtually all ecosystems. Only they can harness the energy of sunlight and by photosynthesis produce the energy-rich sugars that are food for animals. Always with the help of bacteria, they capture nitrogen, the essential component of proteins, and make it available to animals. With the notable exception of forests of pines, spruce, and other conifers (collectively called gymnosperms, from the Greek for naked seed), most of the green plants in terrestrial ecosystems are flowering plants (angiosperms, from the Greek for seed encased by an ovary). Few of the latter plants and the ecosystems they support could survive without insects, which serve them in several important ways. Bees, wasps, butterflies, and even beetles pollinate them, thereby bringing about a union of the sexes that results in the production of offspring, seeds. The seeds of some plants are blown by the wind but others are dispersed by insects, primarily ants. Later on you will see some other ways—possibly surprising to you—in which insects help flowering plants.

The evolutionary paths of the flowering plants and their insect helpers are closely and inextricably entwined. The oldest known fossil of a flowering plant was formed 120 million years ago, during the early Cretaceous period, the age of the dinosaurs. Thereafter new species of flowering plants evolved, and by the end of the Cretaceous, about 65 million years ago, there were tens of thousands of them. Today they are the most important group of plants on the land, numbering 275,000 species—about 85 percent of all green plants. These plants offered new opportunities for insects, and many new insect species appeared as the number and variety of flowering plants increased. Some insects became eaters of plants and others became helpers of plants. Generally speaking, the flowering plants and the insects associated with them coevolved: they reciprocally adapted to each other. For example, many flowers pollinated by bees lure them with their colors and scents and reward them with nectar and excess pollen. The bees, correspondingly, have color vision, are sensitive to odors, and have special structures for carrying pollen and nectar back to their nest.

Insects have other important functions in ecosystems. They are indispensable links in virtually all terrestrial food chains. As herbivores, eaters of plants, they make the nutrients that only plants can manufacture available to animals that do not eat plants. They convert plant tissues to the flesh of their own bodies, which is, in turn, eaten by insectivores such as parasitic and predaceous insects, toads, songbirds, and even mammals as large as bears. They also bridge the size gap between predators and microorganisms that are too small to be profitably eaten by large predators. For example, mosquito larvae, which are large enough to be eaten by fish, feed on microorganisms too small to be eaten by fish.

As parasites and predators, insects regulate plant and animal numbers, forestalling population explosions. Because all organisms produce more offspring than are required to replace themselves, population explosions are always possible if the usual restraints on population increases fail. This has been amply demonstrated by ecological disturbances that are, for all practical purposes, "field experiments." In many instances, populations of formerly uncommon, and economically unimportant, insects have exploded and become disastrously destructive when a crop is sprayed with an insecticide that does not kill the potential pest insect but does kill its parasites and predators. Insects or plants introduced from foreign places, such as the notorious gypsy moth, have become ruinously abundant because their usual complement of enemies was left behind. Introducing their natural enemies, a practice known as biological control, has in scores of cases controlled their populations.

Finally, insects play an important role in the decomposition and recycling of dead plants, dead animals, and excrement, thereby helping to return minerals to the soil and making them available to new life. This was convincingly demonstrated when cattle were brought to Australia in the eighteenth century. Because Australian dung beetles were not capable of coping with the wet, sloppy droppings of cows, dung accumulated in pastures, smothering the growth of plants. But when the proper dung beetles were imported from Europe and Africa, the problem was largely solved. Among the recyclers of organic wastes are termites and certain ants. These burrowing insects, which are ubiquitous and plentiful the world over, are indispensable as tillers of the soil. They dig into the ground and bring to the surface mineral subsoil that mixes with the decaying organic matter on the surface, thereby creating the fertile soil in which plants flourish. On the Great Plains of the United States, just one of the many species of ants present was recently found to have brought to the surface about 1.7 tons of subsoil per acre, and they are still busily engaged in this task.

All insects are to some degree useful members of their ecosystems, but some are more important than others, and not all are necessary to the continued survival of the system in its normal state or some close approximation of it. This is so because there is some redundancy of function in most ecosystems. A particular plant, for example, may be pollinated by several different kinds of insects. If one disappears from the ecosystem, the plant may not be affected if one or more of the other pollinators increases its population to compensate for the missing insect. Much the same argument could be made when an organism's population is regulated by several parasites, predators, or herbivores, or when an assemblage of several scavengers disposes of the dead organisms in an ecosystem. Nevertheless, ecosystems include some species of insects and other organisms that are truly indispensable, the "keystone species." For example, fig trees—of which there are hundreds of different kinds—are keystone members of tropical forests, because they provide resources for many animals and even other plants. Without them, these forests would be drastically impoverished. In turn, all fig trees are totally dependent upon certain tiny wasps for pollination. Without them, figs would become extinct. So the wasps are keystone insects that make possible the existence of keystone plants.

Stephen Alfred Forbes, a brilliant scientist and the first head of the Department of Entomology at the University of Illinois, was far ahead of his time when he formulated the first characterization of an ecosystem in 1887—almost 50 years before the word was coined. Forbes's words in *The*

Lake as a Microcosm are as germane today as they were over a hundred years ago:

> Nowhere can one see more clearly illustrated what may be called the *sensibility* of such an organic complex, expressed by the fact that whatever affects any species belonging to it, must have its influence of some sort upon the whole assemblage. [The observer] will thus be made to see the impossibility of studying completely any form out of relation to the other forms; the necessity for taking a comprehensive survey of the whole as a condition to a satisfactory understanding of any part. If one wishes to become acquainted with the black bass, for example, he will learn but little if he limits himself to that species. He must evidently study also the species upon which it depends for its existence, and the various conditions upon which *these* depend. He must likewise study the species with which it comes in competition, and the entire system of conditions affecting their prosperity; and by the time he has studied all these sufficiently he will find that he has run through the whole complicated mechanism of the aquatic life of the locality, both animal and vegetable, of which his species forms but a single element.

Watching the blossoms of a viburnum shrub at the edge of a forest in June gives us a fleeting glimpse of a tiny fraction of the intricate interrelationships between the organisms of an ecosystem. Many adult insects eat pollen and nectar from the blossoms, a brief but essential fueling stop that they more than pay for by pollinating the blossoms. Among those insects are colorful long-horned beetles, whose larval offspring will feed in rotting wood, returning to the soil minerals that are required by viburnums and all other plants of the ecosystem. A queen yellowjacket wasp that survived the winter in a cavity rotted into an oak sips nectar from the viburnum and will repay the plant for its largesse not only by pollinating it but also by founding a colony with hundreds of workers that will help to rid its foliage, and that of other plants, of leaf-eating insects, which they will feed to their colony's hungry larvae. By its very existence, the wasp increases the chances for survival of a flower fly that also visits the blossoms. The harmless fly is a mimic of this venomous wasp. Its uncanny resemblance to its model warns away hungry birds that have experienced the venomous wasp's painful sting. Blossoms of the viburnum pollinated by flower flies bear small, red fruits that are eaten by birds that disseminate the indigestible seeds far and wide in their droppings.

Our observation of the viburnum acquaints us with only a tiny fraction of the many interacting organisms that constitute the web of life in a very complex ecosystem. Much remains to be learned and relatively little is known about the intricate relationships between the organisms in an ecosystem, but in the following pages I will tell you about many important and often essential functions that insects and some of their relatives perform in terrestrial ecosystems, and will give you a nodding acquaintance with just a few of the hundreds of thousands of insect species that perform these functions, so vital to our own existence.

HELPING PLANTS

1

POLLINATING

On a moonlit night I watched a hawk moth sipping nectar from the white flowers of a honeysuckle, probably attracted to them by the same sweet scent that I could smell. The moth, almost as big as a hummingbird, darted quickly from flower to flower, its wings only a blur as it hovered briefly before each one, never landing but uncoiling its long, flexible "tongue" to poke into the flowers as it sought the nectar hidden deep within them. This was a critical moment for both plant and insect, an indispensable prelude to their ultimate goals, in both cases the production of offspring that would survive them and pass their genes on to future generations.

The hawk moth was playing an essential role in the sex life of the honeysuckle by pollinating its flowers, by transferring sperm-containing pollen grains from the male parts of one flower to the female parts of another. Pollen adhered to the nectar-seeking moth when it brushed against the male parts of the flower. When it visited other honeysuckle flowers, some of this pollen rubbed off on their female structures. In this way it initiated the process of fertilization and the consequent production of seeds. It may have pollinated several hundred flowers during the course of that night.

The sugary nectar secreted by the honeysuckle flowers was equally essential to the hawk moth, the adult stage of the tomato hornworm caterpillar, in this case a female. Metabolizing the sugar in the nectar would release the enormous amount of energy needed to fuel the long, swift flights she would make as she distributed her eggs far and wide. She would lay them one at a time, but only on the few species of plants that her leaf-feeding offspring would accept as food: tomato, tobacco, and related wild plants of the nightshade family. After laying each egg, she would fly off to find another, perhaps distant, plant of this family on which to place her next egg. This behavior is time- and energy-consuming but saves her very large caterpillars from

9

starvation by not burdening these small plants with too many caterpillars with huge appetites.

★ ★ ★ The agent of pollination is most often an animal, usually an insect, but it may be the wind. (A few plants need no outside agent, because their flowers, such as those of our garden peas and beans, are self-pollinating.) A sizable minority of plants are pollinated by the wind, among them pines; many deciduous trees of temperate forests; all grasses, including corn, wheat, and other grains; and some herbaceous plants such as stinging nettles, ragweeds, and the weedy plantains that invade our lawns. Wind-pollinated plants usually have small, inconspicuous flowers without petals but with relatively large female parts that are often multiply branched to better intercept wind-blown pollen grains. They also produce huge amounts of pollen with grains small enough to drift long distances. A hazelnut bush of average size produces, according to Bastiaan Meeuse, at least 600 million pollen grains. But ragweeds may be the champion pollen producers; Robert Bertin reported that one plant can release 8 billion grains of pollen in just one day. It is no wonder that so many people are allergic to wind-borne pollen, especially ragweed pollen.

When most of us think of flowering plants we picture them with large, colorful flowers, attention-grabbing displays that attract the animals which pollinate them. We may even conjure up memories of the fragrances of flowers, which are also attractive to pollinators. Among the animals that pollinate flowering plants are a very few mammals, most of them nectar-drinking bats, and a somewhat larger number of birds, including the sunbirds and honeyeaters of the Eastern Hemisphere and the hummingbirds of the Western Hemisphere. But the great majority of pollinating animals are insects, most of them beetles, flies, moths, butterflies, bees, or wasps.

Although no one has counted how many of the world's almost 275,000 species of flowering plants are pollinated by insects, there is no doubt that insects pollinate the great majority. Of 44 species of crops commonly grown in North America, about 66 percent were identified by Samuel McGregor as being more or less dependent upon insects for pollination. Stephen Buchmann and Gary Nabhan wrote that, of the world's 94 major crop plants, 18 percent are pollinated by the wind, 80 percent by insects (92 percent of these by bees), and about 2 percent by birds. David Roubik listed 1,330 species of plants that are cultivated or harvested from the wild in the tropics, but the pollinators of only 775 are known. About 88 percent of them are pollinated mainly by insects, 5 percent by bats, 1 percent by birds, and 6 percent by the wind. These figures agree with Kamaljit Bawa's earlier estimate that animals pollinate about 98 percent of all wild flowering plants in the lowland tropical rainforests of the world, and that the vast majority of these animals are insects. Bawa's estimate is particularly significant because tropical rainforests harbor so many different kinds of plants and animals. Edward O. Wilson wrote that tropical rainforests cover only 6 percent of the planet's land surface but are home to more than half the species of organisms on earth.

Most flowers reward their visitors with pollen or both pollen and nectar. Most produce large quantities of pollen, enough to fertilize many flowers and a generous excess to feed their pollinators. Pollen is a complete and nutritious food that contains carbohydrates, fats, proteins, vitamins, and minerals. Although nectar may contain small amounts of other nutrients, it is essentially a solution of sugar in water. Some insects thrive on a diet of only pollen, and a combination of pollen and nectar is the only food of all 25,000 known species of bees, but no insect can grow on a diet of only nectar. Some insects, such as hawk moths and many butterflies, eat only nectar as adults, but do not grow and must carry over from the caterpillar stage the proteins, fats, and other nutrients required to produce eggs. But many other adult insects retain few nutrients from the immature stage and must eat a complete diet including protein and other nutrients. Some eat pollen to supply these nutrients, among them the bees, certain flies, beetles, and butterflies.

It is difficult to understand the interactions between plants and pollinating insects without being familiar with the structure of flowers. The example of a bisexual insect-pollinated flower shown here is not specialized to accommodate specific insects. It is a generalized flower much like the wild roses that are accessible to many different kinds of insects including beetles, flies, butterflies, and bees. Its most conspicuous feature is the corolla of five colorful petals, two of them cut away in this drawing. Beneath the corolla is a calyx consisting of five sepals, only two of which are shown, that originally enveloped and protected the developing bud. (Some plants have flowers with more or fewer petals and sepals.) The sepals of this and most other flowers are small, green, and inconspicuous, but in some flowers, such as those of anemones, amaryllis, and orchids, they are large and colorful,

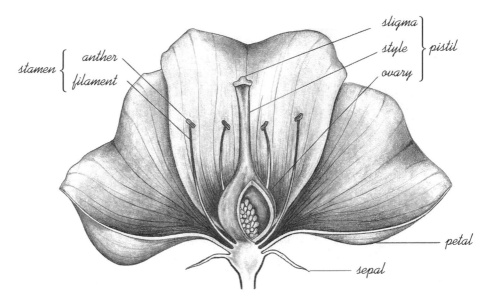

stigma
style } pistil
ovary

stamen { anther
         filament

petal

sepal

A generalized flower with two petals, three sepals, and part of
the ovary wall removed to show the ovules (eggs).

sometimes replacing the showy corolla or combining with it to better attract
the attention of pollinators. Some plants have inconspicuous flowers, but
surround them with large, showy modified leaves known as bracts. Seven
scarlet bracts, which look just like leaves except for their color, attract hummingbirds to clusters of the small, greenish flowers of the poinsettia. The
sexual parts of flowers are the pistils and the stamens. The pistil, the female
part, consists of the stigma, the style, and the ovary. The sticky stigma cap-
tures pollen grains and holds them as they germinate and grow long tubes
that extend through the style to the ovary. The ovary contains ovules (eggs)
that develop into seeds after they are fertilized by sperm that move through
the pollen tubes. Each of the stamens, the male parts of the flower, consists
of a long filament, a stem that bears the pollen-producing anther. In this and
similar flowers, droplets of nectar secreted by special glands may lie in the
bottom of the flower where the stamens and pistils originate, fully exposed
and available to such unspecialized pollinators as beetles and most flies. In
some other plants the nectar is hidden in a special structure such as the spur
described below.

This basic anatomy has been modified by eons of evolution into a marvelously diverse variety of specialized flowers that differ from each other in structure, shape, scent, size, and color. The sexual parts of many are shaped or arranged so as to minimize or altogether prevent self-pollination. Some flowers have special scents, colors, and structures that attract and accommodate just a few coevolved pollinating insects that are, to different degrees, reciprocally specialized to cope with the peculiarities of these flowers. For example, nectar may be accessible only to long-tongued insects, such as hawk moths or certain bees, because it is deep within a long, tubelike flower or at the end of a long spur, a greatly modified petal that extends far beyond the rest of the corolla. Such spurs can be seen on the impatiens and snapdragons that we grow in our gardens. These and other special features encourage reliable pollinators and tend to discourage insects that "steal" nectar without contacting the pollen-bearing anthers and also "fickle" insects that are unreliable pollinators because they flit from one species of plant to another.

Today almost everyone knows that insects pollinate flowers. But less than two centuries ago, virtually no one knew about pollination by insects or other agents. The showy beauty of flowers was attributed to the Creator's wish to please his human creations. According to a French proverb, beauty is "God's handwriting." It was not until 1793 that Conrad Sprengel, in *The Secret of Nature in the Form and Fertilization of Flowers Uncovered,* explained that colorful flowers are lures to attract pollinating insects. Sprengel understood the practical reproductive functions of flowers and the marvelously intricate mechanisms that foster cross-pollination, the fertilization of a plant by pollen from another plant. In 1883, Herman Müller summarized the steps that gradually led Sprengel to his great discovery:

> The inconspicuous hairs which cover the lower part of the petals of the wood cranesbill (*Geranium silvaticum,* L.), and beneath which drops of honey [nectar] lie hid, led Sprengel in the year 1787 to the discovery that most flowers which contain nectar are so arranged that, while insects can easily reach it, the rain is prevented from doing so; and he came to the conclusion "that the nectar of these flowers is secreted for the sake of insects, and is protected from rain in order that the insects may get it pure and unspoiled." Starting from this conception, he next summer studied the forget-me-not . . . and speculated on the meaning of the yellow ring round the mouth of the corolla, which forms a pleas-

ing contrast to the azure-blue of the limb; and he conceived the idea that this might serve to guide insects on their way to the honey. On examination of other flowers he found that coloured dots and lines and other figures occur especially at the entrance to the nectaries, or point towards it, and he was accordingly confirmed in this idea of *path-finders* or *honey-guides* . . . Sprengel could scarcely remain long without perceiving that, as the special colour of one part of the corolla serves to guide the insect after it has settled upon the flower, the bright colour of the whole flower serves to attract the notice of insects while still at a distance. So far, Sprengel had looked upon flowers as contrived simply for the use of insects, but the study of some species of iris, in the summer of 1789, led him to the further discovery that many flowers are absolutely incapable of being fertilised without the aid of insects; and so he concluded that the secretion of honey in flowers, its protection against rain, and the bright colours of the corolla are contrivances of use to the flower itself by bringing about its fertilization by insects.

Just as we need to understand the structure of flowers, we must know a little about the structure, sensory capabilities, and behavior of the insects that pollinate them. Most of these insects—but not quite all of them—belong to four groups, which are, in descending order of importance: bees and wasps, flies, moths and butterflies, and beetles. These groups include at least 200,000 species that pollinate flowers, about 20 percent of all the known insects.

Except for a few species, all adult insects are fundamentally similar in anatomy. Their bodies consist of 18 or 19 segments that are united in three groups to form the head, the thorax, and the abdomen. The head consists of five fused segments, which are distinguishable only in embryos, and it bears the eyes, a pair of antennae, and the organs of ingestion, the mouthparts. Except for the "upper lip" and a tonguelike extension of the "throat," the mouthparts actually consist of three pairs of legs that are grouped close around the mouth and have become highly modified—no longer recognizable as legs—for chewing, as in beetles; lapping, as in flies and bees; sucking, as in moths and butterflies; and piercing and sucking, as in mosquitoes and aphids. The thorax consists of three segments, each with a pair of legs. In most adult insects except flies, the middle and last segments each bear a pair of wings. Flies have wings only on the middle segment. The abdomen consists of ten or eleven segments that have no legs in adult insects, although at

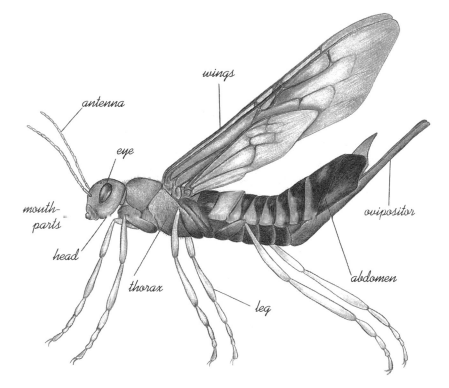

This female sawfly, one of the pigeon horntails, is a nectar feeder
and a primitive relative of the bees, ants, and wasps.

the end of the abdomen there may be an ovipositor (egg-laying organ) in females and genital claspers in males, both of which are actually highly modified legs that are even less recognizable as such than the mouthparts.

All the pollinating insects mentioned above undergo a complete metamorphosis, which results in radical changes in appearance and behavior
between the immature and the adult stages. (Some other insects, such as
grasshoppers, cockroaches, aphids, and termites, undergo a gradual metamorphosis, with egg, nymphal, and adult stages, that does not result in radical changes.) Insects with complete metamorphosis begin life as an egg except for a few that are born live. The larva, the immature stage that hatches
from the egg, does little but eat and periodically molt its skin as it grows.
When fully grown, it molts to the pupal stage, during which the metamorphosis to the adult stage takes place. The differences between larvae and

adults are profound. Larvae never have wings, but almost all adults do. Winged adults but not larvae visit flowers. Some larvae have legs and others do not. In all pollinating insects except beetles, the mouthparts of larvae and adults are different; both larval and adult beetles have chewing mouthparts. Larval flies, bees, wasps, butterflies, and moths have variously specialized mouthparts adapted for eating solid food, but the adults' mouthparts are variously modified for imbibing liquids such as nectar.

Butterflies and moths undergo an especially striking metamorphosis. The larva (caterpillar) that hatches from the egg has up to eight pairs of legs, the usual three pairs on the thorax plus as many as five pairs of stout, fleshy legs on the abdomen. Its strong mandibles are adapted for chewing leaves, the food of most caterpillars, and tiny hooks on its abdominal legs can cling tenaciously to a plant. The adult is so different that it looks like another kind of animal. Caterpillars and butterflies are as different as snakes and birds. The adult's body is compact, has two pairs of broad wings and three pairs of long, spindly legs that act mainly as landing gear. The mouthparts are long and thin like a soda straw for sucking liquids, usually sugar-rich nectar that provides energy for flying as the insect seeks a mate or plants on which to lay its eggs. The caterpillar is an eating machine; the adult is a sex machine devoted to reproduction.

Insects have the same five senses that we do, but rely on vision and their sense of smell to find flowers. Since Darwin's time biologists have agreed that flowers are colorful in order to attract bees and other insects that pollinate them. But in 1910, Carl von Hess, a German ophthalmologist, published experimental results, succinctly summarized by Bastiaan Meeuse, that seemed to contradict this view, because they gave the false impression that honey bees are color-blind. But it was not long before Karl von Frisch, an Austrian biologist who would become a Nobel laureate, showed that von Hess was wrong.

Von Hess knew that totally color-blind people perceive yellow-green as the brightest area of a color spectrum, whereas people with normal color vision see yellow as being brightest. He found that honey bees confined in the dark try to escape by moving to the yellow-green area of a spectrum just as, Meeuse commented, "a color-blind person with the mentality of a bee would have done." Von Hess jumped to the conclusion that honey-bees are color-blind. But he had not taken into account that a bee or any other animal may respond to colors differently depending upon the circumstances. A flower-seeking bee does respond to colors, but it escapes from the dark by moving to what it perceives to be the brightest light regardless of its color.

Von Frisch, seeing this flaw in von Hess's conclusion, experimented with foraging honey bees searching for flowers. On a table near a hive, he placed a checkerboard pattern of small square cards. One card was blue and the others were various shades of gray. He covered the checkerboard with a sheet of glass and placed a small glass dish over each card, all empty except for the one on the blue card, which contained an odorless solution of sugar in water. Soon many bees were coming directly to the dish on the blue card even when he changed its position in the array of gray cards. Then von Frisch did the decisive experiment. He replaced the experimental set-up with a clean one. All of the new dishes were empty, including the one on the blue card, and had no odor that bees might have left on them. Still, the bees trained to come to a blue card ignored the dishes on the gray cards and flew directly to the dish on the blue card, even though it was empty. In another experiment von Frisch showed that bees cannot be trained to come to any of these shades of gray, thereby eliminating the possibility that they had actually been responding to the degree of brightness of the blue card rather than its color, but that this had not been apparent because none of the gray cards in the original experiment happened to match the blue one in brightness. Eventually, von Frisch and others found that honey bees are red-blind but can see ultraviolet, which is invisible to humans; blue, even when it is a component of purple or violet; a narrow spectrum of blue-green; and yellow, even when it is a component of green or orange.

Other insects, including wasps, ants, moths, butterflies, beetles, and flies, can also see ultraviolet, but most of them are red-blind. Sensitivity to colors varies, but it is likely that most insects that visit flowers can see the colors that bees perceive as blue and yellow, because most insect-pollinated flowers are of those colors.

In the Americas, red flowers are scarce except in the tropics, where many flowers are red to attract hummingbirds. In Europe, where there are no flower-visiting birds, the great majority of flowers are blue or yellow. Notable among the few exceptions is the abundant bright red corn poppy, which attracts insects not because it is red but because its petals reflect ultraviolet light. But some exceptional insects, notable among them certain butterflies, are sensitive to red itself. For example, in North America, certain swallowtail and sulphur butterflies visit the scarlet blossoms of the cardinal flower. In

South Africa, according to S. D. Johnson and W. J. Bond, one species of satyrid butterfly is the sole pollinator of at least twenty species of plants with red flowers. This butterfly is so attracted to red that it swoops down on red hats and socks worn by hikers.

A great many flowers have nectar guides, the "honey guides" that Sprengel noted, spots, lines, or patches of color that direct insects to the part of the flower that holds the nectar. On their lower petals pansies have a pattern of radiating lines that converge on the nectary at the center of the flower. The orange flower of the spotted jewelweed has an array of dark spots that lead to the long spur that holds the nectar. In the center of the garden primrose's purple flower is a star-shaped yellow patch that marks the entrance to the nectary. Some flowers have nectar guides that are invisible to us but visible to insects because they reflect ultraviolet. Flowers of the primrose-willow look just plain yellow to us, but when Robert Silberglied photographed them so as to reveal ultraviolet, a large nectar guide was visible in the center of the flower.

Fritz Knoll, a professor at the University of Vienna, was the first to show that an insect, a day-flying hawk moth, takes directions from nectar guides. His results are included in a formidable 645 pages published in the 1920s in *Die Zoologische und Botanische Gesellschaft in Wien*. Some of his experiments are summarized in books by Friedrich Barth and Meeuse. The moth uses its long, thin tongue to feed from the yellow flowers of butter-and-eggs (toadflax), whose nectar is hidden in a long spur. A contrasting orange spot marks the entrance to the spur. When Knoll presented caged hawk moths with these flowers pressed between two sheets of glass, the moths tried to feed from the flowers, pressing the tips of their tongues, which were wet from a previous meal, against the glass. Almost all marks left by the wet tongues were directly over the orange nectar guides—even after Knoll cut them out and moved them to the wrong place on the flower.

The senses of taste and smell are much more important to insects than they are to us. Many insects are deaf, some have no eyes, and the visual acuity of those with eyes is limited; they perceive color, shape, and movement but relatively little detail. Von Frisch found that honey bees can taste the difference between solutions that contain different amounts of sugar, preferring strong solutions to weak ones. According to Mary Wilson and John Ågren, the sugar content of nectar differs among plant species.

Bees and other insects presumably use their sense of taste to choose flowers that provide a good yield of sugar.

Most insects have taste receptors on the appendages surrounding their mouths; some have them on the antennae; and some have them even on their feet, as do bees, flies, moths, and butterflies. It is a great convenience for insects to taste with their feet. Often an insect first touches a flower with its front feet, and thus can make a quick decision to leave or to stay. A simple experiment demonstrates that butterflies do use the taste receptors on their feet. If you immobilize one by clamping its wings together, it will reflexively extend its tongue as if getting ready to drink if you touch one of its front feet with a brush dipped in sugar water. Some moths, fruit flies, bluebottle flies, house flies, and honey bees respond in the same way.

The odors of flowers help pollinators to find them. According to von Frisch, most flowers pollinated by honey bees have a scent discernible to people. But wind- and bird-pollinated flowers are seldom scented. Most birds have a poor sense of smell. For example, as Meeuse noted, honey-suckle flowers pollinated by nocturnal hawk moths have a sweet scent and are white, while another honeysuckle flower that is pollinated by day-flying hummingbirds is orange-red and odorless.

The olfactory organs of insects are usually on the antennae, a useful arrangement for bees, moths, and other insects with long antennae, because these organs of smell can, unlike those of humans, be waved from side to side to locate the source of an odor. Furthermore, some insects, with their antennae spread wide, can ascertain the direction of an odor source by the difference in how strongly each of the antennae is stimulated—just as we can tell the direction of a source of sound by the difference in how loud each ear perceives the sound to be. If one antenna is amputated, insects stimulated by an odor move in circles, turning toward the side with the remaining antenna.

Some day-flying insects spot flowers from a distance by sight and do not respond to their scent until they get closer. Others, including many bees that pollinate orchids, perceive scents from a distance and locate flowers by sight when they get close. The irresistible attraction of the fragrance of a Costa Rican orchid to a native wild bee is made amply clear by Hilda Simon's account of what happened when a man picked a *Catasetum* orchid and carried it back to his house: "The usually shy bees followed him unhesitatingly into the house and into the room to which he took the flower. When he put it away in a drawer to see what the bees would do, he was astonished to see them, in a frenzy of eagerness, attempting to get into the drawer through the slits between the wood." (These bees were all males. You will discover why when I discuss the pollination of orchids.) Night fliers usually locate night-blooming flowers by their scent and, when they get close, use sight to zero in on their usually large, white, and readily visible blossoms. The strong scent of the lovely white flowers of the famous night-blooming cereus cactus of the American Southwest attracts the hawk moths that pollinate it. Meeuse described an experiment that showed that night-flying pine hawk moths can find flowers just by their scent. From 30 feet away, moths flying upwind found honeysuckle blossoms hidden from view in slatted boxes.

When honey bees, and probably some other insects, find a species of flower with abundant nectar, they remain loyal to it for days. This flower constancy benefits both plants and pollinators. The plants profit because they will receive pollen from members of their own species. The pollinators profit by saving time because they have learned how to recognize the flowers and how to gather their nectar. Scents are important to maintaining flower constancy. When a worker honey bee that finds a new patch of flowers returns to the hive, the scent of the flowers still clings to her body. As she (all worker bees are females) does the direction-giving "waggle dance," described below, other workers smell the scent as they crowd close to her and touch her with their antennae. Bees that danced with the scout leave the hive to forage from the newly found flowers, following the dancer's directions and recognizing the flowers by their scent. When marked bees that von Frisch trained to feed from cyclamen flowers went out to forage, they landed on a pot of blossoming cyclamens but ignored an adjacent pot of blossoming phlox. Similarly, marked bees that he had trained to feed from phlox landed on a pot of phlox but ignored the pot of cyclamen right next to it. He repeated this experiment with other flowers and found that the bees were flower-constant only when the flowers had a scent.

🐜 🦟 🪰 The mouthparts of flower-visiting insects are to varying degrees and in different ways adapted for drinking nectar or eating pollen, just as the legs and other parts of their bodies are often adapted for collecting and transporting pollen. Most moths and butterflies do not eat pollen, but—as you will see later—their mouthparts are so specialized for drinking nectar that they are little more than hollow tubes, straws for sipping nectar. Flies have mouthparts marvelously modified for literally sponging up nectar and other liquids. They and the moths and butterflies have lost their mandibles, which in other insects are used for chewing or, as in bees, for manipulating wax and other substances.

Adult bees, the exploiters of flowers par excellence, have mouthparts that, almost like Swiss army knives, serve several functions. As you will soon see, they can both lap and suck up nectar, and the mandibles have many different uses. Depending upon the species of bee, the mandibles are used to handle pollen, to excavate nesting burrows in the soil or even in wood, to cut pieces of leaf that line the bees' nest cells, to gather plant resins to seal cracks in a cavity that contains the nest, or to manipulate wax that may be used to line a cell in the ground or to form the honey pots in bumble bee colonies or the combs in colonies of honey bees. The other five mouthparts, including a long hairy tongue, are adapted for lapping or sucking nectar. When taking nectar that is near the surface of a blossom, honey bees use the tongue to lap it up, but to obtain nectar from deeper within a flower they press together all of the mouthparts other than the mandibles to form a sucking tube. Worker bees carry nectar and pollen back to the colony, where whatever is not consumed immediately is stored for later use. Pollen mixed with a little nectar is stored in cells of the comb. Bees convert the nectar to honey, which is also stored in cells, by evaporating most of its water and adding to it enzymes that convert its sugars to a more digestible form. Nectar is carried back to the colony in the crop, a sac-like part of the digestive system that is often called the honey stomach. Workers carry amazing quantities of nectar—an average of 50 percent of their own weight and a maximum of 92.5 percent, according to Charles Michener, an expert on bee behavior.

When large amounts of pollen are still on the anthers of a flower, honey bees may harvest it with their mandibles, but they usually gather loose pollen caught in the branched hairs on their bodies. (Almost all bees have branched hairs, a characteristic that distinguishes them from wasps, ants, and their other relatives.) However they gather pollen, honey bees transport it in "pollen baskets" on the hind legs, concave areas, surrounded by stiff hairs, on the outer surface of the main segment of each hind leg. Shortly after leaving a flower, the bee—often while flying—uses brushes on her middle and hind legs to sweep pollen from her body past the mouth, where it is moistened with nectar to make it sticky, and then into the pollen baskets. Most species carry pollen in this way, but those of one family, the megachilids, have a pollen basket on the underside of the abdomen. Michener reported that honey bee workers' pollen loads, which are usually bright yellow and readily visible, are bulkier although considerably lighter than their nectar loads, from about 10 to 36 percent of their body weight.

Usually honey bees must visit many flowers to get a full load of pollen or nectar, a fact that enhances their usefulness as pollinators. Ronald Ribbands followed one nectar-gathering worker that visited 1,446 flowers of the same species during one 106-minute foraging flight. But the average number of visits is lower. Depending upon the species of flower and climatic conditions, a worker visits from 50 to 1,000 blossoms on each nectar-collecting trip. A full load of pollen can sometimes be gathered in far fewer visits. According to Ribbands, a single flower, a poppy for instance, may contain more than enough pollen to constitute a load. But bees had to visit from 66 to 178 nasturtium flowers or 585 tiny white clover blossoms to accumulate a full load. Each year, an average colony of honey bees harvests, according to Thomas Seeley, about 44 pounds of pollen and 265 pounds of nectar. The pollen is used in raising about 150,000 larvae, and the nectar, most of it converted to honey, is eaten by both larvae and adults; it is the stored honey that tides the colony over the cold winter. To accomplish the monumental task of gathering sufficient food, the 20,000 or more workers that populate a colony at any given time must cooperate to utilize flowers as efficiently as they can.

Cooperation and efficiency are fostered by the symbolic dance language that foragers use to direct each other to productive patches of flowers. When a successful scout returns to the hive, she performs a waggle dance on the vertical surface of a comb, describing a figure 8, whose crossbar is a straight-line run during which she waggles her whole body from side to side and emits a high-pitched buzz. The waggle run is the most informative part of the dance. Its duration indicates the distance to the flowers—the slower the run the greater the distance. The direction is indicated by the angle of the run with respect to the vertical. Just as we follow the convention that the top of a map is north, honey bees understand that the top of a comb represents the direction of the sun. If the flowers are in the direction of the sun, the waggle run goes straight up the comb; if they are directly away from the sun, it goes straight down; if they are 60 degrees to the left of the line from the hive to the sun, it goes 60 degrees to the left of the vertical. As a bee dances in the darkness of the hive, other foragers crowd close and touch her with their antennae as they follow her moves. After a few rounds with the dancer, these new recruits fly to the indicated flowers. If they find them to be productive, they recruit more foragers by repeating the waggle dance when they return to the hive. But if the flowers are not productive, they do not dance upon their return. In this way the attention of the colony is focused on one or more patches of productive flowers.

Karl von Frisch and his coworkers painstakingly deciphered the waggle dance by charting the dances of marked bees that returned to an observation hive from sugar water feeders at known distances and directions from the hive. After years of experimentation, von Frisch demonstrated his new understanding of the waggle dance by watching bees in an observation hive and then following their directions to find a hidden feeder:

> A feeding station was set up whose location was unknown to me. I watched dances from 9 to 10 A.M. and again from 11 A.M. to 12 P.M. After I had derived a distance of about 330 m[eters] from the distance curve and had learned the compass direction from the curve relating the direction of dancing and the azimuth, I set out and had to look around for a few minutes at the presumed distance; then I discovered the feeding station hidden behind a bush. The direction deduced from the dances in the morning led 2° left past the goal; the line found in the afternoon ran 4° to the right of it.

Unlike bees, adult moths and butterflies have no mandibles, and must use their only mouthpart, a long "tongue," to reach the nectar in the deepest of flowers. The tongue is a thin, flexible tube, an extreme modification of just a part of one of the three pairs of mouthparts that most insects have. The tongue is usually shorter than the length of the body, although in some hawk moths it may be longer. Suction is provided by a muscular pump in the head. It would be awkward for a butterfly or a moth to fly or even walk unless it could get its long tongue out of the way. When the tongue is not being used, it is coiled up under the head like a watch spring. This is its natural and involuntary resting position. If you stretch out the tongue of a freshly killed moth or butterfly, it will coil back up most of the way when you release it. The tongue also contains tiny muscles that help to tighten the coil. When the insect is ready to drink nectar, blood under pressure is forced into two hollow channels in the tongue's walls, causing it to uncoil.

True flies—not dragonflies, butterflies, or other insects imprecisely called flies—are virtually unique among the insects in having only one pair of wings, which are on the middle segment of the thorax. Diptera, the scientific name of this order, means "two wings" in Greek. Among the flies are such familiar species as house flies that invade our homes, mosquitoes, horse flies, and black flies that suck our blood, and less familiar ones such as gnats, midges, robber flies, bee flies, and fruit flies.

Many different kinds of flies, including even mosquitoes and horse flies, drink nectar from flowers or eat pollen. The hover flies, also known as flower flies, are probably the single most important group of flower-visiting flies. Many are large, colorful, and easy to see when they are on flowers. But people often do not recognize them as flies, because many of them have a superficial but very deceptive protective resemblance to stinging wasps, bumble bees, or honey bees.

Hover flies (family Syrphidae) and other flies with similar mouthparts of the sponging type can drink nectar and other liquids, even thin films, and can also liquefy and ingest soluble solids such as dried-up nectar or honeydew, the sugary and fluid excreta of aphids and related insects. Flies collect liquid nectar with the spongelike labella at the tip of their mouthparts. It flows through tiny capillary channels on the surface of the labella to the tube that ends between them and is then sucked up into the digestive system. Flies consume dried nectar or honeydew—or in the case of the house fly, sugar in a bowl on your table—by dissolving it in their saliva and regurgitant. The fly presses its labella against the dry substance and then wets it. When the taste receptors on the labella sense that enough sugar has dissolved in the saliva and regurgitant, the fly sucks the substance up. Hover flies also use the labella to scoop up pollen, which contains nutrients they need to produce eggs or sperm.

Few of the 350,000 known species of beetles pollinate flowers; those that do are relatively scarce in temperate areas but more abundant in the tropics. Beetles that visit flowers are often large, tend to be clumsy, and have strong, stout mandibles used for biting and gnawing. With few exceptions, their mouthparts are not adapted for taking nectar from any but the shallowest of flowers. Among these exceptional species are blister beetles of the genus *Nemognatha*, which have a long, nectar-sucking proboscis—in some tropical species even longer than the body.

But the great majority of flower-visiting beetles can cope only with shallow flowers in which the food reward, nectar, pollen, or another edible tissue, is easily reached. These may be large, open flowers such as those of magnolias and water lilies, or small open flowers massed together in large, flat inflorescences, such as those of Queen Anne's lace and elderberry. Beetles are likely to injure flowers as they blunder about, pollinating by "mess-

Wasting as little of her short lifespan as possible, a female soldier beetle continues to eat the pollen of a dogbane blossom as a male inseminates her.

ing and soiling." Accordingly, many beetle-pollinated flowers protect their precious ovules by embedding the ovaries in protective tissues or by massing flowers close together to cover the ovaries.

Pollination, by insects or other means, does not always result in fertilization. If self-pollinated, some plants prevent self-fertilization with a physiological mechanism that inhibits their own pollen from germinating on their stigmas, preventing sperm from getting to the ovary. Others let the pollen grain germinate but prevent sperm from moving down the style to the ovary. These plants are immune to their own pollen. "Genetic self-incompatibility of this sort," wrote Robert Wyatt, "is probably the most widespread and effective mechanism for promoting outcrossing." Although Charles Darwin did not know about these mechanisms, he vividly described their visible effects:

> It is an extraordinary fact that with many species, even when growing under their natural conditions, flowers fertilised [pollinated] with their own pollen are either absolutely or in some degree sterile; if fertilised with pollen from another flower on the same plant, they are sometimes, though rarely, a little more fertile; if fertilised with pollen from another individual or variety of the same species, they are fully fertile . . . but if with pollen from a distinct species, they are sterile.

Darwin's experiments and others done since show that most plants produce superior offspring if they are cross-pollinated rather than self-pollinated. This condition is called hybrid vigor by plant breeders. The opposite effect of self-pollination is called inbreeding depression. The superior performance of hybrid corn (maize) seed is a well-known example of hybrid vigor. Inbred corn plants, the offspring of self-pollinated parents, produce a low yield of grain. But if two inbred lines are crossed, the hybrids produce much more grain than did their parents. Hybrid corn yielded an average of about 140 bushels per acre in the 1990s, while the nonhybrid varieties planted in the 1930s yielded an average of little more than 35 bushels per acre. About half of this improvement is due to the use of hybrid seed.

Darwin demonstrated hybrid vigor with a rigorously controlled experiment in which he cross-pollinated some flowers of a butter-and-eggs plant and self-pollinated others on the same plant:

> For the sake of determining certain points with respect to inheritance, and without any thought of the effects of close interbreeding, I raised close together two large beds of self-fertilised and crossed seedlings from the same plant of *Linaria vulgaris*. To my surprise, the crossed plants when fully grown were plainly taller and more vigorous than the self-fertilised ones. Bees incessantly visit the flowers of the Linaria and carry pollen from one to the other; and if insects are excluded, the flowers produce extremely few seeds; so that the wild plants from which my seedlings were raised must have been intercrossed during all previous generations. It seemed therefore quite incredible that the difference between the two beds of seedlings could have been due to a sin-

gle act of self-fertilisation; and I attributed the result to the self-fertilised seeds not having been well ripened, improbable as it was that all should have been in this state, or to some other accidental and inexplicable cause.

Darwin, meticulous scientist that he was, tested two plants that readily self-fertilize when insects are excluded, thinking that by so doing he was increasing the odds against hybrid vigor. He wrote: "some flowers on a single plant of both species were fertilized with their own pollen, and others were crossed with pollen from a distinct individual; both plants being protected by a net from insects. The crossed and self-fertilized seeds were sown on opposite sides of the same pots, and treated in all respects alike." Even these plants, which readily self-pollinate, profited from cross-pollination: "the crossed seedlings were conspicuously superior in height and in other ways to the self-fertilized." Large size is important because big plants are likely to produce more seeds and thus more offspring.

🐜 🐜 🐜   Self-pollination is more prevalent and often more useful ecologically and genetically than Darwin and other early students of pollination thought it to be. A few plants have adaptations that facilitate or even ensure self-pollination and many are capable of both self- and cross-pollination. Among the latter are violets, which in early spring bear showy, insect-pollinated flowers on tall stems, and in late spring have only short-stalked, green flowers that must be self-fertilized because they never open. Some plants fall back on self-pollination if cross-pollination, their first option, fails to occur— in a cold spring or in excessively rainy climates, pollinators may not fly for days on end, and during long droughts, they may be scarce. According to Peter Bernhardt, the beautiful woodland bloodroot of early spring is pollinated by wild, solitary bees on warm days. But when the weather is too cold for the bees, as often happens, self-pollination occurs when the stamens of an aging flower collapse onto its stigmas. Although self-pollination has obvious short-run benefits, it is doubtful, as Robert Wyatt wrote, that any exclusively self-pollinating species can survive centuries or eons of natural selection, because it will lack the greater genetic variability that results from having genes from two parents.

As odd as it may seem, insects help in the self-fertilization of tomato flowers. Some varieties of tomatoes, including the cultivated ones, are fertilized mainly by self-pollination within a single flower. But only if sufficient pollen falls from the anthers to the stigma will a full-size fruit with the normal number of seeds result. The problem is that little pollen falls from the anthers unless the flower is adequately shaken by the wind or an insect. Most insects that land on tomato flowers are not good shakers, but bumble bees are. They are "buzz pollinators" that vibrate and shake their bodies as they collect pollen from the flower. When tomatoes were first grown in greenhouses, the flowers were shaken with a mechanical vibrator, but now domesticated colonies of bumble bees maintained in the greenhouse do the job.

Most plants have evolved various ways of facilitating cross-pollination by insects or even making self-pollination virtually impossible. Perhaps the simplest way is for the pistils and stamens of a flower to become mature at different times. Thus a flower with a mature pistil can be fertilized only by pollen from another individual with mature stamens. In some species, such as fireweed, the stamens mature first, and in others the pistils mature first. Hogweed, like Queen Anne's lace, bears an umbrella-shaped head that consists of scores of small white flowers. Self-fertilization is minimized because all the flowers of an inflorescence switch from the male to the female role more or less synchronously. The stigmas mature only after the anthers have shed their pollen. Honeysuckles employ a different tactic. A stigma on a long style and four long stamens protrude from the tubelike flower. On the flower's first evening, it is functionally male because the stamens are above the drooping stigma. On the next and several succeeding evenings, the style bends upward so that the stigma is above the stamens. The flower is then functionally female. The insect-pollinated avocado, that delicious neotropical fruit introduced to the world by the Indians of Central America, has a more complex way of preventing self-pollination by timing the functions of male and female parts. As Bastiaan Meeuse explains in *The Story of Pollination,* the numerous varieties of this plant belong to one of two coexisting classes. The stigmas of one class are receptive only in the morning, and their anthers produce pollen only in the afternoon. The reverse is true of the other class. This arrangement guarantees cross-pollination. Avocado growers plant trees of both classes in their groves. Otherwise, bees—the plant's major pollinators—cannot accomplish pollination and thereby initiate fruit set.

🦗 🦗 🦗   Other plants have gone a different route. The structure of their blossoms is physically modified to assure or at least increase the probability of cross-pollination. These modifications range from the relatively simple, as in the garden primrose, to the intricately complex, as in several wild arum lilies, the giant water lily of Brazil, and many species of orchids. Primrose flowers are of two types. Those of some plants have a long style that raises the stigma to the top of the long, tubelike corolla, while their short stamens are in the tube well below the stigma. In flowers of the other type, the style is short and the stigma is well below the stamens, which are at the top of the tube. In *Insects and Flowers,* Friedrich Barth explained how this arrangement makes cross-pollination almost inevitable. When a butterfly probes for nectar in a flower with deep-set stamens, its long tongue becomes coated with pollen. If it next visits a plant of the other type, its long tongue may deposit pollen on the stigma deep in the corolla. The short tongues of bees cannot reach the deep-set stamens of primrose flowers with long styles; they can acquire pollen only from high stamens, but since their tongues are too short to reach the deep-set stigmas of these flowers, they can deposit pollen only on the raised stigmas of flowers with long styles.

Some plants ensure cross-pollination by trapping and holding insects. Among them are the giant Amazon water lily *(Victoria amazonica)* and some other plants discussed below. This water lily is famous for its gigantic floating leaves with upturned edges. It is said that a leaf can support the weight of a small child. Bastiaan Meeuse recounted boyhood experiences with this plant in a botanical garden in Java. The leaves would not support him because he was too big, but he noticed that when the flowers opened in the evening they were white and gave off a strong fragrance. The next morning they were closed, but they reopened in the evening, at which time they were purplish and gave off little odor. Meeuse eventually learned the answer to this puzzle. In the plant's native habitat in Brazil, the newly opened flowers attract many plant-feeding scarab beetles, which are trapped when the petals fold up at night. They feed voraciously on the inner walls of the flower, but at least some of them earn their keep by bringing in pollen from another flower. Only during that first night are the stigmas receptive to pollen. The following evening the stamens dust the trapped beetles with pollen just before the flower reopens. As Meeuse put it, the pollen-covered beetles "no longer held spellbound by sweet odors, take to the wing to drop in on another fresh and fragrant *Victoria* bloom whose stigmas are receptive."

Some plants of the arum family with bisexual flowers have remarkable ways of attracting insects and ensuring cross-pollination. What you might take to be the "flower" is actually an inflorescence that consists of a spadix, the large stalk that bears the true flowers, and a spathe, a bract that wraps around the spadix on all sides, enclosing it at the bottom, but leaving a wide opening at the top. The inflorescence of a North American arum, the jack-in-the-pulpit, is a familiar example of this unusual structure. The protruding club-shaped spadix is Jack the preacher and the overarching spathe is his canopied pulpit.

In 1926, Fritz Knoll published a detailed study of the inflorescence of and the insects attracted to the bisexual black arum, which grows in central Europe and along the Mediterranean coast. The spadix extends upward from the base of the spathe, and ends in a thick, clublike prolongation. The spathe is divided into upper and lower chambers by a barrier of long, bristle-like sterile flowers that encircle the spadix below the club. Below this barrier is a broad band of male flowers that are just short-stalked stamens, and below it a ring of bristle-like sterile female flowers. The bottom of the spadix is circled by a broad band of female flowers that are just ovaries with a stigma. The club of the spadix and the inner surface of the spathe are smooth and covered with a film of oil, so slippery that insects which land on them slip and fall down into the spathe to land on the uppermost ring of bristles, which guards the lower chamber, acting as a sieve that lets small insects fall through to be temporarily trapped in the lower chamber but keeps out large and potentially damaging insects, which soon fly away. The lower ring of bristles keeps trapped insects from crawling up the spadix, but is not wide enough to stop falling insects by obstructing the entire width of the chamber.

The black arum, like other arums, uses deceit to attract insect pollinators. On the day the inflorescence opens—but not before or after—the club at the top of the spadix emits a strong odor of feces. The odorous substance evaporates and disperses through the air, its evaporation hastened by the large amount of heat produced by the club. Six clubs from another species of arum tied to the mercury bulb of a thermometer raised the temperature to almost 108°F, about 38°F above the air temperature.

As is to be expected, black arums attract mainly flies and beetles that feed on or breed in dung. Knoll's field notes, translated from the German, capture the flavor of doing research in the field:

May 16: On the ground in a large, overgrown depression. A beautiful day, very windy in the morning, later almost calm. Sunshine. A normally formed inflorescence, the spathe 16 cm [6 inches] long, at its opening 4.2 cm [1.6 inches] wide. The club light cinnamon brown. The spathe stood completely unobstructed among shorter grasses, easily reached from all sides. At dusk I cut off the inflorescence and killed the insects that were found in the flower chamber.

In three inflorescences Knoll found a total of 227 insects, all but 5 of them dung feeders. The 5 exceptions were all rove beetles that live in dung and prey on dung-feeding insects. Of the dung feeders, 171 were small black dung flies, and 51 were beetles, 49 of them small dung-feeding scarabs, and the other 2 dung-feeding species of the water scavenger family.

The inflorescence of the black arum lasts for only 2 days. It opens during the night of the first day, and in the morning attracts and traps insects, which it holds captive until the morning of the next day. The trapped insects crawl on the female flowers as they attempt to escape, and will cross-pollinate them if they had previously been trapped in another inflorescence of black arum. They survive by drinking a sweet liquid the stigmas exude. During the second night, the male flowers produce pollen for the first time and shed enough of it to thoroughly dust the trapped insects. Self-pollination is impossible, because by this time the stigmas of the female flowers have withered. The next morning, the pollen-coated insects escape by climbing up the spadix; by that time the bristles have begun to wither and the insects can pass through them. The spadix, unlike the club, is not slippery. If these newly escaped insects happen to be trapped by another and newly opened inflorescence of black arum, they will cross-pollinate that plant.

Amorphophallus titanum, the giant arum lily of the jungles of Sumatra, is famous for its immense inflorescence, which may be from 8 to 10 feet tall and over 30 inches wide. Translated from Greek, the generic name means deformed penis, a reference to the spadix that sticks straight up from the spathe. The overpowering stench of carrion released by the huge heat-producing spadix attracts pollinating carrion beetles that are temporarily trapped in the flower chamber. On rare occasions, this plant has flowered in greenhouses in various parts of the world. In 1999, the eleventh one to flower in the United States was being viewed and smelled by hundreds of people lined up at a greenhouse at the Huntington Botanical Garden in San Marino, California.

The flowers of most plants, including those considered so far, contain both the male and females organs. In other words, they are bisexual. But some plants have unisexual flowers, some with only stamens and others with only

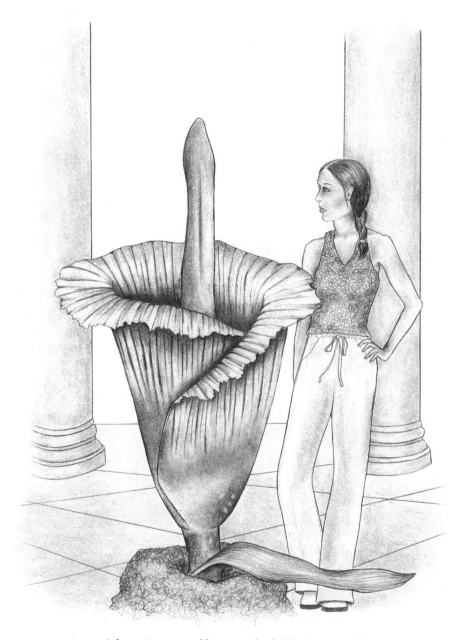

A rare sight, a giant arum lily *(Amorphophallus titanum)* in bloom
at the Huntington Botanical Garden in California.

pistils. That unisexual flowers evolved from bisexual ones is indicated by the presence in some plants, such as the shrubby cinquefoil, of vestigial stamens in female flowers and vestigial pistils in male flowers.

Some species of plants with unisexual blossoms have both male and female flowers on the same individual and are said to be monoecious, from Greek roots that mean one house. In other species, some individuals have only male flowers, while others have only female flowers. Botanists say they are dioecious (two houses). Dioecy is the usual condition in animals but is uncommon in plants. Those plants that are dioecious benefit, though, because that condition guarantees cross-pollination, while cross-pollination is not necessarily inevitable in monoecious plants. Some dioecious plants are wind-pollinated, among them the cottonwoods, chestnuts, hazels, willows, and date palms. The ancient Mesopotamians grew dates, and had a working knowledge of pollination. By the eighth century B.C.E. and probably earlier, they were hand-pollinating dates by cutting male flowers and shaking them over female flowers.

Many dioecious plants are pollinated by insects. They must have dependable pollinators that will visit plants of both sexes. A dioecious arum that grows in the Nilgiri Hills of southern India has an inflorescence that consists of a spathe that surrounds and forms a canopy over a protruding clublike spadix that bears either male or female flowers. In 1934, Edward Barnes published in the *Journal of the Bombay Natural History Society* a fascinating article that describes how this plant is pollinated. The spadix in both male and female flowers gives off an unpleasant odor that attracts fungus gnats. Gnats that come to a female inflorescence are likely to have visited a male inflorescence first and therefore to be loaded with pollen. Male plants outnumber female plants by about nine to one. As the gnats move down the spadix of a male inflorescence, they crawl over the flowers, are dusted with pollen, and escape through a small opening at the bottom of the spathe in male but not in female inflorescences. A gnat that next enters a female inflorescence moves down the spadix to a group of flowers with stigmas so tightly packed that the fly will surely be trapped. It will ultimately die, but as it struggles to escape it coats the stigmas with pollen. Female inflorescences soon become jammed with dead gnats.

Some monoecious plants evolved ways to prevent female flowers from being fertilized by pollen from male flowers on the same plant.

The female flowers of many monoecious plants have, like some bisexual plants, physiological ways of preventing self-fertilization if pollen grains from male flowers on the same plant land on their stigmas. Other monoecious plants can avoid or minimize self-pollination in the first place. Some wind-pollinated monoecious conifers bear female flowers on their upper branches, where they are not likely to be dusted by falling pollen from the male flowers on the lower branches. Another tactic is for male and female flowers on the same plant to become sexually mature at different times. Individuals of the opposite sex must, of course, be present when they are required; consequently, some individual plants will have mature flowers of one sex at the same time that others have mature flowers of the other sex.

Approximately a thousand species of figs occur in subtropical and tropical regions. They may be trees or vines and include, in addition to cultivated edible figs, wild species such as banyan trees and strangler figs. Most figs are monoecious, although about 128 closely related species, including the edible fig, are functionally dioecious. To understand how figs are pollinated we must first understand the peculiar structure of their inflorescence. A fig fruit is not the product of a single flower as are most other fruits, including strawberries, apples, and tomatoes. A fig fruit is a complex inflorescence, a hollow, fleshy, flask-shaped structure. Before it ripens, its inner walls are lined with hundreds of tiny flowers out of sight in the inflorescence. (The Chinese character for a fig signifies "a fruit without a flower.") The only opening into the inflorescence is the "eye," a small hole covered by flexible scales at the end of the fig opposite its stalk. Ripe figs are eaten by mammals and birds that disperse the tiny, still viable seeds in their feces.

All figs, whether monoecious or functionally dioecious, are pollinated by tiny wasps of the family Agaonidae, commonly known as fig wasps. The relationship between fig and wasp is mutually beneficial, and neither fig nor wasp could survive without the other. The fig can be pollinated only by the wasp, and the wasp's larvae can feed and grow only in the ovary of a fig. This relationship between fig and wasp seems to be almost universally species-specific: a fig hosts only one species of wasp and is pollinated only by that species. According to Georges Valdeyron and David Lloyd in their article on the common edible fig, this holds true in the great majority of both Old and New World figs. Seldom is a fig pollinated by two species of wasp. William Ramirez reported that only 2 of the 40 species of figs that he studied in South and Central America and Florida were pollinated by more than one wasp, and in both cases only 2 wasps were involved.

The flowers in the inflorescences of both monoecious and dioecious figs are "stripped-down models." A female flower is just a pistil with only one ovule in its ovary. There are two types, one with a long style and another with a short style. Male flowers consist of little more than stamens, from one to five per flower, depending on the species.

An inflorescence of a monoecious species contains functional flowers of both sexes, but the female flowers mature first and the male flowers mature only after the female flowers are no longer receptive to fertilization. Thus an inflorescence goes through two phases, a male and a female phase, that are separated in time. There is no self-fertilization because all inflorescences on a tree are in the same phase at the same time.

Cross-pollination is achieved when a female wasp bearing pollen enters a female-phase inflorescence by pushing through the scales that cover its eye. The wasp pollinates both short-styled and long-styled flowers, but she thrusts her needle-shaped ovipositor down through the styles of short-styled flowers to place a single egg in the ovary. Short-styled flowers produce no seeds, because the wasp larvae feed within the ovary. The full-grown larvae stay in the ovary until after they metamorphose to the adult stage. The female wasp cannot, however, lay eggs in long-styled flowers because her ovipositor is too short to reach the ovary, so these flowers do produce seeds.

By the time the wasp's offspring attain the adult stage and are ready to emerge from the inflorescence, the inflorescence is in the male phase. Male wasps, which are wingless and blind, emerge before the females. After chewing their way out of the ovary that confines them, they search for females, which have yet to emerge. A male mates with a female, which is likely to be his sister, through a small hole that he chews through the wall of the ovary in which she is still trapped. He ultimately "commits suicide" by leaving the fig via a burrow he gnaws through its wall. After mating, the female enlarges the hole the male made in the ovary wall and crawls out. She will ultimately leave the inflorescence through a burrow made by a departing male, but only after she has acquired pollen from the male flowers, which are then mature.

The species of figs that occur in the Western Hemisphere belong to two different groups (subgenera), and as Ramirez pointed out, the wasps that pollinate them differ in the way they acquire pollen before leaving the fig. One of these groups is pollinated exclusively by wasps of the genus *Tetrapus*. The stamens of the figs in this group shed their pollen spontaneously, and the female wasps become thoroughly dusted with it before they leave the inflorescence. The figs in the other group are pollinated only by wasps of the genus *Blastophaga*. These wasps use their mandibles and legs to gather pollen, which is not spontaneously shed, and place it in storage cavities on their bodies. Female wasps of both genera, laden with pollen, leave their natal inflorescence through tunnels made by males, and, probably guided by an odor, fly to another tree of the same species with inflorescences in the female phase. The pollen-laden females enter the inflorescence through its eye, place pollen on the stigmas, and lay eggs in some of the ovaries.

Some trees of the edible Smyrna fig bear inflorescences with only female flowers, all of which have long styles. In the cultivated species it is these inflorescences that ripen to become the figs we eat. Other trees of the same species bear inflorescences with both male flowers and female flowers. The latter have short styles and seldom produce seeds because their ovules are destroyed by the larval wasps that feed within them. In their book on pollination ecology, Knut Faegri and Leendert van der Pijl wrote that this species is "functionally dioecious," because inflorescences on some trees produce pollen but only rarely do ovaries survive to produce seeds, while inflorescences on other trees produce seeds and never any pollen. Growers call the female trees "figs," and the functionally male trees "caprifigs." In the past, some people thought that "figs" are the cultivated form of the species and that the inedible "caprifigs" are the wild form. But this view is incorrect. Both forms occur in the wild and in cultivated orchards.

The Smyrna fig has its own specific pollinator, a fig wasp whose scientific name is *Blastophaga psenes*. The generic name is derived from two Greek words, *blastos*, a germ or seed, and *phagein*, to eat. *Psenes* is the ancient Greek name for the fig wasp. This wasp behaves and reproduces much like other fig wasps, with one exception: it can reproduce only in a caprifig. The cycle of reproduction begins when a pollen-laden female enters a caprifig through its eye and lays one egg each in all or most of its short-styled flowers. The larvae develop in the ovaries and emerge as adults a few weeks later, the males outnumbering the females about ten to one. After mating, the inseminated females leave the ovary in which they developed, become coated with pollen from the then mature stamens, and fly away from the caprifig to start another cycle of reproduction. A female may enter either a caprifig or a "fig." Entering a "fig" is a dead end for the female. She pollinates the flowers but cannot lay eggs in them because her ovipositor is too short to reach the ovary through the long style. She is an efficient pollinator. Ira Condit re-

ported that 21 figs that had each been entered by only 1 wasp bore an average of 850 seeds each.

Edible figs introduced as cuttings from Europe have been cultivated commercially in the United States, primarily in California, since the nineteenth
century. Until the beginning of the twentieth century, all figs grown in California and elsewhere in the country were varieties that produce fruit without being pollinated and therefore do not generate the seeds that give the
delicious nutty taste to the Smyrna fig. Consequently, cuttings and seeds
of the Smyrna fig were planted in California. The trees thrived, but their
fruits invariably fell from the tree before they ripened. The problem was that
there were no caprifigs in California and therefore no wasps to pollinate the
Smyrna figs.

For centuries fig growers in the Old World knew that Smyrna figs need to
be pollinated. They grew caprifigs, and when the female "fig" trees were
ready for pollination, they hung in each tree a basket containing a few
caprifigs, which produced enough wasps to pollinate the female inflorescences. But many botanists thought that this practice of "caprification" was
useless, no more than the superstition of ignorant peasants. Condit reported
that as late as 1898 a botanist "was ridiculed by scientists in Italy for his belief in the necessity for caprification." In 1887, Dr. Gustav Bisen was hooted
down in Fresno, California, when he recommended the importation of
caprifigs and figs wasps. But reason eventually prevailed, and by the beginning of the twentieth century caprifigs and the fig wasp had been introduced
into California, and ever since large commercial crops of various varieties of
the Smyrna fig have been grown there.

Orchids probably have at least as many different and often bizarre ways of inducing insects to pollinate them as do all other flowering
plants together. No other group of plants can match the incredible diversity
and variability of this, the largest plant family, with over 25,000 species.
Most of them grow in the tropics of the Old or New World, but a respectable
number occur in the north and south temperate zones. Some live rooted in
the soil, as do the lady's slippers of North America and Europe. Many tropical species, including the cattleyas, the "corsage orchids," are epiphytes that
cling to the bark of trees for support but derive no nourishment from them.
They are watered by rain and absorb nutrients from debris that collects
about their roots. In Australia, ghostly white subterranean orchids that can-

not photosynthesize because they have no chlorophyll survive only as parasites of other plants. Some of their flowers never break through the surface of the soil, but others that just barely do are, Kingsley Dixon and his coauthors report, pollinated by termites—probably the only plants pollinated by these insects.

Orchid flowers, including the familiar cattleyas, are greatly modified and often highly specialized versions of the simple flower I described earlier. In the cattleyas, not the most specialized of the orchids, three large and colorful petals and three equally large and colorful sepals form the perianth, the showy part of the flower. The lower petal of the perianth is enlarged to form the lip, which in this and other species is a landing platform for insects who come for a drink of nectar. In other orchids the lip is variously modified to perform some surprising functions—anything from acting as a trap to mimicking an insect. The sexual organs of all orchids are highly modified. In the cattleyas and most other species, the filaments of the stamens and the style of the pistil are fused to form a single long, thin column, usually with the anthers at its tip and the stigma just behind them. In most orchids the numerous sticky pollen grains are clumped to form two tightly packed bundles called pollinia. In *The Forgotten Pollinators,* Stephen Buchmann and Gary Nabhan describe a pollinium as looking like a tiny bright yellow egg yolk attached to a pliant but leathery brown stalk. The flowers of most orchids are bisexual, but a few species are dioecious.

Nectar may be exposed and easily reached, or it may be hidden at the end of a long spur, or corolla tube. Spurs vary in length. In his book on the fertilization of orchids by insects, Charles Darwin noted that the flower of the orchid *Angraecum sesquipedale,* found only on the island of Madagascar, has an extraordinarily long spur that contains the nectar. "In several flowers sent me," he wrote, "I found the nectaries [spurs] eleven and a half inches long, with only the lower inch and a half filled with nectar." He speculated that "in Madagascar there must be moths with [tongues] capable of extension to a length of between ten and eleven inches." A hawk moth with that long a tongue—much longer than the body—has since been discovered on the island, but as far as I know it has yet to be caught in the act of taking nectar from this orchid. But L. Anders Nilsson and his colleagues did prove that in Madagascar another hawk moth with a tongue 5 inches long visits orchids with a spur about 5 inches long.

Some orchids provide nectar, edible tissues, or even perfumes useful to the pollinator. Pollen is usually inaccessible and thus not available as food.

Other orchids, between 8,000 and 10,000 species according to Nilsson, trick their pollinators into performing their duties, and offer them no reward.

Orchids force pollination by trapping insects in several ways. Among the trappers are the dozen or so lady's slippers found in North America. Although they are considered to be the most primitive of the orchids, the way they force bees to pollinate them is quite sophisticated. The lip of the flower closes upon itself to form a large bladder-like pouch that resembles a shoe, inspiring the plants' common name. Attracted by the flower's fragrance, a visiting bee, usually a member of some native species rather than the imported honey bee, enters the "shoe" through a large opening near its base. When the bee tries to leave through the same opening, it can't climb out because the inner surface of the shoe is too slippery. After several futile attempts to escape by this route, it gives up but is soon attracted to the far back of the shoe, where light comes through transparent "windows" near the only possible exit. The departing bee must squeeze under the column, first brushing against the stigma and then the anthers, which smear it with sticky pollen grains. (These orchids do not form discrete pollinia.) If the bee enters another lady's slipper, pollen will rub off on the stigma.

The orchids of the South American genus *Coryanthes*, commonly known as bucket orchids, trap the bees that pollinate them—believe it or not—in a bucket of water. The lip forms a deep bucket with a spout below its rim. "Faucet glands" secrete a thin liquid, probably water, that fills the bucket up to the level of the spout. Hilda Simon noted that a bee that comes to nibble on the fleshy, nourishing tissues on the bucket's rim "invariably loses its footing on the smooth rounded surface [and] falls into the liquid." The bee cannot fly out of the bucket because it is soaking wet. Its only escape is to crawl through the lip of the spout. Since the tip of the column all but blocks the spout, the escaping bee is forced to contact the two pollinia on the column, which stick to the upper side of its body. After it dries off it may fly to another bucket orchid, where it will leave the pollinia on the sticky stigma as it is once again forced to make its escape from a bucket.

Charles Darwin wrote that certain plants of the genus *Catasetum* must "be considered the most remarkable of all orchids." He was right, although he knew only part of the story, the role of the orchids; he did not know the even more remarkable role of their pollinators. *Catasetum* plants are visited almost exclusively by male bees of various species of the subfamily Euglossinae, commonly known as orchid bees. These orchids are extraordinary because they reward their pollinators with droplets of perfume rather than food, because they forcefully "shoot" their pollinia at visiting bees, and because they are dioecious.

The brilliant metallic golden, blue, and green orchid bees, native only to the New World tropics, have been called flying jewels of the rainforest. Both sexes forage for pollen and nectar at flowers other than orchids, but only males visit *Catasetum* plants to collect droplets of perfume secreted by both male and female flowers. Attracted by its fragrance, a male enters the narrowed center of a male flower's lip to reach the gland that secretes the perfume. Friedrich Barth succinctly described what happens next. As the bee vigorously rubs the surface of the scent glands with his front legs, mops of finely branched hairs on his feet absorb droplets of the perfume. From time to time the mopping operation is interrupted for a few seconds as the bee hovers just above the flower, "kicking the legs rapidly in a manner reminiscent of the way honeybees pack pollen into the pollen basket." As the orchid bee hovers over the flower, he transfers the perfume from his front legs into two large chambers in the greatly swollen main segments of his hind legs. The middle leg on each side scrapes the perfume from the corresponding front leg and moves it to the opening of the "perfume flask" in the corresponding hind leg. Euglossine males also collect other odorous substances from a variety of sources, including flowers other than orchids, decaying wood, rotting fruits, and feces. These substances form a fragrant bouquet whose chemical composition is distinctive enough to identify the species of orchid bee from which it came. How the males use their "perfume" remains a mystery, but T. Eltz and three coauthors suggest that they release it only during the brief and rarely observed courtships at the small territories males establish near treefalls in the forest.

How do these bees pollinate the orchids? As a male enters a flower to reach the perfume-secreting gland, he cannot help touching two slender appendages of the column referred to as "antennae," thereby triggering the sudden release of two tautly stretched pollinia. When this tension is abruptly released they shoot forward, hit the bee, and stick tightly to his body. In Darwin's words,

> When certain definite points of the flower are touched by an insect, the pollinia are shot forth like an arrow, not barbed however, but having a blunt and excessively adhesive point. The insect . . . flies sooner or later away to a female plant, and, whilst standing in the same position as before, the pollen-bearing end of the arrow is inserted into the stigmatic

cavity, and a mass of pollen is left on its viscid surface. Thus, and thus alone, can the five species of *Catasetum* which I have examined be fertilised.

Some species of *Catasetum* may be visited by several species of orchid bees, and a bee may visit several different species of these orchids. How then is proper pollination achieved? Buchmann and Nabhan wrote that orchids have evolved ways of "ensuring that their pollinia get transferred *only* to another flower of their own species and are not 'wasted' by being left in flowers of an altogether different species." In the catasetine orchids, the male flowers of different species are so constructed that their pollinia adhere to different parts of a bee's body. Robert Dressler found that in different species of *Catasetum* and related genera pollinia may be attached to the top of the bee's thorax, the "elbow" of its foreleg, its head, or the tip of its abdomen. Female flowers are constructed so that they force bees to enter them in such a way that the stigma will contact only pollinia that were properly placed on the bees by male flowers of their own species.

The most bizarre way in which plants achieve cross-pollination is to trick male insects into having brief sexual flings with their blossoms. Bastiaan Meeuse wrote that this is "so incredible that I would not accept it myself if it had not been reported by at least a dozen reliable investigators on three continents." Certain Eurasian, South American, and Australian orchids subject male bees and wasps to the frustration of trying to copulate with a dummy female, actually a flower that is—at least to the insect—a convincing olfactory, visual, and tactile mimic of a female of its own species. Some of these orchids minimize competition from female wasps by blossoming when male wasps, which emerge from their pupae first, are abundant, but female wasps, which emerge later, are still scarce. The unfortunate male gains nothing from being seduced by the orchid, which secretes no nectar, but when he leaves, he carries on his body pollinia that will fertilize the next flower that vamps him.

Among the orchids pollinated by such floral sexual deceits are several European members of the genus *Ophrys*. The traditional common names of these European species, bee orchid, fly orchid, and the like, derive from the visual resemblance of their flowers to insects. To the human eye, the mimicry is discernible but far from perfect, although it is apparently good enough to fool insects, whose visual acuity is not sufficient to resolve much detail. Flowers of the looking-glass orchid of southern Europe and north Africa

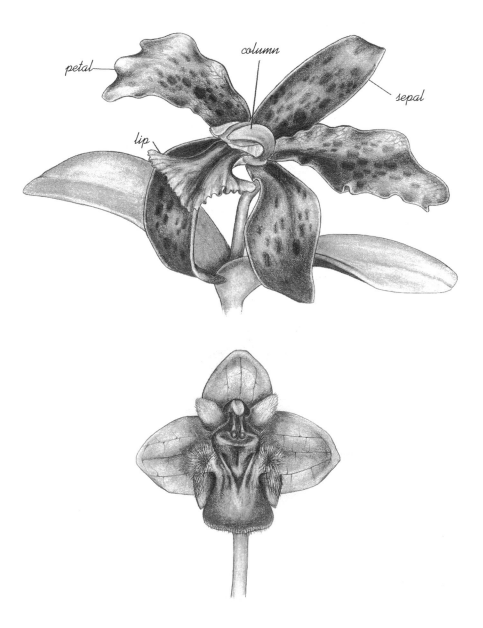

The *Ophrys* orchid below the showy *Cattleya* looks just enough like a female
bee to fool a myopic male lured by the flower's imitation sex pheromone,
a mimic of the chemical signal that female bees use to attract males.

mimic a female wasp whose males are the orchid's only pollinators. The plant is named for the large, convex, bright blue area on the lip of the flower, whose mirror-like metallic look imitates the blue sheen of the crossed wings of a wasp. The thick fringe of red hairs that circumscribes the edge of the lip suggests similar hairs on the abdomen of the wasp, and its antennae are nicely imitated by the thin, dark upper petals.

Experiments and observations made by A. Pouyanne in Algeria early in the twentieth century, summarized by Michael Proctor and Peter Yeo in their book on pollination, showed that male wasps are attracted to the look-ing-glass orchid from a distance by its odor, which we now know mimics the sex-attractant pheromone, a chemical signal recognized by other members of the species, emitted by the female wasp. Pouyanne found that a few spikes of orchid blossoms held in the hand soon attracted male wasps, "sometimes several hustling one another on the same flower, and apparently quite oblivious of the observer." Several males were attracted to the odor of blossoms that were hidden from sight by a wrapping of newspaper. But the right visual stimulus is also required to fool the wasp. Male wasps tried to mate with detached flowers laid face up on the soil, but when the re-semblance of the flowers to a wasp was obscured by putting them face side down, the wasps were attracted by the odor but had difficulty finding the flowers. Males sit lengthwise on the lip of the flower with their heads just beneath the column, thrust the end of the abdomen into the fringe of long red hairs at the lower end of the lip, and make "tremulous, almost convul-sive" movements that are exactly the same as the movements they make when copulating with a female. Before the male leaves, two pollinia be-come attached to his head. The males of this species are not known to ejacu-late on orchid flowers, but Edith Coleman found that male wasps of other species deceived by female-mimicking Australian orchids do ejaculate on the flower.

An ant-mimicking Australian orchid that, as R. Peakall and his two coau-thors reported, achieves cross-pollination by seducing amorous flying male ants is among the very few plants of any kind pollinated by ants. With the exception of this Australian orchid, most ant-pollinated plants are prostrate species whose blossoms hug the ground and are thereby readily accessible to crawling worker ants. Although ants are of little importance as pollinators, they are, as you will see, involved in the lives of plants in other significant ways. Preeminent among them is the all-important dissemination of seeds, the subject of the next chapter.

2

DISPERSING SEEDS

All plants disperse soon after they are "born," ending up anywhere from inches to many miles away from their parents. The great majority of the higher plants—as opposed to the lower or more primitive plants such as mosses, horsetails, and ferns—disperse as seeds. Most of the species in the plant kingdom are higher plants, the plants that are most familiar to us, consisting of the gymnosperms, such as cycads and cone-bearing trees such as firs and pines, and the angiosperms, such as oaks, cactuses, daisies, and orchids. A seed is a whole individual, an embryo accompanied by a store of nutrients that, after germination, sustain it as it grows to become an established seedling. Enclosed in a tough shell, or seed coat, it is protected against injury and desiccation. The seeds of some plants, such as those of certain willows and orchids, can survive for only a few weeks; but many survive for years or decades, and lotus seeds have remained viable for hundreds of years.

Some higher plants have an alternative method of reproduction and dispersal, a form of cloning, the dispersal of vegetative parts other than seeds that can take root and grow. A twig shed by a streamside willow may float downstream, take root on a sandbar, and grow to become a tree. The cholla cactus of the Sonoran desert is often called the jumping cholla, because its spiny segments are so loosely attached that they break off at the slightest touch, seemingly jumping out at you as you brush past. A cholla segment may cling to a passing deer or other animal and fall off to take root in some place far from its parent.

Dispersal has several advantages, three of which are especially important: First, young plants that "leave home" do not have to compete with their parents—or a crowd of siblings that might have stayed put—for essential resources that may be in short supply, such as water, soil nutrients, and "a

44

place in the sun." Second, they may escape from the plant-eating insects and disease-causing microbes and fungi that often become established in the stand of plants in which their parents grow. Finally, plants may disperse and colonize distant habitats not yet occupied by their species or, rarely, as after a volcanic eruption or a landslide, by any other species. In 1883, the island of Krakatoa, which is near Java in the East Indies, was denuded of all life by an immense volcanic eruption. After only three years, scattered individuals of eleven species of ferns and fifteen species of flowering plants were growing on Krakatoa, probably colonizers from the closest island, about 25 miles away. Only 25 years later, the once barren island was almost totally covered with green. There was a much greater variety of plants than before, and 263 species of animals, mostly insects, had returned.

Closer to home, the bare soil of an abandoned crop field is soon covered by a community of grasses and herbaceous plants, most new invaders but some developing from seeds that had lain dormant in the soil. Over the next few decades, the field will be colonized by a succession of plant communities, culminating in eastern North America with a "climax" forest community of oaks, hickories, and other trees. In mature forests in New England, you sometimes come upon tumbled stone walls that once bounded crop fields, mute evidence of the ecological succession that began early in the nineteenth century, when many farms in New England were abandoned after the Erie Canal made it possible to ship grain from the more fertile areas bordering the Great Lakes.

The universal importance of dispersal is indicated by the many different ways plants have evolved to disseminate their seeds, nicely summarized by Leendert van der Pijl in *Principles of Dispersal in Higher Plants*. Most animals are mobile at some time during their lives and can disperse—as most of them must—by swimming, flying, or walking. But, with few exceptions, seeds cannot move under their own power. The exceptions are a few "creeping" seeds with bristles that can push or drag them for short distances over the ground as the bristles twist and turn when they alternately become dry or moist as the humidity decreases and increases. Some plants disperse their seeds by "shooting" them away. When the two walls of some beanlike seed pods dry, they split apart and curl up, gripping the slick seeds and putting enough pressure on them to shoot them off just as you can shoot a slippery watermelon seed by pinching it between your thumb and forefinger. Some other plants that shoot their seeds are violets; wood-sor-rels, which can shoot a seed for more than 6 feet; and the amazing sandbox tree of South America, whose fruit bursts explosively, hurling seeds as far as 150 feet.

🐜 🐜 🐜 But most seeds are transported by some external force: wind, flowing water, or wandering animals. Among the many windborne ones are the winged seeds of maples and ashes, which twirl like helicopter blades as they are blown away. The plumed seeds of cottonwoods, milkweeds, dande-lions, and many other plants may travel for miles as they float on the breeze like parachutes. The tumbleweeds—several species in different families—disperse seeds when the dead parent plant breaks loose from its roots and scatters its seeds as the wind rolls it over the ground. The common tumble-weed of the American West, a "trademark" of the plains, sometimes piles up against fences and buildings in a huge mass that obscures the structures from view. In the Amazon and Orinoco basins of South America, the seeds of some land plants are disseminated by water currents when their habitat, the wide floodplain of a river, is inundated. When the flood recedes, many seeds are left behind on the wet soil. But some seeds float off and are carried out to sea and washed up as far away as the shores of England and Scandi-navia.

Animals disperse the seeds of many plants. Most of these mobile seed disseminators are ants, birds, and mammals, but a few fish and reptiles also spread seeds. Burdocks, tick trefoils, beggar ticks, and many other plants of various families have sticky or spiny seeds or fruits, the latter by definition the bearers of seeds, that get caught in the fur of a mammal and later fall to the ground. (You have probably found some of these hitchhikers clinging to your pants after a walk in the country.) In South Dakota's Custer State Park, my daughter Susan and I found a wad of hair entangled with several burs hanging on a shrub, no doubt shed by a passing bison that may have ac-quired the burs miles away. Many plants, including wild cherries, mountain ashes, and mulberries, enclose their seeds in fleshy fruits that are swallowed whole by animals, mainly birds. The pulp of the fruit is digested, but the seeds are passed in the droppings. Because most birds locate food by sight, these fruits are usually conspicuous, often bright red, orange, or yellow.

Few insects other than ants regularly disseminate seeds. I know of only a few. Several Old World scarabs, the dung beetles—such as the sacred scarab of ancient Egypt—both transport and plant seeds. They make balls of cattle dung, which may contain undigested seeds, and roll them away from the "mother lode" of dung to bury them in the soil with one of their eggs. Seeds not destroyed by the grub that hatches from the egg eventually germinate. Certain Brazilian scarabs provide their grubs with palm seeds rather than dung. They bury the seeds and lay eggs on them, and those not eaten by the grubs germinate. The dung-burying scarabs of Amazonia, seemingly insignificant insects, play essential roles in their ecosystems. As Kevina Vulinec has pointed out, if they were not there, feces would accumulate, the soil would not be sufficiently turned over and aerated, and seeds deposited in the dung of vertebrates would remain on the surface, vulnerable to rodents, fungi, and seed-eating insects.

In the twelfth chapter of *The Origin of Species,* Darwin noted:

> Now, in parts of Natal it is believed by some farmers, though on insufficient evidence, that injurious seeds are introduced into their grassland in the dung left by the great flights of locusts which often visit that country. In consequence of this belief Mr. Weale sent me in a letter a small packet of the dried pellets [excreta of the locusts], out of which I extracted under the microscope several seeds, and raised from them seven grass plants, belonging to two species, of two genera. Hence a swarm of locusts, such as that which visited Madeira, might readily be the means of introducing several kinds of plants into an island lying far from the mainland.

And Henry Ridley has described how termites disperse seeds with the help of aardvarks. These large mammals, inappropriately called anteaters, subsist mainly on termites, using powerful claws to rip open large termite mounds and a long, sticky tongue to lick up the insects. Ridley quoted a Mr. Burtt-Davy, who described how the seeds of a certain grass are dispersed in the Transvaal of South Africa: the termites "collect fragments of the grass and its seeds, and store them in the upper chambers of the nest. After the nest is broken down by the attacks of the Aardvark . . . the seeds germinate and produce a patch of the grass. This grass only occurs in open spots, pathways, etc., in the veldt, being, from its low growth, unable to compete with the taller grasses, so that it chiefly occurs on the bare sites of the destroyed nests, covering them with a dense mat."

Finally, there is a dung beetle found in southern Africa that rolls irregularly shaped dung fragments over the ground. B. V. Burger and W. G. B.

Petersen discovered that this beetle also rolls seeds of the spineless monkey orange tree for long distances, often over a mile, and at an astonishingly rapid speed. The seeds, covered with a thin layer of soft flesh and about an inch in diameter, are in a hard-shelled fruit about the size, shape, and color of an orange. The seeds fall to the ground after the fruit is cracked open by a monkey who eats the edible flesh that surrounds the seeds but leaves them unharmed. The aroma of fruit and seeds is pleasantly sweet to humans, but it contains odorous chemicals that are also found in the dung of mammals. Burger and Petersen showed that these chemicals are attractive to the beetles, presumably deceiving them into accepting the seeds in lieu of dung.

Of the animals that disperse seeds, only birds and mammals are more important than ants. In his book on ant-plant mutualism, Andrew Beattie noted that the seeds of 35 percent of herbaceous and woody flowering plants are disseminated by ants.

All of the many ant species are social, and most nest in the soil. Their colonies consist of a queen and, depending upon the species, dozens, thousands, or even millions of her daughters, which are all sterile and constitute the worker and soldier castes. The vast majority of ants are primarily carnivores that, as Kathleen Keeler puts it, "consume animals smaller or slower than themselves." Indeed, Bert Hölldobler and Edward. O. Wilson in their Pulitzer Prize–winning *The Ants* argue that ants are descended from carnivorous wasps. How is it, then, that they are important disseminators of seeds? The answer is twofold. A few types of ants, especially desert species, are avid hunters of insects but have broadened their diet by also harvesting and storing seeds. They eat most of the seeds they collect, but some survive undamaged because workers often drop a seed they are carrying to the nest or discard an uneaten seed from their underground granary. Harvester ants are incidental seed dispersers, but other ants are, in the words of Hölldobler and Wilson, "purposeful" seed dispersers. They collect and carry to their nests certain seeds that bear fleshy, disposable, edible appendages known as elaiosomes, which loosely translated from its Greek roots means "edible oily bodies." The ants relish these edible appendages and benefit from eating them. The plants benefit because the ants disseminate their seeds, usually discarding them intact in an aboveground trash pile after consuming their fleshy appendages.

Ants are better seed dispersers than other insects because they move the seeds a significant distance and do not necessarily harm them. Almost all ants carry seeds back to the nest, rather than eating them on the spot as do other insects, often dropping some in the process. They also leave at least some seeds unharmed, the few that are lost or discarded by harvester ants and all or almost all of those collected by ants for their elaiosomes.

Only a few ants, mostly species of the grasslands or deserts, are seed harvesters. But they are a diverse lot from several subdivisions of the ant family, which indicates that this life style evolved independently several times. In North America fewer than 50 of the 750 species of ants are harvesters. But in their chosen habitats, harvester ants are sometimes overwhelmingly important both as destroyers and disseminators of seeds. Fritz Went and two coworkers found that harvester ants are the most abundant ants in Death Valley, California, and that in a given area the total weight of these ants is about equal to the total weight of all rodents.

Western science's awareness of harvester ants has a convoluted history. People of ancient civilizations were well aware that ants harvest and store seeds for future use, and many of their writers recognized this phenomenon. The Bible (Proverbs 6:6–8) advises:

> Go to the ant, thou sluggard;
> Consider her ways and be wise:
> Which having no chief,
> Overseer, or ruler,
> Provideth her bread in the summer
> And gathereth her food in the harvest.

In one of his fables, Aesop told of ants drying damp seeds in the sun. A grasshopper approached and begged for a few seeds, complaining that it was starving. The busy ants stopped working momentarily and asked the grasshopper what it had been doing all the past summer and why it hadn't collected a store of food for the winter. The grasshopper replied that it was so busy singing that it didn't have the time. Aesop did not know that, even if given plentiful food, the grasshopper could not have survived the cold of winter, as do ants in their cozy underground nests. The grasshopper had been singing to attract a mate, which would lay dormant eggs that would survive the winter to perpetuate the species. But although Aesop knew little about grasshoppers, his fable shows that the Greeks were aware of harvester ants at least 2,500 years ago.

In 1873, J. Traherne Moggridge, an English naturalist, wrote that the Greeks' ancient knowledge of harvester ants "had begun to be called in question" by European naturalists in the middle of the eighteenth century. By Moggridge's time, authorities on the subject were "opposed to the belief that European ants ever do systematically collect and make provision of seeds, and [believed] that instances of such occurrences in tropical climates remain as isolated facts . . . which it is difficult to explain." They even suggested that ant pupae had been mistaken for seeds. (If an ant nest is disturbed, streams of fleeing workers carry away the tiny cocoons enclosing the pupae.) This difference of opinion came about because eighteenth- and nineteenth-century European naturalists, most of whom lived far from the Mediterranean, knew the ants of their northern areas, which are not harvesters, but were unfamiliar with ants that live in far southern Europe or the tropics, many of which are harvesters. When Moggridge visited Mentone, a French city on the Mediterranean coast, he confirmed that harvester ants exist in Europe when he found ants carrying seeds to their nest, "a long train of ants, forming two continuous lines, hurrying in opposite directions, the one with their mouth full, the others with their mouths empty."

In a prescient passage in his book, *Harvesting Ants and Trap-Door Spiders,* Moggridge recognized that harvester ants, although they eat most of the seeds they collect, are also disseminators of viable seeds:

As the ants often travel some distance from their nest in search of food, they may certainly be said to be, in a limited sense, agents in the dispersal of seeds, for they not unfrequently drop seeds by the way, which they fail to find again, and also among the refuse matter which forms the kitchen midden in front of their entrances, a few sound seeds are often present, and these in many instances grow up and form a little colony of stranger plants.

Almost all harvester ants utilize seeds from a variety of plants. Steven Rissing and Jeanette Wheeler reported that an ant in Death Valley harvests seeds from at least 29 species of plants, ranging from grasses such as fescue to herbaceous species such as plantain and brittle bush. Rissing later reported finding the seeds of 36 plants in the refuse piles of these ants. Elizabeth Davison found seeds of 20 species of plants—ranging from grasses to herbaceous species—when she excavated nests of an Australian harvester ant. Moggridge said that the harvesters of the Mediterranean coast of France "collect almost indiscriminately from any fruiting plant that falls their way." One species he observed collected seeds of 54 species of plants from at least 18 different families. Harvesters are, however, a little more discriminating than Moggridge thought. Recent investigations show that they will not collect certain seeds even when they are abundant. And a few Australian harvesters, Alan Andersen has shown, are exceptionally fussy about the seeds they eat, collecting them from only one or a few species of grass.

Ants are efficient harvesters. Various reports indicate that, depending on the species and the size of the colony, they collect anywhere from a few thousand to hundreds of thousands of seeds per colony per year. Workers of at least one species vary their harvesting technique according to the abundance and distribution of seeds. If seeds are scarce and widely scattered, workers forage alone and disperse widely. If they find a seed they bring it to the nest and then go out alone to search for another seed. But if a worker finds a place where seeds are plentiful, she collects a seed and marks her return path with a trail pheromone, a chemical marker, that leads other ants to the seeds. Ants that find a seed at the indicated site reinforce the pheromone trail as they return to the nest, and soon a large force of workers is exploiting the bonanza. The trails may be long and followed by many workers. Moggridge saw his ants collect seeds from as far as 72 feet from their nest. A century later, Wheeler and Rissing found a column of about 17,000 ants that extended about 130 feet from the nest to the source of seeds.

Data on the quantity of seeds stored by harvesters are scarce, because it is always difficult and sometimes impossible to completely excavate ants' nests, which often extend several feet or even yards down into the soil. Moggridge, with perseverance, excavated an entire nest of harvesters and exposed the special chambers in which they stored seeds. These granaries were of various sizes and shapes, but he said that the average one was "about as large as a gentleman's gold [pocket] watch." He made a careful estimate of the quantity of seeds stored in this nest:

During the spring of 1873 I removed with but very little loss the contents of two granaries from a very extensive nest of Atta [Messor] structor, consisting principally of seeds of clover, fumitory, and pellitory. These seeds, when perfectly clean and freed from earth, weighed in the one case 4 sc. 4 grs., and in the other 4 sc. 8 grs. [A scruple (sc.) is 20 grains (grs.) and a grain is about 0.002 ounce.] Now there cannot have been less than eighty such granaries in this nest, so that, if we take five scruples as the average weight of the seeds in each granary, and thus, allowing for loss in collection, which we may fairly do, we should have a total weight of more than sixteen ounces, or one pound avoirdupois weight of seeds contained in the nest.

Much more recently Walter Tschinkel made an actual count and found that as many as 300,000 seeds may be stored in a large nest of the Florida harvester ant.

Moggridge noted that traditional Jewish law deals with the ownership of harvester ant stores found in grain fields. The existence of such a law shows that these stores of grain are not trivial but are large enough to have engendered disputes. The law specifies that if a store of seeds is found when the field is being reaped, it belongs to the farmer. If it is discovered after the field has been reaped, it is to be divided between the farmer and the poor people who glean the field.

The ways in which harvester ants prevent stored seeds from germinating or spoiling are an amazing example of how evolution has shaped behavior. Moggridge and others after him confirmed the reports of ancient Greek and Roman writers that the ants prevent seeds from germinating by biting off the part of the embryo that will become the root. If a seed the ants fail to deactivate germinates in a granary, the ants discard it in a trash pile on the soil outside the nest. Ants prevent seeds from spoiling by keeping them dry. They locate their granaries in dry sites and "waterproof" their floors, walls, and ceilings. Moggridge reported that the harvesters he studied paved the floors of their granaries, presumably to hold back moisture from the soil. "The texture of the floor usually differs markedly from that of the surrounding soil, and the fine grains of silex [sand] and mica which are selected for its construction are more or less cemented together."

If seeds stored in the nest do get wet—as might happen during heavy rains—the ants carry them to the surface to dry in the sun. Aesop knew this, but the first modern observer to report ants drying seeds in the sun was a British officer, Lieutenant Colonel W. H. Sykes, who in the early nineteenth century was stationed in India. In 1836, Sykes reported his observations in the first issue of the *Transactions of the Entomological Society of London*, which is still being published today. An excerpt from his diary records his observations:

Poona, June 19, 1829.—In my morning walk I observed more than a score of little heaps of grass-seeds *(Panicum)* in several places on uncultivated land near the parade-ground; each heap contained about a handful. On examination I found they were raised by . . . [ants] hundreds of which were employed in bringing up the seeds to the surface from a store below: the grain had probably got wet at the setting in of the monsoon, and the ants had taken advantage of the first sunny day to bring it up to dry. The store must have been laid up from the time of the ripening of the grass-seeds in January and February. As I was aware this fact militated against the observations of entomologists in Europe, I was careful not to deceive myself by confounding the seeds of a panicum with the pupae of the insect.

Sykes's observation that harvester ants dry wet seeds in the sun was confirmed by Moggridge, and just a year later Gideon Lincecum, a physician-naturalist, observed that a harvester ant in Texas "subsists almost entirely on small seeds, great quantities of which [these ants] store away in the granary-cells to supply food for winter," and that "during the rainy seasons in the autumnal months it happens right often that . . . water penetrates their granaries . . . In this emergency they bring out the damaged grain the first fair day and [expose] it to the sun until near night." On one occasion he saw "on a flat rock as much as a gallon of wheat sunning." Since then many other observers have substantiated Sykes's observation.

✻ ✻ ✻ Harvester ants are a distant second as disseminators of seeds to the many more species of ants that gather seeds with elaiosomes. In addition to oil, elaiosomes contain protein, starch, sugar, and vitamins. Attracted by chemicals in the elaiosomes, ants collect the seeds, carry them to their nests, eat the elaiosomes or feed them to their larvae, and then discard the seeds, which are usually uninjured, because they are too hard for the ants to crack.

In Moggridge's time, elaiosomes had yet to be discovered, but their presence on violet seeds explains what was to him a mystery. Moggridge concluded, quite correctly, that none of the common English ants harvests seeds because they did not gather seeds he threw in their way. But he noted an exception: "I was gathering some fresh capsules of the common sweet violet in a garden at Richmond, near London, and in pouring the seeds out of my hand into the paper bag made to receive them, a few were spilled to the ground. In a short time afterwards I was greatly surprised to see some of

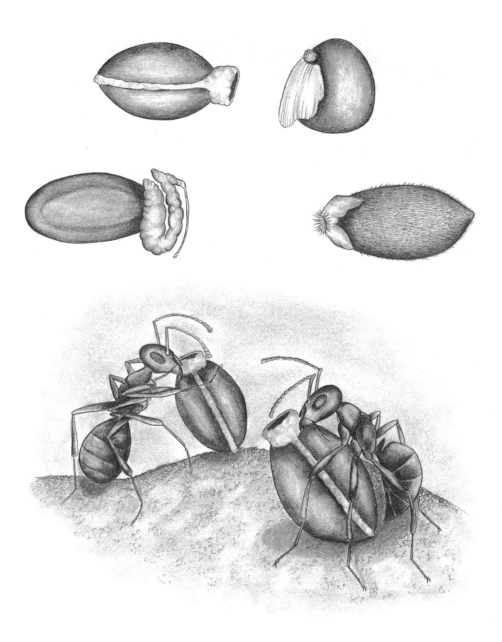

Seeds with elaiosomes, clockwise from the upper left: violet, golden corydalis, milkwort, and acacia. Below ants nibble on the elaiosomes of violet seeds.

these spilled seeds in motion, being carried by the common black ant . . . into its nest. On seeing this I hastened to get some more fresh violet seeds, and also a quantity of seeds taken from ant's granaries at Mentone, and scattered these where the other seeds had lain. After watching for half an hour a few of the violet seeds were carried in, but not one of the granary seeds was removed, though these were examined with some curiosity." The solution to the mystery is that violet seeds bear elaiosomes whereas the seeds from the harvester ant granary lacked elaiosomes, because they had never had them or because the harvester ants had eaten them.

Many plants do not have seeds with elaiosomes, but we know that at least 3,100 species do—1,500 from Australia, 1,300 from South Africa, and 300 from elsewhere—and there are surely many other plants not yet known to have this trait, especially in tropical areas not thoroughly explored by botanists. Plants of at least 70 families, many not related or only distantly related to each other, bear elaiosomes, in Australia alone at least some plants in 23 families. Among the North American families that include at least some members with elaiosomes are the lily, poppy, violet, purslane, birthwort, buttercup, spurge, and daisy families.

The great diversity of plants with elaiosomes shows that this useful trait evolved independently many times. The multiple origins of elaiosomes are also indicated by the variety of their shapes and the variety of plant tissues and organs from which evolution has molded them. In violets, the elaiosome is a modification of the short stem that attaches the newly formed seed to the wall of the ovary. In spurges it consists of the enlarged lips of the tiny opening in the seed coat through which sperm cells enter. In other plants elaiosomes are modifications of the wall of the ovary that surrounded the seed, or of the petals and sepals of the flower that produced it. Different species of plants usually have elaiosomes of different shapes: girdles that ring the seed, sheaths that more or less cover it, or caps or finger-shaped bodies at one of its ends. According to Andrew Beattie, the elaiosomes of different plants are adapted to the prevailing climate. In moist areas, they tend to be soft and rot within a week or two. In arid areas, they are generally tough and so resistant to decay that they may last for several months. Because an ant may try to eat a seed after consuming its elaiosome, the seeds are generally difficult for the mandibles of an ant to pierce. Some are so round and smooth that the mandibles slip off. Others are covered with coats too hard for the ants' mandibles to pierce.

In an article on seed dispersal by ants, Steven Handel and Andrew Beattie

pointed out that some plants that have evolved elaiosomes have greatly altered their architecture to make it easier for ants to find their seeds. Among
them is a trillium that grows in western North America. Most trilliums bear
near the top of a tall stalk a showy flower with three petals and just below it
a whorl of three almost stemless leaves that attach directly to the stalk. But
in this western trillium, the flower is very close to the ground, and thus the
seeds are within reach of ants crawling along the forest floor. As in other trilliums the leaves originate below the flower, but, unlike those of other trilliums, these leaves have long stems that raise them high above the ground-
hugging flower to increase their exposure to the sun.

Plants with ant-dispersed, elaiosome-bearing seeds, first discovered early
in the twentieth century, occur on all the continents except Antarctica but
are not evenly distributed among habitats. According to Andrew Beattie, the
seeds of plants of many different growth forms and life styles bear elaiosomes: small trees, shrubs, vines, herbaceous annuals and perennials, cactuses, grasses, and sedges, and epiphytes that grow on the bark of tropical
trees. Some of these plants blossom under the trees in moist forests in the
north temperate zone early in spring, before the forest floor is shaded by the
leaves of trees. Among this group of approximately 300 plants are such familiar North American species as bloodroot, violets, and spring beauty. It
was not until 1975 that Rolf Berg reported the discovery in Australia of another 1,500 plants with elaiosomes, most of them shrubs of dry areas with
nutrient-poor soils. A few years later Mark Westoby and others discovered
yet another 1,300 plants with this trait in a similar habitat in South Africa, a
community of shrubs—known as the fynbos—that grow in nutrient-poor
soils.

Westoby and his colleagues investigated the intriguing fact that in Australia there are far more plants with elaiosome-bearing, ant-dispersed seeds on
infertile soils than on fertile soils. The percentage of species with seeds dispersed by animals such as insects or birds is similar for the two soil types:
about 48 percent on fertile soils and about 52 percent on infertile soils. But
the dissimilar plant communities that grow on these two soil types differ
greatly in the numbers of species with ant-dispersed or bird-dispersed seeds.
On fertile soils, about 33 percent of the plants attract birds by covering their
seeds with fruits that birds like to eat, but only about 15 percent attract
ants by producing seeds with elaiosomes. But the percentages are reversed
on infertile soils, where only about 5 percent of the plants bear fleshy fruits,
while 47 percent have seeds with elaiosomes. Why do elaiosomes replace

fleshy fruits on plants in infertile soils? Westoby speculated that the reason is that a large fleshy fruit is more "expensive" to make than a much smaller elaiosome. Most plants on infertile, mineral-poor soil can't afford to "spend" the large quantities of nitrogen, potassium, phosphorus, and other minerals needed to produce fruits big enough to attract such large seed dispersers as birds. They can, however, afford to produce elaiosomes, which, because they are far smaller than fruits, are less expensive to produce but still large enough to attract the much smaller seed-dispersing ants.

The ants that disperse elaiosome-bearing seeds are a varied lot that represent the four major subfamilies of the ant family Formicidae ("formica" means "ant" in Latin). This family is divided into 11 subfamilies, 7 minor ones and 4 major ones that include most of the over 8,800 ant species. As far as I know, no one has counted the seed-dispersing species, but they number at least in the hundreds.

Ants may move seeds only a few inches or feet, but sometimes carry them much farther. Certain harvesters, we saw, may transport seeds to a nest from as much as 130 feet away. Seed-carrying ants carry elaiosome-bearing seeds even farther. Henry Ridley quoted a Swedish entomologist who found that such seeds may be carried for as much as 230 feet, and according to Beattie, they are sometimes moved almost 330 feet. Long-range dispersal is often advantageous to the plant, but even short-range dispersal can be beneficial if competition from nearby parents or siblings is not critical.

Seed dispersal may even be a "cooperative effort" between two different insects or other animals. The journey of a strangler fig seed in Borneo may, as Tim Laman discovered, begin with a fruit-eating animal, perhaps a bird such as a rhinoceros hornbill, and end when an ant that nests high in a tree picks it out of a bird dropping that landed on a branch and carries it off to its nest in the crotch of a branch. These ants eat many of the fig seeds they collect, but some survive to germinate in the organic debris in and around their nest. Strangler figs survive only if they germinate high up in the crown of a tall tree because as seedlings they cannot survive in the dense shade of the forest floor. Thus, as Laman wrote, the strangler fig takes a "shortcut to essential sunlight" by beginning its life in the top of another tree. It ultimately sends roots down to the soil and becomes a huge tree that surrounds and eventually smothers its host tree.

Ants benefit seeds even when they discard them in the soil of trash piles that surround the entrance to the nest, soil fertilized by feces, seed chaff, and other discarded wastes. Seeds of four species collected by a desert harvester ant were, Steven Rissing reported, more than three times as numerous in ant trash piles as in nearby sites of equal size. Furthermore, seedlings grow exceptionally well in trash piles. Rissing found that plants growing in soil with ant trash produced many more seeds than plants growing in less fertile sites. For example, plantains growing close to harvester ant nests in soil fertilized by trash bore an average of 174 seeds per plant, whereas those growing nearby in ordinary desert soil bore an average of only 27 seeds per plant.

The invasion by the Argentine ant of the brushy fynbos plant community of South Africa's Cape Province is an ecological disaster but also a revealing demonstration of the importance of ants as dispersers of seeds. Anthony Milewski and W. J. Bond found that among the more than 60 plants in the fynbos community, the seeds of 29 percent bear elaiosomes and are dispersed by ants. The continued survival of these plants, including some proteas with magnificent blossoms, is threatened, because the indigenous ants are being replaced by the Argentine ant, a native of South America that has spread to virtually all the subtropical areas of the globe and is probably the most pernicious ant in the world. The native ants eagerly collect seeds with elaiosomes soon after they mature and carry them for considerable distances to their underground nests. By moving seeds into their burrows, the native ants not only disperse them but also rescue them from destruction by rodents and other seed eaters and protect them from the fires that periodically burn the fynbos. The Argentine ants, Bond and P. Slingsby report, are slower and less adept at finding these seeds, and after eating the elaiosomes abandon the seeds on the surface of the soil within 2 inches of where they found them. They are of no help to the seeds. On sites still occupied by native ants, seedlings of one of the ant-dispersed plants were 50 times more abundant than on sites where the native ants had been driven out by the Argentine ant.

A plant that germinates from a seed cannot survive without the necessary nutrients. Most plants grow on fertile soil and do not lack nutrients, but others grow in places where nutrients are in short supply, some on infertile soil and others, epiphytes, clinging to the bark of trees. As you read on you will become acquainted with some of the ways in which insects literally feed plants.

3

SUPPLYING FOOD

After dispersal, seeds germinate and seedlings grow if the soil has the required nutrients. But a few plants can, with the help of insects, grow in soil in which nutrients are almost or virtually absent. Some plants, as you just read, are fertilized by the organic trash discarded by ants. But insects also feed plants in other ways. Their bodies become plant food if they fall prey to insect-eating plants. Some epiphytic plants of the Old World tropics have internal chambers that house ant colonies which discard their feces and other nutrient-rich trash in special nutrient-absorbing chambers. And some epiphytic plants of Central and South America grow on the surface of arboreal ant nests, whose walls are composed of fertile organic debris.

When I was still a budding naturalist, I found a small patch of strange plants that I had never seen before. They were in a gravel pit in Trumbull, Connecticut, growing on ground from which the soil had been scraped away down to a soggy layer of sand and gravel barely above the water table. The plant was a rosette less than three inches across. Radiating from its center were several round, bright green leaves on short stems, each leaf bedecked with numerous upright hairs topped with a tiny drop of liquid that glittered in the sun. When I looked more closely, I saw that insects were stuck to several leaves by those droplets. A few days later, I described these plants to Aretas Saunders, my high school biology teacher. He explained that they are sundews, named for the little droplets that sparkle in the sun, and that they are trappers and digesters of insects.

The 90 known species of sundews are not the only insect-eating plants; worldwide there are 538 carnivorous species in eight families, all green flowering plants, just a few of the 275,000 known flowering plants. They

are, however, as Thomas Givnish wrote, "one of the great anomalies of the natural world." They have turned the evolutionary table on insects. Whereas plants are usually the "prey" of insects, insects are the prey of these plants. Although this is an amazing inversion of the usual food chain, it has evolved independently at least eight times.

Carnivorous plants are widely distributed geographically. They occur on all the continents except Antarctica, but not at the high latitudes of the Arctic, not in extremely dry deserts, and not in the seas. Most grow in moist places with abundant sunshine, but their ecological distribution is limited. Almost all grow on nutrient-poor soils: acid bogs, peat, sand, or a sterile layer of sand and gravel like the one where I found sundews in Connecticut. They can exist in such habitats, where most plants cannot grow, because their insect prey supply nutrients lacking or deficient in the soil.

In addition to oxygen, hydrogen, and carbon, plants require 14 mineral nutrients to thrive. Seven are called macronutrients, because plants require large amounts of them. For example, to grow 100 bushels of corn—the yield of a half acre of good Illinois soil in a good year—it takes 160 pounds of nitrogen, 125 pounds of potassium, 75 pounds of sulfur, 50 pounds of calcium, 50 pounds of magnesium, 40 pounds of phosphorus, and 2 pounds of iron. The remaining mineral nutrients, manganese, boron, chlorine, zinc, copper, molybdenum, and nickel, are micronutrients, required only in very small or even trace amounts.

Hollywood movies sometimes show a man walking through the jungle being snapped up by a huge carnivorous plant. But as Francis Lloyd wrote in his book on carnivorous plants, "the man-eating tree of Madagascar must at present be excluded [from the list of carnivorous plants], since the evidence of its existence is elusive." Lloyd wrote with tongue in cheek: there is no carnivorous plant that can trap a large animal, and he knew it. But on rare occasions small vertebrates such as mice and frogs fall into a pitcher plant's pitfall trap. Insects and the occasional spider are the usual prey of carnivorous plants. Some of the trapped insects are plant eaters that obtain all of their nutrients from their food plants, perhaps from plants that grow on nearby fertile soil. Others are themselves carnivores that eat other insects, which are more often than not plant eaters.

The movement of minerals from the soil through a four-part food chain, noncarnivorous plant to plant-feeding insect to carnivorous insect to carnivorous plant, is just part of a complex web of interacting organisms that contribute to the survival of carnivorous plants. The first member of the food chain, the noncarnivorous plant, may owe its existence to an insect that pollinated its mother's flowers, to a fruit-eating bird or an ant that carried it away from its parent when it was still a seed, and to one or more predators that ate caterpillars that could have eaten its leaves. The plant-feeding insect might not have survived the winter if it had not found a snug shelter under the fallen leaves of oaks, pecans, and other trees in a nearby woodland. The carnivorous insect, an adult robber fly for example, that ate the plant feeder spent its larval life in a rotting log of a maple eating beetle larvae. The carnivorous plant would not exist if an insect had not carried pollen from its father to its mother. (Sundews and some other insect-eating plants avoid trapping their pollinators by bearing their flowers at the top of a tall, bare stalk, well away from the deadly leaves.)

Charles Darwin was the first to show that carnivorous plants profit from their animal diet. A species of sundew that occurs on both sides of the Atlantic, the same one that I found in Connecticut, was the subject of experiments he did with his son Francis:

> My experiments were begun in June 1877, when the plants were collected and planted in six ordinary soup-plates. Each plate was divided by a low partition into two sets, and the *least* flourishing half of each culture was selected to be "fed," while the rest of the plants were destined to be "starved." The plants were prevented from catching insects for themselves by means of a covering of fine gauze, so that the only animal food which they obtained was supplied in very minute pieces of roast meat given to the "fed" plants but withheld from the "starved" ones. After only 10 days the difference between the fed and starved plants was clearly visible: the fed plants were of a brighter green and the tentacles of a more lively red. At the end of August the plants were compared by number, weight, and measurement with . . . striking results.

Plants fed roast meat weighed three and a half times as much as starved plants, produced almost twice as many seed capsules, and almost two and a half times as many seeds. Their total seed crop weighed almost four times as much as that of starved plants. A few years later, in 1883, M. Büsgen found that sundews fed aphids benefited almost as much as did Darwin's meat-fed sundews.

We now know that carnivorous plants do not require prey if the soil in which they grow is fertilized with nitrogen, phosphorus, and other minerals. Eating insects is necessary only if the soil is short on minerals. Insects are sometimes so scarce that a plant cannot produce seeds. In an article on pitcher plants, Carmen Chapin and John Pastor suggested that traps may be "a gamble to obtain an abundance of insect nutrients for storage in years when massive outbreaks [of insects] occur." Yolanda Heslop-Harrison described just such an occurrence in England: A dense 2-acre patch of sundews trapped about 6 million butterflies that landed after a migratory flight from Europe; 4 to 7 butterflies adhered to each plant.

🐜 🐜 🐜 The traps of all carnivorous plants are leaves variously modified to catch insects, to digest them, and to assimilate the nutrients released by digestion. There are six different kinds of traps. Each one evolved independently of the others and catches insects in its own way. Some traps are passive. They can catch insects without moving anything. They are the pitfall traps of the pitcher plants, the lobster pot traps of certain aquatic plants, and the passive flypaper of the Portuguese sundew. Others are active trappers that move some part of the trap in capturing insects: the active flypaper of the sundews, the snap trap of the Venus flytrap, and the suction traps of the aquatic bladderworts. Victor Albert and two coworkers recently showed, through molecular studies, that the flypaper traps had five separate evolutionary origins and that the pitfall traps had three.

🐜 🐜 🐜 Pitcher plants are a diverse and widespread group including dozens of species in three families. Some species of one family or another occur in Australia, the Old World tropics, and North and South America. Independently of each other, the ancestors of these three families evolved essentially the same method for catching insects, the pitfall trap. The pitfall, the pitcher of the pitcher plants, is the simplest of traps—no different in principle than the kind of pit people dig to trap unwary animals. Each pitcher borne by a plant is a large leaf curled upon itself to form a watertight, funnel-shaped tube open at its top and partially filled with liquid. In most species the interior is shielded by a flap that overhangs the opening like an awning.

Pitcher plants of all three families have much in common. The top of

The hanging trap of this tropical African pitcher plant is an extension of a leaf.

the pitcher is usually brightly colored—often red—seemingly simulating a flower; the rim of the pitcher's opening usually secretes nectar and a scent that attracts insects; and the upper part of the pitcher's inner wall is exceedingly slippery. The way pitcher plants of the Old World tropics *(Nepenthes)* capture prey was described by Adrian Slack and by Barrie Juniper and J. K. Buras. Insects are attracted by the colorful awning to nectar secreted by glands on the awning's inner surface and on the slippery rim of the pitcher. As they feed, many lose their footing and plunge into the liquid at the bottom of the pitcher. If they try to escape by crawling up the pitcher's wall, most of them will not gain a foothold on its waxy, slippery surface, but if they do make it to the top, their escape is blocked by long, down-curved barbs that line the rim. Trapped insects drown. They are usually digested by enzymes secreted by the pitcher. But there are exceptions. The northern pitcher plant, the most familiar and widely distributed of the fourteen species of pitcher plants that occur in North America, does not secrete digestive enzymes, but depends upon bacteria to digest its prey. In either case the digested nutrients are absorbed by the pitcher's wall.

Pitchers may be safely occupied by certain specially adapted creatures. Some spiders spin a web just inside a pitcher's mouth, thereby intercepting insects that could have been caught by the plant. Some caterpillars feed on the inner wall of the pitcher above the level of the liquid. The water in the trap of the North American pitcher plant is host to a miniature ecosystem that includes bacteria, protozoa (relatives of amoebas), rotifers (tiny aquatic creatures known as wheel animalcules), tiny crustacea (relatives of crabs and crayfish), and three different kinds of flies, including a mosquito. Insects caught by the plant are the nutritional foundation of this ecosystem. The maggots of a flesh fly feed mainly on newly drowned insects at the water's surface. The larvae of a midge feed on long-dead insects at the base of the pitcher, and the larvae of pitcher plant mosquitoes swim about and filter bacteria and protozoa from the water.

Aquatic plants of the genus *Genlisea* have traps that, as Charles Darwin put it, are "a contrivance resembling an eel-trap, though more complex." As with an eel trap or a lobster pot, a creature moves into the trap by pushing through a circle of inward-slanting spines but cannot go back "against the grain" of the same spines.

There are at least sixteen species of *Genlisea*, found in tropical South

America, the West Indies, tropical Africa, and Madagascar. They are floating, rootless plants with two types of leaves: broad, green, photosynthetic ones and peculiar Y-shaped trap leaves as much as 6 inches long in some species. Each arm of the Y is a narrow tunnel formed by a long, loosely spiraling, ribbon-like blade. The spiral gapes open at each wind of the ribbon, allowing small insect larvae or other aquatic animals to enter. It is not known whether the plant attracts prey by a water-borne chemical or some other cue. Prevented from leaving the tunnel by the inward-pointing spines, the captured creatures can move only to the junction of the arms, where they enter the long stem of the Y, a tunnel with solid walls and still more inward-pointing spines. They finally enter the "stomach," a bulbous swelling of the stem, where nutrients released by digestion are absorbed. Little more is known about *Genlisea* plants. According to Givnish, they grow in nutrient-poor habitats. Each plant is submerged except for a tall, flower-bearing stalk that extends well above the water surface. The flowers are presumably pollinated by flying insects.

Only three plants have passive flypaper traps, two in Australia and one, the Portuguese sundew (not to be confused with the "true" sundews), in Portugal, Spain, and Morocco. The Portuguese sundew has very long and almost needle-like leaves with rows of closely spaced, hairlike, stalked glands, each with a glistening drop of mucilage at its red tip. Below them are many tiny, stalkless glands that secrete digestive enzymes scattered over the surface of the leaves. The leaves radiate from a common point, roughly resembling a tuft of pine needles—hence this plant's Portuguese name, "herba piniera orvalhada," dewy pine. It grows mainly in poor soils in the coastal hills and, as Adrian Slack noted, "differs from the vast majority of carnivorous plants in favouring soils which are normally rather dry, and often alkaline."

Darwin pointed out that the Portuguese sundew is a very efficient trapper of insects:

Mr. Tait informs me that it grows plentifully on the sides of dry hills near Oporto, and that vast numbers of flies adhere to the leaves. This latter fact is well known to the villagers, who call the plant the "fly-catcher," and hang it up in their cottages for this purpose. A plant in my hot-house caught so many insects during the early part of April, although the weather was cold and insects scarce, that it must have been in some manner strongly attractive to them. On four leaves of a young and small plant, 8, 10, 14, and 16 minute insects, chiefly Diptera [flies, mosquitoes, and midges], were found in the autumn adhering to them.

These plants probably do lure insects. The drops of mucilage emit a honey-like scent that very likely does attracts insects, although as far as I know, this has not yet been proved; but Portuguese sundew, as you will see in detail later, is also visually attractive, at least to humans and presumably to some insects.

It would be difficult to improve on Darwin's description of how insects are trapped on Portuguese sundews: "when a small insect alights on a leaf, . . . the drops [of mucilage] adhere to its wings, feet, or body, and are drawn from the gland; the insect then crawls onward and other drops adhere to it; so that at last, bathed by the viscid secretion, it sinks down and dies, resting on the small [stalkless digestive] glands with which the leaf is thickly covered." Digestion and absorption proceed rapidly. Darwin found that cubes of egg white placed on digestive glands had largely liquefied after 24 hours. Others reported that a mosquito is completely digested and absorbed in 26 hours.

🐝 🐜 🪰 The leaves of my Connecticut sundew and most other true sundews (not to be confused with the unrelated Portuguese sundew) are active flypaper traps. Darwin found one or more insects trapped on 31 of the 56 leaves of a random sample of these plants collected in the wild in England, and as I said earlier, sundews of the same species collected elsewhere in England trapped from 4 to 7 large butterflies per plant. Their success suggests that they have evolved some way of attracting insects. According to Slack, these plants secrete no nectar and have no odor we can detect. But, he suggested, they may have an alluring scent that insects but not humans can smell; or the glistening drops of clear mucilage on the red tips of the stalked glands that clothe the leaves may attract the insect prey.

The mucilage of the true sundews is much stickier than that of the Portuguese sundew. An insect that lands on a leaf is instantly stuck and cannot escape. The stalked glands (tentacles) move surprisingly quickly. According to Darwin, if a bit of raw meat is put at the tip of a tentacle, it begins bending in

10 seconds and is strongly curved inward in 5 minutes. He described the concerted action of the tentacles:

> When an insect alights on the central disc [of a leaf] it is instantly entangled by the viscid secretion, and the surrounding tentacles after a time begin to bend, and ultimately clasp it on all sides. Insects are generally killed, according to Dr. Nitschke, in about a quarter of an hour, owing to their tracheae being closed by the secretion. If an insect adheres to only a few of the glands of the exterior tentacles, these soon become inflected and carry their prey to the tentacles next succeeding them inwards; these then bend inwards, and so onwards, until the insect is ultimately carried by a curious sort of rolling movement to the centre of the leaf. Then, after an interval, the tentacles on all sides become inflected and bathe their prey with their secretion, in the same manner as if the insect had first alighted on the central disc. It is surprising how minute an insect [such as a mosquito] suffices to cause this action.

The glands that top the tentacles secrete not only the sticky mucilage, but also digestive enzymes that release the prey's nutrients. They also—with the help of minute glands on the tentacle's stem and the leaf blade—absorb the products of digestion. When digestion is complete, all that remains are the insect's wings and body wall.

The Venus flytrap, the first carnivorous plant discovered, is the most famous of the insect-eating plants and has interested people more than any of the others. According to Francis Lloyd, the first written account of the Venus flytrap is in a 1760 letter written by Arthur Dobbs, then governor of the North Carolina colony. Dobbs called the plant "the great wonder of the vegetable kingdom" and compared its trap leaf to an "iron spring fox trap." Over a century later, Darwin wrote that this plant "is one of the most wonderful in the world." The Venus flytrap amazes us because of the fast snap of its trap leaves, which operate much like a spring trap with steel jaws that snap shut if its trigger is touched.

It is a small plant, a rosette of about a dozen leaves, seldom more than 3 inches in diameter. A leaf consists of a long, winged, green stalk and the blade that forms the trap. The two lobes of the blade, joined at the midrib and bearing a row of long, stiff spines at their edges, are the jaws of the trap.

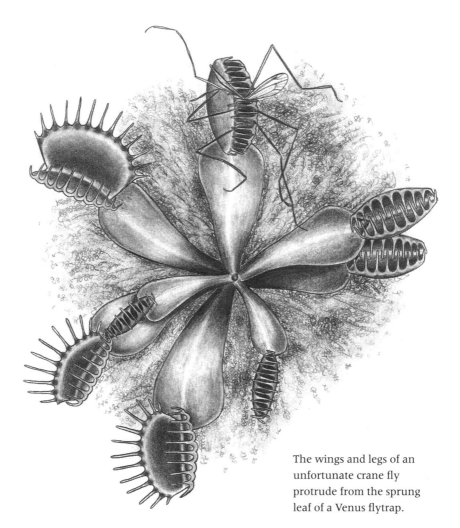

The wings and legs of an
unfortunate crane fly
protrude from the sprung
leaf of a Venus flytrap.

The upper surface of each lobe secretes abundant nectar; the lobe may be
green, yellow, or red, and it bears three long, thin hairs, the triggers that
spring the trap. If an insect lands on a trap and touches a trigger, the trap
snaps shut in as little as a second. The spines on the lobes interlock, forming
a cage from which only a very large insect can escape. For several hours, the
jaws of the trap press closer and closer together, sometimes crushing the
trapped insect. Once the trap is tightly sealed, acid and digestive enzymes
pour out from glands on its inner surface.

The Venus flytrap, Slack reports, prefers savannas with damp, sandy soil with a small admixture of peat. Its range is limited to a small coastal area of North and South Carolina that is a semicircle with a 65- to 75-mile radius extending inland from Wilmington, North Carolina. With one exception in Florida, attempts to introduce the plant to apparently similar habitats outside of its natural range have failed. This marvelous plant was once threatened with extinction by commercial collectors and is still threatened by human encroachment on its habitat.

The suction trap of the bladderworts, probably the most sophisticated of the plant traps, is even more amazing than the snap trap of the Venus flytrap. Although they occur worldwide and are common in North America and Europe, bladderworts have attracted little popular notice because they are usually small and very inconspicuous. Some of the 250 known species are aquatic. Others grow in soils wet enough to support potential prey: small, aquatic, swimming organisms. All or most of the rootless plant is out of sight beneath the surface of the water or the soil, except when tall stalks with small but conspicuous white, yellow, or purple blossoms at their tops poke high above the surface.

The bladders are the traps. They vary in size among species; in some the bladder is only a hundredth of an inch long but in others it is as long as a fifth of an inch. The smaller plants trap microorganisms such as protozoa, but the larger species, such as the greater bladderwort, can trap mosquito larvae and other small aquatic insects. The details of the traps' structure vary. But because all function in essentially the same way, the trap of the greater bladderwort, widespread on both sides of the Atlantic, serves to illustrate the basic principle.

The wide mouth of the bladder, normally closed by a valve, is surrounded by long, thin, branched hairs that guard the shorter trigger hairs at the rim of the mouth. The longer hairs form a loose funnel that may steer passing creatures to the hairs that spring the trap. When the trap is set, the walls of the bladder are concave and compressed, greatly decreasing its inner volume. If some creature brushes against a trigger, the walls of the bladder suddenly spring outward, increasing the bladder's volume and thereby producing a partial vacuum that sucks water and prey into the bladder through the now open valve. This can take as little as 27 milliseconds, 27 thousandths of a second. The prey is digested in the bladder. Lloyd quotes Mary Treat, who in the

1870s commented, "I was forced to the conclusion that these little bladders are in truth like so many stomachs, digesting and assimilating animal food."

Insect-eating plants attract prey much as flowers attract pollinators, with their colors, scents, and the secretion of nectar. Some have suggested that insects are attracted to true sundews and the Portuguese sundew because they mistake the drops of mucilage for nectar or dew. This may be. Water is scarce in the microhabitats of most insects, and some drink drops of dew from leaves.

Many carnivorous plants are marked with red. But we know that most insects other than butterflies are red-blind. So butterflies may be attracted by the red on carnivorous plants, but what attracts the many red-blind insects that become trapped? The answer is probably that they are attracted by ultraviolet light reflected by these plants, just as insect pollinators are attracted by the ultraviolet reflected by the flowers of many plants. Daniel Joel and his colleagues were the first to show that carnivorous plants both reflect and absorb the ultraviolet part of the spectrum, thereby creating colors and patterns that are invisible to us but not to insects. Portuguese sundew, the true sundews, the Venus flytrap, and the pitfalls of many pitcher plants are marked with red but also reflect ultraviolet. For example, the old dead leaves of Portuguese sundew reflect ultraviolet and thus provide a bright background for the rosette of green leaves they surround. The green leaves, which absorb ultraviolet, appear to be darker to insects.

Some ants feed plants without becoming their prey. These ants, and no other insects, supply certain plants with mineral nutrients much as gardeners feed their flowers with fertilizer. These ant-fed plants are all epiphytes. They cling to the bark of trees—sometimes high above the forest floor—but take no nourishment from their host tree. Therefore, except for those fed by ants, epiphytes depend upon nutrients in debris trapped on the bark and in the crotches of a tree. If the forest grows on infertile soil, the debris caught in the trees is likely to be infertile too. Thus plant-feeding ants facilitate the growth of epiphytes in environments where they might not otherwise be able to grow. But these are not one-sided relationships; the ants also benefit.

These ant-plant mutualisms are of two types. First, there are "ant gardens." Some ants of the New World tropics that nest in trees literally plant the seeds of certain epiphytes in the fertile organic debris they use to build the walls of their nest. In return, the plants furnish the ants with nectar and other food. In relationships of the second type, certain epiphytes, called myrmecophytes (meaning ant plants), accommodate ants by forming within themselves hollow chambers in which ant colonies nest. Some of them also feed the ants. The ants reciprocate by discarding their nutrient-rich feces and trash in special nutrient-absorbing chambers in the plant.

Ant gardens, beautifully described by Sally Kleinfeldt, are dense clusters of epiphytes that grow on the walls of the nests of certain tree-dwelling ants but are almost never found elsewhere. These ants build their nests of "carton," their equivalent of papier maché. Carton is composed of soil and organic debris, often including bird or mammal dung, all glued together by the honeydew the ants collect from sapsucking insects. The relationships between these ants and the few species of plants they favor are specific and mutualistic. The ants are attracted to the seeds of these plants, probably by their odor, but not to seeds of other plants. They collect the favored seeds and "plant" them on the wall of their nest, an ideal site for germination and growth. Nourished by the fertile carton, the plants flourish and ultimately cover the nest. Almost never do these plants grow in sites other than ant nests, and they also grow much better on the nests than elsewhere. The plants reciprocate by providing food for the ants: fruit pulp, elaiosomes on their seeds, and extrafloral nectar secreted by glands on stems and leaves rather than by flowers. In addition, their roots may strengthen the wall of the nest. The plants in the ant garden may harbor sap-sucking insects, such as treehoppers, scale insects, or aphids that the ants tend and "milk" for honeydew.

Only a few species of ants plant gardens. These ants all have three characteristics that make them successful gardeners. First, they build free-standing nests of carton that provides nutrients for the plants. Second, they depend on foods offered by the plants. And finally, they form colonies consisting of several separate nests connected by carton-covered runways, which facilitate the multiplication of ant gardens, because worker ants can easily carry seeds from gardens on established nests to new satellite nests.

The German biologist Ernst Ule was the first to recognize the ant gardens in the upper Amazon basin for what they are. In 1902, he wrote, "We know for a fact that ants sow flowering plants on shrubs and trees and grow and care for them as protection for their nests. Thus they establish true hanging gardens that I have named ant gardens [Ameisengärten] . . . The ants plant and cultivate these epiphytes, which could not otherwise survive. In return these plants make it possible for them to build nests in trees, supporting the nests against violent rain storms and protecting them against the scorching rays of the sun" (translated from the German). He thought the plants contribute mainly to the architecture of the nest.

Ule's report drew the fire of William Morton Wheeler, in his time the foremost American authority on ants. Wheeler disagreed with Ule, and after citing several passages from Ule's papers, wrote in 1921, "I have cited these passages because they may be regarded as a classical example of the uncritical mixtures of observation, inference, assertion and speculation, which abound in the work of observers in the tropics and constitute the only foundation on which some of the closet naturalists of Europe and the United State have been building their specious hypotheses." Although Ule was overly enthusiastic and tended to overstate his case, he did not deserve such scathing criticism. Wheeler contended that the ants do not collect and sow seeds in their gardens. But as he admitted, he made his own observations when no seeds were available because the plants in question had not yet borne fruit. He could not, therefore, directly confirm or refute Ule's observations but presented several circuitous arguments to prove him wrong. Relatively little research was done on ant gardens for the next fifty years, possibly partly because of Wheeler's negative comments. But in the 1970s Ghillean Prance and Miramy Macedo demonstrated convincingly that the seeds of ant garden plants are sown by the ants that tend the gardens.

In 1988 Diane Davidson found that ant gardens were common in a rainforest in Peru. Along a 7.5-mile trail through the forest, she counted 879 of them, an average of about one every 15 yards. Even those occupied by the same species of ant varied in shape and size. Including the plants that covered them, some were smaller than a baseball while others were globes more than 3 feet in diameter. The great majority of the plants in the ant gardens belonged to ten species of seven different families. These plants grow exclusively or almost exclusively in ant gardens. Seven species were found only in ant gardens and three were only rarely found growing elsewhere. Davidson found, as had Auguste Forel 90 years earlier, that the great majority of ant gardens were occupied by the same two species of ants living together in harmony. One of them, *Camponotus femoratus,* is large and very aggressive, while the other, *Crematogaster limata,* is not as aggressive and is less than half the size of its partner. Of the 879 ant gardens along the trail, 758 were still inhabited by ants. They were usually occupied by both of these ants. Rarely a nest was inhabited only by the small ants, but Davidson never found large ants nesting independently of the small ones.

The two ants share the same nest by putting up partitions that divide it into separate chambers, but they rub shoulders in the surrounding territory and use the same foraging trails. They also share food resources. The large ant usurps high-quality foods but relinquishes lower-quality ones to the small ant. The ants probably benefit each other. The small one benefits itself and the large ant by building protective carton shelters over aggregations of sap-sucking insects from which both obtain honeydew. The large ants probably protect the nest from birds and mammals. Wheeler found them to be effective defenders: "touching a leaf or twig on which a few of the [large ants] are engaged in collecting their sweets is an experience to be remembered. The large workers, without a moment's hesitation, bury their mandibles in one's skin, and curving the [abdomen] forward drench the wound with formic acid."

The large ants carry the seeds, which are too heavy for the small ants to haul, to the nest and plant them in the carton. When Davidson offered the ants a "buffet" of different kinds of seeds, they carried away only seeds of the plants that regularly grow on their nests, rejecting all others. They probably recognize the favored seeds by their odor. As Ule thought, the plants may give the nest physical protection, but there is no doubt that the ants regularly exploit foods provided by the plants, which together probably constitute a well-rounded diet: the pulp of their fruits; the fat- and protein-containing elaiosomes on their seeds; the solid, fat-containing "pearl bodies" on leaves and shoots; the sugary extrafloral nectar; and the honeydew excreted by scales and other sap-sucking insects.

Some plants provide internal chambers, hollow stems, or other forms of housing for ants of various species. These accommodations are called domatia, from "domus," the Greek for house. In Southeast Asia, there are many epiphytes that house ants in domatia. The ants repay their hosts by fertilizing them with their trash, which consists mostly of their feces, dead ants, and fragments of prey insects. Domatia-bearing trees in the tropics of Africa and the Western Hemisphere are repaid by their ant guests in a different way, as we will see later.

In Southeast Asia there are, according to Andrew Beattie, at least 200 species of epiphytes that provide domatia for ants. Most occur in open forests in rainy areas with infertile soil. The plants need the otherwise scarce nutrients provided by the ants, and the ants depend upon the domatia for housing, because these areas have few dry cavities not preempted by termites. Some of these epiphytes provide extrafloral nectar or other foods for their tenants, which are primarily carnivorous scavengers and drinkers of honeydew. In addition to feeding the plant that houses them, some ants disperse its seeds and may also protect it against plant-eating insects. These associations tend to be specific; the same kinds of ants always or usually occupy the same kinds of plants, but ants of some species associate with more than one species of plant and vice versa. A colony may spread over several plants. The egg-laying queen is in one of them, and as the colony's population increases, satellite nests for housing and raising her offspring are established in domatia in nearby plants.

The strategy of housing ants to get their mineral-rich feces and trash evolved independently several times, as is attested by the fact that the domatia of different plant genera are modifications of different plant parts: leaves, roots, hollow stems, bulblike swellings on stems, or rhizomes (underground stems). The domatia and their entrances are formed solely by the plant whether or not ants are present. The ants do not enlarge them or change them in any way.

The domatia contain organs for absorbing minerals from trash left by the ants, fine roots that extend into the domatia or absorptive "warts" lining separate chambers for trash disposal. But how is the trash digested—how are its complex constituents broken down to simpler soluble forms that plants can absorb? Not much is known about this process, but Camilla Huxley proposed that organisms that live in the moist trash piles do the digesting. She wrote of these trash pile inhabitants, "the fungi, mites, dipteran [fly] larvae, nematode worms, and microorganisms must form a miniature ecosystem in the warted chambers which cycles minerals, releasing them in soluble form . . . This [may be] an important link in the transfer of minerals, especially from insect fragments, into the plants."

These coinhabitants of domatia are helpful to the plants. But other associates are not helpful. In Sarawak, the Malaysian area of Borneo, Daniel Janzen observed that domatia-bearing plants are sometimes parasitized by other epiphytes not equipped to house ants. Virtually all of these other plants are closely associated with ant-housing epiphytes, and almost always many of their roots penetrate into the trash piles in the host plant's domatia to steal nutrients.

Studies with radioactive tracers have shown that, as many observers have inferred, the epiphytes that house ants do in fact absorb nutrients from the ants' trash piles. In three different experiments, Camilla Huxley fed ants radioactive nutrients, phosphate, sulfate, and methionine (an amino acid). These radioactively "labeled" nutrients were traced as they passed through the food chain. Radioactivity ultimately appeared in the stems and growing leaves of plants occupied by the labeled ants. The radioactivity could have come only from the feces and dead bodies of radioactively labeled ants deposited in the trash piles.

🐜 🐜 🐜 In Sarawak, Janzen studied the natural history of several ant-housing epiphytes. Two of these plants, *Hydnophytum formicarium* (from Greek and Latin, "ant-housing truffle plant") and *Myrmecodia tuberosa* (from Greek and Latin, "tuberous ant house"), are particularly well adapted to associate with the ants that occupy them, *Iridomyrmex cordatus*. The two epiphytes are similar. Both are shrubs—with several leafy stems in *Hydnophytum* but only one in *Myrmecodia*. Both develop a huge swelling, commonly known as a tuber, at the base of each stem. Short, thick roots extend from the tuber, apparently functioning as holdfasts that anchor the epiphyte to the bark of the host tree rather than as absorbers of nutrients. The tubers are riddled with cavities, the domatia, with openings to the outside. The tubers vary greatly in size. In *Hydnophytum* some weigh only a little more than 2 ounces and their cavities have a capacity of only two-hundredths of a quart. Others may weigh 6.6 pounds and have a capacity of three-quarters of a quart. *Myrmecodia* tubers have some smooth-walled cavities, in which the ants nest, and others with walls covered with nutrient-absorbing "warts," in which the ants dump trash. A *Hydnophytum* cavity serves both functions, having a smooth-walled area in which the ants nest and a separate warty area where they discard trash.

The maximum population of *Iridomyrmex* in a colony depends upon the amount of living space available, which varies with the number and capacity of the domatia it "owns." Janzen censused a colony spread over nine *Hydnophytum* tubers and one *Myrmecodia* tuber. It included 12,640 workers, one wingless egg-laying queen, 46 nonreproducing winged queens, and 357

A *Myrmecodia* plant with roots grasping a branch and with part of the
"tuber" cut away to show the domatia in which ants nest.

winged males. This colony was ready to found new colonies. The winged
queens and males would soon fly off and mate with partners from other colonies. The males would die after mating, but the females would search for a
domatium in which to found a new colony and snap off their wings after
they settled down.

The behavior of *Iridomyrmex* is well adapted to nesting in domatia. These
ants propagate their host plant by planting seeds in the carton with which
they surround the base of its tuber or in the protective cover that they place
over colonies of scale insects. They chew open the fruits of their host to collect the seeds and will even retrieve seeds that have been passed in the droppings of birds. This ant species would not be a satisfactory partner for its host
plant if it did not differ from most other tree-dwelling ants in leaving its
trash in the nest rather than throwing it out.

The workers discard a variety of prey fragments in their trash. Among the insect fragments that Janzen found were heads of the colony's own workers and of other ants, and pieces of beetles, parasitic wasps, termites, and crickets. There was an occasional plant seed but no other vegetation and no dirt or sand. The foragers brought in "conspicuously old" parts of insects that had apparently been dead when the ants found them. They also removed body pieces from large insects that were dying. After Janzen put a large, crippled, bright green walkingstick insect on a *Hydnophytum* tuber, the ants cut it into tiny pieces and carried them into the tuber. "Two days later, the bright green . . . fragments were conspicuous in at least three different debris deposits in this tuber."

A plant must survive the onslaughts of its enemies if it is to complete its life cycle by producing seeds. Flowering plants are forage for nearly half a million known species of animals, from tiny, flattened, leaf-mining caterpillars that feed as they burrow between the upper and lower epidermal layers of thin leaves to deer that browse on the leaves. The great majority of these animals are insects. Most plants fight back, some with spines, thorns, or dense growths of hairs on their leaves, and almost all with noxious deterrent chemicals. But some plants have, as you will see next, enlisted stinging ants and other insects to help them defend themselves.

4

PROVIDING DEFENSE

In 1965, I learned—through an exceptionally painful experience—that ants can be very effective defenders of plants. While collecting insects on a hot day in southern Mexico, I happened to brush lightly against a branch of a small tree. Almost instantly a horde of small ants swarmed onto my bare arm and stung me ferociously. The stings burned and were painful for hours. An encounter with a tree protected by these ants has been likened to walking into a large stinging nettle plant, but to me it felt much worse. I remember thinking at the time that these ants would surely drive away a deer, cow, or any other mammal that tried to browse on a plant they occupy. I know now that the ants were ready for me even before I brushed against the tree. Bert Hölldobler and Edward O. Wilson wrote that these vicious stingers become alert at the mere smell of a cow or a person and are poised to attack even before the plant on which they live is touched. In the year following my painful experience, Daniel Janzen published an article on this ant-plant association. I learned from it that the plant was a bull's-horn acacia, and that it is regularly occupied by ants that defend it against browsing mammals, herbivorous insects, and even competing plants that grow nearby—"weeds" from the acacia's point of view. Later in this chapter, you will hear much more about the complex ant-acacia association, one of the great wonders of nature.

 Many different kinds of plants have enlisted insects to help defend themselves against their plant-feeding enemies. Preeminent among the defenders are many species of ants, but there are also predators other than ants, parasitic wasps and flies, and even predaceous mites. Associations between plants and defenders are mutually beneficial but differ in complexity and in how intimately the participants are committed to each other. In simple associations, including many with ants and virtually all with other predators or parasites, defenders and plants are not permanently associated, and the plant rewards defenders only with the sugary exudate of extrafloral nectaries. The commitment between plant and defender is not essential; either could survive without the other. But in complex associations, almost always with ants, plant and ant are strongly committed to each other, sometimes so inextricably that neither could survive without the other. The ants are permanent residents of the plants, which provide nesting cavities (domatia) and food that includes not only nectar, but also fats and proteins. Unlike most simpler ones, these associations are often specific, with a plant hosting only one or two kinds of ants.

Many plants, one or more species of at least 68 different families, can secrete nectar even when they have no blossoms, because they bear extrafloral nectaries on stems, leaves, petioles (leaf stems), or other structures. These plants usually occur where ants are abundant, most in the tropics but some in temperate areas. Among those of northeastern North America are various plums, cherries, roses, hawthorns, viburnums, poplars, oaks, elderberries, trumpet creepers, and catalpas. Like floral nectar, extrafloral nectar consists mainly of water with a high content of dissolved sugars and, in some plants, small amounts of amino acids. The extrafloral nectaries of some plants are known to attract ants and other insects, but the natural history of most plants with these nectaries is unknown. Nevertheless, most ecologists believe that all extrafloral nectaries attract insects that will defend the plant.

Ants, probably the most frequent and certainly the most persistent defenders of plants, are well suited to this role; their characteristics predestined them to become protectors of plants. For one, the highly active workers require a great deal of energy; plants exploit this need by providing extrafloral nectar that supplies ants with abundant energy. Barbara Bentley and M. H. Way described how ants return this favor. They guard the nectaries, driving away or killing intruding insects that might compete with them for nectar, and many capture interloping insects to feed their protein-hungry larvae. Many of these intruders are herbivorous enemies of the plant. Also ant scouts soon recruit a contingent of other workers from their colony to a newly discovered nectar-producing plant, increasing the forces that guard the plant. Furthermore, ants occur in almost all terrestrial habitats, and—as picnickers know—they are persistent and abundant, often enormously so. Bert Hölldobler and Edward O. Wilson reported that on the savannas of Africa's Ivory Coast there are over 2,800 ant colonies per acre. In the Amazon rainforest, about 30 percent of the animal biomass, the combined weight of insects and all other animals, consists of ants.

Ants often have an obvious impact on plants they guard. In 1904, O. F. Cook reported that an avidly predaceous ant protected cotton planted by the Indians of Alta Vera Paz in Guatemala from the destructive boll weevil. Several ants were on almost every plant, attracted by extrafloral nectar but doing double duty as hunters of insects. Cook noted that plants visited by these ants were not damaged at all or were only slightly damaged, while the few that had no ants were heavily damaged. The Indians "give the ant a special name, *kelep*, not applied to any other species." Gary Ross explained that in Ketchi, the language of these Indians, *kelep* means "cotton-protecting ant." In Zanzibar, F. L. Vanderplank found that an ant which nests in coconut palms more than triples the trees' yield by killing large bugs that suck sap from growing nuts. A large group of palms with very few or no ant nests produced only 18 nuts per tree, but another group with large ant nests yielded 61 nuts per tree.

Biologists once thought that the secretion of extrafloral nectar has some purely within-plant physiological function, and that plants "have no more use of their ants than dogs do their fleas." This view and the opposing "protectionist" hypothesis that ants defend plants had been disputed for over a hundred years when, in 1910, a skeptical William Morton Wheeler commented on the controversy. He called for proof of the protectionist view, which was then lacking, quoting A. W. F. Schimper, who had written that sufficient proof requires demonstrating that "visitations of the ants confer protection on the plants . . . and that in the absence of the insects a much greater number would perish or fail to produce flowers or set seeds, than when the insects are present."

We now have an abundance of the proof that was called for. Barbara Bentley reviewed the relevant evidence in 1977, and since then many more observations and experiments have provided more proof that ants benefit plants. Two marvelous ecological stories, both uncovered by Kathleen Keeler, tell how ants attracted to extrafloral nectaries protect two morning glories against attacking insects in different ways. One of them, *Ipomoea leptophylla*, of the high plains of North America, grows in short-grass prairies that receive only 15 inches of rain per year. The other, *Ipomoea carnea*, occurs in Central and South American forests that have dry and wet seasons and get about 70 inches of rain per year. Keeler showed that these two closely re-lated plants, adapted to radically different habitats and climates and associ-ated with different species of ants, use essentially the same tactic to thwart completely different enemies.

The principal insect enemies of the North American morning glory feed mainly on its flowers or fruits rather than its leaves. Grasshoppers feeding on flowers indirectly block pollination and the production of seeds by de-stroying the stigma or even by destroying the corolla. Without their colorful corollas flowers do not attract pollinators and are not fertilized. An adult grasshopper can consume a large corolla, about 2.5 inches long, in an hour. A caterpillar and a seed beetle affect seed production directly. The caterpillar devours the ovaries in a bud or flower, and seed beetle larvae eat seeds as they burrow in developing fruits.

Extrafloral nectaries at the base of each sepal attract several kinds of in-sects, but 96 percent of them are ants, several different species of them. When buds are still small, less than a quarter of an inch long, the sepal nectaries are already present and producing nectar. They continue to do so as the flower develops and while the fruit matures. The leaves also have nectaries on their undersides, but I will discuss them later. Keeler's observa-tions leave little doubt that ants protect flowers and fruits from the com-bined enemy force of grasshoppers, caterpillars, and seed beetles. She com-pared the seed production of 6 plants that grew where there were no ants with that of 17 plants that grew elsewhere and were occupied by ants. The unprotected plants bore only 45 seeds per plant, but the plants occupied by ants bore 211 seeds per plant.

Keeler recently told me that although the ants are not big enough to kill or seriously injure grasshoppers, they drive them away by nipping at their feet. Seed beetles are more vulnerable because they are much smaller than grass-hoppers. The ants prey on the adult beetles, disturb females as they lay their eggs on developing fruits, and eat many of the eggs they do manage to lay. She did not describe interactions between ants and the ovary-eating cater-pillars.

The Central American morning glory also has extrafloral nectaries that re-cruit defending ants, but their exact locations are slightly different. Those as-sociated with a flower are on its supporting stem just below the sepals rather than on the sepals themselves. Those on leaves are on the leaf's petiole rather than on the underside of its blade. The Central American species lives in an ecosystem radically different from that of its North American cousin, and is threatened by insects that inflict different injuries. Its leaves, present only in the wet season, are fed upon by insects, and its flowers, present only in the dry season, are plagued by nectar-robbing bees.

The nectar robbers are solitary (nonsocial) carpenter bees, which bore nesting galleries in solid wood. Too large to get into a morning glory blossom, they crawl down the outside of the corolla to its base and, ignoring the nearby extrafloral nectaries, get at the floral nectar by gnawing a hole through the corolla. They do not pollinate blossoms, because they contact neither stamens nor pistils. They "steal" nectar and give nothing in return. But ants defending nectaries on the flower stem often come to the rescue and drive the bees away before they do any damage. In an area with few ants, Keeler found, 58 percent of 69 flowers had been robbed, but in an area with many ants, only 31 percent of 116 flowers had been robbed.

If the leaves of the Central American morning glory come into contact with an acacia defended by ants, the acacia ants, intent on removing competing plants from the vicinity of their host tree, try to destroy the morning glory leaves by severing their petioles. Keeler found that morning glories next to an acacia are subject to severe damage. But *Solenopsis* ants on morning glories greatly limit the damage by confronting the acacia ants. The acacia ants run off when a pugnacious *Solenopsis* approaches. In the rare event that an acacia ant does not retreat, the *Solenopsis* attacks it and drives it off.

Many other observations and experiments show that the raison d'être of extrafloral nectaries is to attract a defensive phalanx of ants and—to a lesser extent—other insects. The following examples show that these defenders can be effective. In Costa Rica, Dennis O'Dowd found that extrafloral nectaries on the undersides of leaves of balsa saplings attract ants that defend the leaves against chewing insects. Saplings frequented by ants lost at most about 5 percent of their leaf area, while those not protected by ants lost as much as 23 percent.

As Charles Pickett and W. Dennis Clark have shown, the extrafloral nectaries of a prickly pear cactus in southern Arizona attract ants that kill or drive away sap-sucking bugs that injure flower buds. On plants without ants, 62 percent of 101 buds were aborted, but on plants visited by ants, only 19 percent of 115 buds were aborted.

The familiar catalpa tree, planted widely as an ornamental, is sometimes totally stripped of its leaves by catalpa sphinx caterpillars, which will feed only on catalpas. Nectaries on the undersides of the leaves attract five kinds of ants that attack the caterpillars. In his research, Andrew Stephenson found ants to be abundant on catalpas: In June and early July an average of 547 per hour, all of them carrying nothing, climbed up the trunk of a tree; when they came back down, some carried eggs or small catalpa sphinx caterpillars. The ants, with some help from ladybird beetles and a parasitic wasp, considerably reduced the injury done by caterpillars. By comparing branches that ants visited with branches from which they were barred by a band of sticky material, Stephenson showed that the ants significantly decreased the leaf area eaten by the caterpillars, with the ultimate result that trees protected by ants produced 50 percent more fruits than unprotected trees.

There is no free lunch—not for people, not for other animals, and not even for plants. Everything that an organism is or does entails the expenditure of energy, which ecologists measure in calories. Natural selection favors economical organisms, those that are sparing in the use of resources. Every calorie saved elsewhere can be allocated to producing offspring and thereby increasing the individual's evolutionary fitness. Plants with extrafloral nectaries could profit by minimizing the production of nectar, secreting it only when the ants or other insects it attracts can do it some good. Several studies, recently summarized by Anurag Agrawal and Matthew Rutter, show that some plants use their extrafloral nectar frugally, and it is likely that most if not all plants do the same.

Some plants initiate or increase their production of nectar when herbivores are damaging them. In his research on catalpas, Stephenson noted that the average daily secretion of the sugars in nectar by 25 undamaged leaves was only about 1.65 milligrams, while that of 25 partially damaged leaves was 8.05 milligrams. Agrawal and Rutter recounted several instances where leaf damage induced the flow of extrafloral nectar, which brought ants to the defense of the plants. Another parsimonious tactic is to secrete nectar only on otherwise poorly protected structures. According to my friend and colleague May Berenbaum, a leading expert on plant-insect interactions, systematic observations and experiments have shown that young leaves usually lack or have a lower concentration of the built-in poisons that protect older leaves against herbivores. Keeler observed that the nectaries on the undersides of the leaves of the North American morning glory secrete nectar only when they are very young. More than sixteen times as many nectar-feeding insects, including many ants, arrived on young leaves with flowing nectar than on old leaves that no longer secreted nectar. Young morning glory leaves probably secrete ant-attracting nectar because they are not as well protected by poisons as the old leaves.

Plants can also economize by not secreting nectar in habitats in which ants are rare or absent. Barbara Bentley found that a species of *Bixa* which is widespread in the New World tropics behaves differently at different altitudes. When growing at low elevations, where ants are abundant, it secretes extrafloral nectar, but plants growing at 3,600 feet, where ants are scarce, do not even have extrafloral nectaries. The plant has switched to an alternative defense. "Instead," Bentley wrote, "the buds are covered by thick, leathery sepals, which give greater protection to the developing bud than is provided by the papery sepals of the lowland form."

David Tilman reported an interesting observation of the North American wild black cherry and its major defoliator, the eastern tent caterpillar. He found that in early April, when the leaves are unfurling and the tent caterpillars—newly hatched from egg masses that survived the winter—are still small enough for ants to handle, the leaves secrete abundant nectar and ants are numerous on the trees. But by early May, the caterpillars are too large for the ants, the secretion of nectar has greatly decreased, and ants have all but disappeared from the trees.

Ants are attracted to some plants that do not secrete extrafloral nectar by the sweet honeydew excreted by sap-sucking members of the order Homoptera, aphids, treehoppers, leafhoppers, mealybugs, scale insects, and others. Honeydew, being essentially a solution of sugars in water, is an analogue of extrafloral nectar and an important food for many ants. In a single day, a homopteran may produce an amount of honeydew equal to several times its own body weight. Ants solicit honeydew by stroking the hind end of a homopteran with their antennae. The homopterans are cooperative. If not tended by ants, they squirt out the honeydew to get the sticky stuff well away from their bodies, but when stimulated by an ant, they raise their abdomens and gently ease out a droplet. Some ants tend colonies of aphids or other Homoptera much as people tend herds of cattle. They kill or chase away intruding insects of all kinds, inadvertently including some injurious to the plants, as they try to keep away honeydew robbers and predators that would eat their "cattle." There is chemical communication between

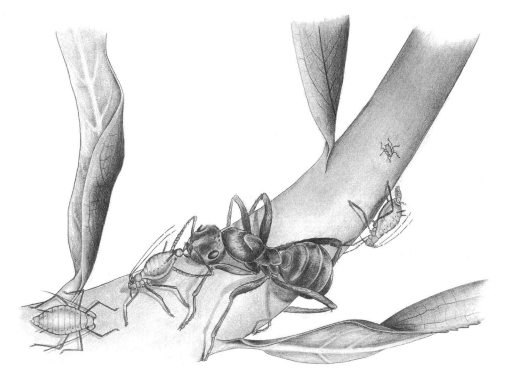

An ant solicits honeydew from an aphid on a goldenrod stem
as a nearby aphid gives birth to a nymph.

homopterans and their attending ants. For example, ants respond to a volatile alarm pheromone that disturbed aphids release to warn other aphids in their group that they risk attack. When ants smell this pheromone they rush to the rescue, helping to drive off intruders.

Harboring Homoptera that attract ants is a tradeoff for the plant. It must spend nutrients and energy to support the Homoptera, but the ants attracted by the Homoptera may more than repay this cost by deterring other herbivores whose depredations would cost the plant far more.

Consider the three-way mutualistic relationship between a North American goldenrod, honeydew-producing treehoppers, and ants, well described by Frank Messina. The growth of the goldenrods can be severely stunted by two leaf beetles that sometimes completely defoliate a plant. Eggs of these beetles, which survive the winter in the soil at the base of a goldenrod plant, hatch in early May and the larvae immediately climb up the plant to feed on its leaves. Some time later adult treehoppers lay their eggs on the undersides of the goldenrod leaves, but only on leaves of plants closely associated with an ant mound. Ants guarding the treehopper colonies to protect their source of honeydew attack not only enemies of their treehoppers but also adult and larval leaf beetles. When Messina compared goldenrods with ants to nearby goldenrods without ants, he found that plants profit greatly by supporting treehopper colonies that attract ants. Plants not occupied by ants grew only 27 inches tall and bore only 536 seeds per plant, while plants with tree-hoppers guarded by ants grew to a height of 36 inches and produced 3,585 seeds per plant.

Little is known about how extrafloral or floral nectar affects the level of protection against herbivores from parasites and predators other than ants. Although parasites and predators of insects (several hundred thousand are known worldwide) greatly outnumber the known species of ants (less than 9,000 worldwide), they are far less numerous at extrafloral nectaries. There are two reasons for this. First, in most ecosystems ants are so numerous that their total population will most likely greatly outnumber and outweigh the combined population of all parasites and predators. Second, parasites and predators are only transient visitors to extrafloral nectaries, while ants usually occupy and guard a plant against herbivores as long as it continues to secrete nectar. Parasites and predators are likely to attack herbivores present when they visit a plant, but they soon move on to search for hosts or prey elsewhere.

Andrew Stephenson found that although nectaries on catalpa leaves are frequented mainly by ants, they are also visited by a parasitic wasp and two species of predaceous ladybird beetles. He saw ladybirds attack the eggs and newly hatched caterpillars of the catalpa sphinx, and saw hundreds of these caterpillars that had been parasitized by the wasp *Apanteles congregatus*. Using her sharp, spearlike ovipositor, the female *Apanteles* inserts an egg into the body of a caterpillar. Then something amazing happens. At the very beginning of the wasp embryo's development, when it is in the four-cell stage, it divides and redivides until it has split into several dozen identical replicas of itself—a process much like twinning in humans. By the time the caterpillar has grown large, the parasitic larvae are ready to molt to the pupal stage. Then all of them simultaneously chew their way through the caterpillar's skin and spin tiny, white cocoons on its back. You may have seen cocoons of this same parasite on tomato hornworms, close relatives of the catalpa sphinx, in your tomato patch. Don't destroy them. Let the adult parasites emerge to destroy more hornworms.

At four elevations from 1,960 feet to 5,400 feet in the mountains of Costa Rica, Suzanne Koptur made painstaking observations of two mimosa-like trees of the genus *Inga*. At the lowest elevation, ants of several species were abundant on the trees—guarding the extrafloral nectaries. Parasites seldom visited the nectaries. The ants were aggressive defenders, frequently carrying away or driving off herbivorous katydids, beetles, leafhoppers, and caterpillars—the last the most damaging of the trees' insect enemies. Leaves from which ants were excluded by barriers of Tanglefoot, a sticky substance, suffered four times as much damage as leaves to which ants had free access. But ants became less abundant as the elevation increased. At 1,960 feet ants were abundant on the trees, at 4,300 feet they were scarce, and at 4,900 and 5,400 feet they were virtually absent. But the trees secreted extrafloral nectar at all levels, and there was a "changing of the guard" at higher elevations: as ant numbers declined, the number of parasites that came to the trees' extrafloral nectaries increased, and the number of caterpillars they parasitized increased.

🐝 🐜 🪰 Some parasites that attack herbivorous insects—perhaps many or even most—locate a suitable host by orienting to its food plant's odor, a tactic that often leads the parasite to hosts that would otherwise be too far away to find. The odds of finding a host are improved if parasites respond only to the odor of damaged plants. For example, as H. C. J. Godfray noted, parasitic wasps fly upwind to damaged thistles or artichokes (the latter are just big thistle flower buds). If the damage had been caused by their host insect, as was often the case, they were in luck. The odds are improved even more if a parasite responds only to the odor of plants whose injuries have been caused by herbivorous insects. In one study, Vincent Nealis found that a parasite of the imported cabbage worm (a caterpillar) orients to leaves damaged by caterpillars but not to artificially damaged leaves.

Some plants "cry for help" when they are under attack by herbivorous insects. The "cry" is not a sound, but a volatile odorous chemical that the plant releases into the air to attract parasites of its attackers. Working in the laboratory, Ted Turlings and two coworkers showed that seedling corn plants whose leaves had recently been damaged by armyworm caterpillars released volatile terpenoids that attracted a wasp that parasitizes armyworms. (Terpenoids are chemicals responsible for the characteristic odors of many plants, including cloves, lemons, and lavender.) The volatiles are released only as a result of an interaction between a corn plant and a caterpillar. When leaves were artificially injured with a razor blade, few parasites were attracted, but if artificially injured leaves were smeared with caterpillar regurgitant, they were as attractive as leaves damaged by caterpillars. Uninjured leaves smeared with caterpillar regurgitant attracted few parasites. Turlings and James Tumlinson later showed that these volatiles are released by both injured and uninjured leaves on the same plant. H. T. Alborn, Turlings, and their coworkers isolated and identified volicitin, the chemical component of caterpillar regurgitant that elicits the release of the plant volatiles that attract parasites. The application of just volicitin alone to artificial injuries on leaves elicited the release of the plant volatiles.

🐜 🦟 🐝 In *The Naturalist in Nicaragua,* published in 1874, the English mining engineer Thomas Belt recorded his observations of the amazing relationship of certain ants with bull's-horn acacias. His account firmly established the existence of the acacia-ant association, and to this day his outline of its major aspects remains unchallenged. He wrote:

> Clambering down the rocks, we reached our horse and mule, and started off again, passing over dry weedy hills. One low tree, very characteristic of the dry savannahs, is a species of acacia growing to a height of fifteen or twenty feet. The branches and trunk are covered with strong curved spines, set in pairs, from which it receives the name of the bull's-horn thorn, they having a very strong resemblance to the horns of that quadruped. These thorns are hollow, and are tenanted by ants, that make a small hole for their entrance and exit near one end of the thorn, and also burrow through the partition that separates the two horns; so that the one entrance serves for both. Here they rear their young, and in the wet season every one of the thorns is tenanted; and hundreds of ants are to be seen running about, especially over the young leaves. If one of these be touched, or a branch shaken, the little ants . . . swarm out from the hollow thorns, and attack the aggressor with jaws and sting. They sting severely, raising a little white lump that does not disappear in less than twenty-four hours.

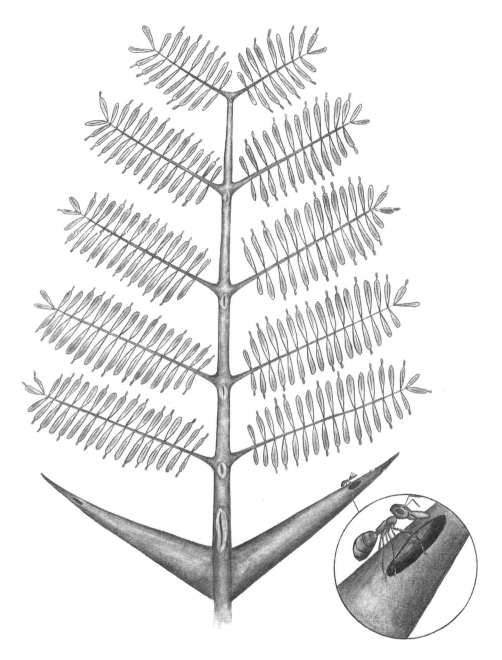

A branch of an ant acacia with extrafloral nectaries along the stem, Beltian
bodies, consisting of protein and fat, at the tips of the leaflets, and
an ant entering the hollow thorn that houses its colony.

Belt goes on to arrive at what was then a startling conclusion: the association between the ants and the acacias is mutually beneficial. He next explained just how ants and plants benefit each other.

These ants form a most efficient standing army for the plant, which prevents not only the mammalia from browsing on the leaves, but delivers it from the attacks of a much more dangerous enemy—the leaf-cutting ants. For these services the ants are not only securely housed by the plant, but are provided with a bountiful supply of food. The leaves are bi-pinnate [featherlike, with a long row of leaflets on either side of the leaf stem]. At the base of each pair of leaflets, on the mid-rib, is a crater-formed gland, which, when the leaves are young, secretes a honey-like liquid. Of this the ants are very fond; and they are constantly running about from one gland to another to sip up the honey as it is secreted. But this is not all; there is a still more wonderful provision of more solid food. At the end [of each leaflet] . . . there is, when the leaf first unfolds, a little yellow fruit-like body united by a point at its base to the end of . . . [the leaflet]. Examined through a microscope, this little appendage looks like a golden pear. When the leaf first unfolds, the little pears are not quite ripe, and the ants are continually employed going from one to another, examining them. When an ant finds one sufficiently advanced, it bites the small point of attachment; then, bending down the fruit-like body, it breaks it off and bears it away in triumph to the nest. All the fruit-like bodies do not ripen at once, but successively, so that the ants are kept about the young leaf for some time after it unfolds. Thus the young leaf is always guarded by the ants; and no caterpillar or larger animal could attempt to injure them without being attacked by the little warriors. The fruit-like bodies are about one-twelfth of an inch long, and are about one-third of the size of the ants; so that an ant carrying one away is as heavily laden as a man bearing a large bunch of plantains. I think these facts show that the ants are really kept by the acacia as a standing army, to protect its leaves from the attacks of herbivorous mammals and insects.

The colonization of an acacia begins when a lone fertilized ant queen lands on it, snaps off her now useless wings, and burrows into a thorn in which she will lay eggs and raise a small group of infertile daughters, the colony's first workers. In Belt's words, "The thorns, when they are first developed, are soft, and filled with a sweetish pulpy substance; so that the ant, when it makes an entrance into them, finds its new house full of food. It hollows this out, leaving only the hardened shell of a thorn." When the first workers mature, they assume all the chores of the colony, including raising the larvae. Thereafter the queen does nothing but lay eggs, so prolifically that the colony may eventually expand to occupy all the thorns on a tree and even thorns on nearby acacias. Daniel Janzen once found a colony of over 12,000 workers. The queen stayed on one branch but workers carried eggs and small larvae to other branches and trees occupied by the colony.

The acacia provides the ants not only with domatia but also with a diet that contains all the nutrients required by both adults and larvae. Only rarely do the ants feed insects to their larvae. The nectar secreted at the base of each compound leaf is eaten mainly by adults and provides the energy they need to fuel their vigorous activity. The pearlike bodies at the tip of each leaflet of a compound leaf—now known as Beltian bodies in Belt's honor—consist largely of proteins and fats and are the food of the rapidly growing but inactive larvae. The nectaries and the Beltian bodies apparently provide all the required nutrients, since the ants can survive and reproduce on a diet of nothing else.

Acacias make yet another provision for their ants. Species of acacias not protected by resident ants drop their leaves in the dry season, when they blossom, and remain leafless until the wet season, but many ant acacias have a few green leaves throughout the dry season. Ant acacias are usually the last to drop their leaves but quickly replace them with small tufts of interim leaves on their flower-bearing shoots. These dry-season leaves, which persist until the wet season, have nectaries and Beltian bodies. The products of these leaves, as Janzen wrote, "are essential to keep the ant colony from starving to death." The tree needs its ants even in the dry season to protect the new flowering branches from herbivores and its woody parts from rodents and from the egg-laying parents of wood-boring beetle larvae. And of course it is advantageous to have a resident ant colony in place and ready to expand when the wet season and the main growth period begin.

Janzen has clearly demonstrated that the ants protect the tree that houses them from herbivorous insects and competing plants. There is as yet no direct proof that they protect it from browsing mammals. But it is likely that they do. Consider, for one thing, their response to encroaching people, whom the ants perceive as a threat, probably because they cannot differentiate between them and leaf-eating mammals such as deer or cows. If a person comes within a yard of their tree, the ants gather on the part of the tree closest to the intruder and will attack en masse if the person brushes against the tree. This is a painful experience for a person, as I found out, but just imagine how much more painful it would be for a deer that thrusts its tender nose and lips into the foliage.

Many different kinds of herbivorous insects, including sucking bugs, tree-hoppers, and caterpillars, land on ant acacias or are placed on them as eggs by their mothers, but the defending ants make short shrift of these potentially damaging intruders. They patrol the tree day and night, spotting intruders by sight during the day and perceiving them by touch at night. They move in quickly to bite or sting intruders, often ganging up on lone invaders. With this strategy the ants are able to kill or chase away most herbivorous insects, but a few can survive and feed despite the ants. Among them are a scarab beetle whose body armor is too tough for the ants to penetrate, a caterpillar that can physically throw off attacking ants, and another caterpillar that the ants simply ignore. My guess is that this last caterpillar emits a substance that mimics the pheromone by which the ants recognize each other.

The ants also destroy vegetation encroaching on their tree, such as climbing vines and the branches of neighboring plants of other species. They also "weed" the area beneath their tree, destroying all plants that sprout there, leaving a large area of bare earth that may be more than 3 feet in diameter.

Janzen was the first to show experimentally that their ant guests benefit bull's-horn acacias. He freed some acacia shoots of ants and compared them with other acacia shoots occupied by ants. He eliminated the ants by clipping all the thorns off plants, by cutting plants down and observing the ant-free shoots that sprouted from the stumps, or by using an insecticide. The results were striking. Herbivorous insects were much more numerous on shoots without ants than on shoots occupied by ants. In daylight, herbivorous acacia-feeding insects were 22 times more numerous on the shoots without ants than on shoots with ants. The ants were less efficient in the dark, but even then acacia-feeding insects were 12 times more numerous on shoots without ants.

The acacia shoots profited greatly from this protection. Janzen observed that during a period of almost 10 months about 57 percent of the ant-free shoots died but only about 28 percent of the occupied shoots died, and surviving shoots without ants grew much less vigorously than ones with ants. At the end of this period, shoots with ants had more than twice as many leaves and thorns as did the others, and they had gained an average of over 20 ounces, while ant-free shoots gained an average of less than 2 ounces.

And during a 70-day period, shoots without ants increased in length by less than 10 inches, but those with ants grew almost 25 inches.

The ant-acacia association is virtually obligatory for both partners; neither is likely to survive without the other. Janzen observed that acacia plants not protected by ants eventually die without producing seeds. The ants involved in this particular association, *Pseudomyrmex ferruginea*, have been found nesting only in bull's-horn acacias, the one that Janzen studied, *Acacia cornigera*, and a few other species similarly hospitable to them.

Many plants other than the New World bull's-horn acacias maintain standing armies of ants by providing them with nesting sites and food. The protection the ants offer these plants varies: all deter herbivorous insects, some destroy competing plants, some may drive off browsing mammals, and one species of ant eats the spores of fungi that infect its host plant. Among the best known of these plants are species of four genera: *Cecropia* and *Piper* of the New World tropics and *Macaranga* and *Leonardoxa* of the Old World tropics. The African members of the genus *Acacia* can be added to this list, but they house their ants in hollow swellings of the stem rather than in swollen thorns. All of these plants are small to medium trees except for *Piper*, which is a vine. In every case, their domatia are stems rather than modified thorns. These stems may be naturally hollow or have a pith that is soft enough for the ants to remove. All these plants except *Cecropia* have extrafloral nectaries, and in some of them the main or an additional source of sugar is honeydew secreted by scale insects, which some plants house in special protective chambers. All but *Leonardoxa* have food bodies, analogous to Beltian bodies. The ants that inhabit *Leonardoxa* depend upon insect prey for nutrients other than sugar. Ants on some of the other plants also rely on prey to some extent, and some even make their scale insect herds do double duty, much as we do with cattle, not only milking them but also killing some to feed to their larvae.

HELPING ANIMALS

5

GIVING SUSTENANCE

At the edge of a bog in northern Michigan, a colony of aphids, each one clothed by a secretion of white waxen threads, sucks sap from a branch of an alder shrub. They are all daughters of one mother who conceived parthenogenetically, without being fertilized by a male, and gave live birth to numerous offspring, perhaps as many as seven per day. Several of them are eaten by an aphidlion, the larval form of a lacewing, that had not been driven away by ants guarding their "herd" of honeydew-excreting aphids—the ants do not recognize the aphidlions as a threat to their "livestock" because, like wolves in sheep's clothing, the aphidlions disguise themselves with wax threads they pluck from the aphids. The aphidlion soon becomes an adult and flies off to find a mate. But it doesn't get far before it is snatched up and eaten by a cruising dragonfly. The dragonfly would have used nutrients from the lacewing's body to develop eggs to lay in a nearby pond, where the nymphs that hatched from them would have preyed on aquatic insects and occasionally even small fish. But later that day, an olive-sided flycatcher, recently returned from its winter home in South America, swooped down on the dragonfly from a perch at the top of a tall tree and fed it to one of the young waiting hungrily in its nest in a black spruce.

This food chain intermeshes with many others to form a complex food web, one of many that together constitute the community of life in any ecosystem. The alder-flycatcher food chain, as do virtually all others, begins with a green plant. Among the green plants are certain algae, mosses, horsetails, ferns, and gymnosperms, but except in the seas, the majority of them are angiosperms. Green plants are the foundation of food chains because they are the ultimate source of all the nutrients—from vitamins and minerals to proteins and carbohydrates—required by animals from the tiniest mite to the gigantic elephant. Only green plants—with the exception of some

bacteria and other microorganisms—can capture the energy of the sun and make it available to animals through photosynthesis, synthesizing energy-rich sugars from water and carbon dioxide, using sunlight absorbed by chlorophyll as the source of energy. Plants, with the help of nitrogen-fixing bacteria, also synthesize amino acids, the building blocks of proteins. Complex many-celled animals, such as insects and mammals, cannot synthesize amino acids and must obtain them directly or indirectly from green plants. Finally, life as we know it is made possible by an elegantly mutualistic interaction between the plant and animal kingdoms. Green plants also produce, as a byproduct of photosynthesis, almost all the oxygen in the atmosphere. Both plants and animals require this gas of life to obtain energy by "burning" (oxidizing) the sugars made by plants. Happily, the oxidation of sugars produces the carbon dioxide that green plants use to photosynthesize more sugars.

The second link in virtually all terrestrial food chains is some animal that eats plants, usually an herbivore, which eats only plants, but sometimes an omnivore, which feeds on both plants and animals. Plant-eating animals, the great majority of them insects, play an indispensable and pivotal role as intermediaries in food chains by making the nutrients synthesized by plants available to animals that do not eat plants. These nutrients pass on to succeeding links of the food chain when herbivores or omnivores are eaten by predators, parasites, blood feeders, or carrion eaters. Insects also bridge the size gap between large predators and unicellular plants or animals too tiny to be profitably eaten by a large animal. The aquatic nymphs of mayflies, for example, eat microscopic single-celled algae they filter from the water. The mayflies grow to be suitably large "packages" of food for predators such as small fish that eat them when they are still nymphs and for birds that feast on the adults after they fly from the water. When pollution virtually wiped out the mayflies of western Lake Erie, fish populations dwindled. When the pollution was reduced, the mayflies returned and fish populations returned to normal levels.

Among the vertebrate herbivores are many mammals, such as giraffes, antelopes, bison, rabbits, lemmings; some birds, such as the grass-eating Canada geese, fruit-eating waxwings, and seed-eating sparrows; and a few reptiles, the leaf-eating iguanas and some turtles. But the great majority of herbivores are insects. According to Arthur Weis and May Berenbaum, nearly half the known species of insects, or about 450,000 species, feed on plants, attacking all their parts in various ways. Thus about 37 percent of the

1,200,000 known animals are insects that feed on green plants. Most of them are grasshoppers and katydids; stick insects; thrips; true bugs, such as the squash bug; aphids, leafhoppers, and scale insects; beetles; flies; moths and butterflies; and sawflies, ants, bees, and their relatives.

The eaters of insects are a heterogeneous mix that includes mammals as small as shrews and bats and as large as bears; many different birds, such as woodpeckers, flycatchers, swallows, and warblers; many reptiles, such as lizards and some small snakes; amphibians, such as frogs, toads, and salamanders; most arthropods—segmented animals with paired, jointed legs and a "skeleton" that is the hardened wall of the body—other than insects, such as spiders, scorpions, and centipedes; and of course a huge variety of insects, including among others antlions, damselflies, dragonflies, mantises, tiger and ladybird beetles, robber flies, parasitic flies, almost all ants, and both parasitic and predaceous wasps.

Insects, indispensable members of almost every ecosystem on land or in fresh water, are far more numerous and ubiquitous than most of us imagine. Nevertheless, our experiences do hint at their abundance: choruses of chirping crickets and katydids all around us as we walk in the country on a summer night; flurries of grasshoppers we flush when we stroll through a meadow; even the many insects that splatter our windshield after a drive through the country. But these signs of abundance are just the tip of an iceberg. For every insect we see, there are tens of thousands that we do not see, because they are small or hidden beneath the soil, within a plant or an animal, under a rock, or in some other "crack or crevice" of the environment. Martin Speight and his two coauthors concluded that there may be over 1,390 *Culex* mosquito larvae per square foot in the water of a Korean rice field, and almost 4,800 arthropods (mostly insects and mites) per square foot in the ground litter in a forest in India. In *Six-Legged Science*, Brian Hocking estimated that in East Africa the summed per-acre weight of just two species of ants equals that of the dense population of hoofed mammals such as zebras, gazelles, and wildebeest.

Because of their diversity and enormous numbers, insects are extraordinarily important as converters of plant tissue to animal tissue. But just how important are they, and how do they stack up as converters against the much larger and usually far more visible herbivorous birds and mammals? Information pertinent to this question is difficult to collect and not much is available, but there is enough to establish that insects are the most important of the herbivores in terms of both numbers and biomass.

We can get some idea of the importance of herbivorous insects by looking at the population size of some of the more abundant species. Consider, for example, periodical cicadas. The immature cicadas live in the soil sucking sap from the roots of trees and shrubs for 13 or 17 years, and then in their last spring emerge synchronously from the soil as adults, a bonanza of food for insect-eating creatures, actually much more than they can eat. Henry Dybas and D. Dwight Davis estimated that in the Chicago area in 1956 about 1,500,000 adult cicadas, weighing approximately 1.2 tons, emerged per acre in lowland forests. This is about 533 tons per square mile, and the emergence encompassed hundreds of square miles.

The termite *Macrotermes natalensis* builds huge mounds that dot the African landscape. They sometimes tower to a height of 16 feet, can be as much as 16 feet in diameter at their base, and may harbor 2 million or more workers. These termites are agriculturists that live on a fungus they cultivate on a mulch of vegetation collected on foraging trips away from the mound. In the New World tropics, leafcutter ants cultivate fungus on a fertile mulch they prepare by masticating pieces of fresh leaves and flowers they snip from a wide variety of green plants. Columns of foraging workers may be more than 100 yards long. Those in an outgoing file are "empty-handed," but their sisters, by their sides in a returning file, carry in their jaws large pieces of leaves or flowers to their huge underground nests. Leaf-cutting ants, wrote Bert Hölldobler and Edward O. Wilson, are the dominant herbivores of the New World tropics, consuming much more vegetation than any other comparably diverse group of animals, including mammals and other plant-feeding insects. Forest-dwelling leafcutters harvest between 12 and 17 percent of all the leaf material produced by the trees and other plants. Leaf-cutters are impressively abundant. For example, there may be twelve colonies per acre of a Brazilian grass-harvesting species, each colony including a million or more workers. Their yearly harvest of grass varies from about 400 to 2,000 pounds per colony. If we take 1,200 pounds as the average, they collect about 7 tons of grass per acre each year!

Few scientific studies compare the relative ecological importance of insect and vertebrate herbivores. To this day the most useful comparisons are those in a few research papers summarized by Richard Wiegert and Francis Evans in 1967. Done in abandoned crop fields in the process of "returning to nature," these studies showed unequivocally that herbivorous insects greatly outnumbered and outweighed the most numerous vertebrate herbivores, savannah sparrows and mice. Eugene Odum and two coworkers found that in summer the average population of 10 sparrows and 10 mice per acre in a field in Georgia was vastly outnumbered by the over 18,000 herbivorous grasshoppers and crickets per acre. Astonishing as it may seem, the biomass of the grasshoppers alone was more than seven times that of the much larger sparrows and mice combined. A later study of the same field by Frank Golley and J. B. Gentry also included harvester ants and other herbivorous insects in the field and therefore shows an even greater difference in biomass between vertebrate and insect herbivores. The biomass of the sparrows and mice was less than seven-tenths of a pound per acre, while the biomass of the grasshoppers, crickets, harvester ants, and other herbivorous insects was close to 6 pounds per acre. The herbivorous insects outweighed the herbivorous vertebrates by a factor of almost nine.

But the ultimate bottom line is not numbers and biomass. It is growth, how much new animal tissue, food for animals that do not eat plants, is produced by herbivores. According to Wiegert and Evans, the sparrows and mice in the abandoned field in Georgia produced an average of only a little more than seven-tenths of a pound of new tissue per acre per year. The insects, more fecund and more efficient converters of plant tissue into animal tissue, produced 20 pounds of new tissue per acre per year. In other words, the herbivorous insects produced close to thirty times more food for animal-eating creatures than did the much larger and more visible vertebrates.

Many arthropods, including hundreds of thousands of insects and tens of thousands of insect relatives such as scorpions and spiders, depend upon insects for sustenance. They are by far the most important eaters of insects in number of species, number of individuals, and biomass. Some insect eaters are predators, which, like cougars, kill their prey immediately and soon eat all or most of it. Others are parasites, which live in or on the body of a host animal, and absorb nutriment from its blood, but do not necessarily kill it. According to R. R. Askew's estimate in *Parasitic Insects*, 15 percent or more of the known insects, about 135,000 species, are parasites of other insects, and at least as many plus other arthropods are predators of insects. Insects that parasitize other insects are often called parasitoids because most of them ultimately kill their host. A parasitic wasp larva, for example, at first acts like a true parasite, absorbing nutriment from its host's blood but doing it little or no harm. But when nearly full grown, it acts like a predator by devouring all or most of its host.

Of the approximately 85,000 known terrestrial arthropods other than insects, 45,000 or more are predators that eat mainly insects. All but about 4,000 of them subdue their prey by injecting them with venom. This capability evolved independently at least four times: in centipedes, scorpions, pseudoscorpions, and spiders. Each of these groups has a different way of injecting venom. In centipedes, the first pair of legs are not used in walking, but are highly modified as fanglike, venom-injecting "poison jaws" that protrude forward under the head and look more like mouthparts than legs. The first pair of legs (pedipalps) of pseudoscorpions resemble the front legs of a crab or lobster in that they bear strong pincers. The pincers inject venom. Scorpions also have formidable pincers on their pedipalps, but a bulbous and sharply pointed stinger at the end of the abdomen injects the venom. Finally, spiders use fangs on their paired mouthparts, the chelicerae, to inject their prey with venom.

Centipedes are readily recognizable because the body behind the head consists of from 15 to over 100 similar segments, each of which bears a single pair of legs. Most are of moderate size, but some tropical giants are a foot long and have been seen capturing birds and other small vertebrates. Most centipedes are insect eaters that spend the night prowling for prey. A centipede that J. L. Cloudsley-Thompson kept in captivity "even ate bees and wasps which she caught in mid-air, rearing up the fore part of her body to snatch them with her poison claws as they flew past." He observed that "she dropped them quickly and waited until her poison had had time to take effect."

The adroitness of these creatures is contrary to an anonymous bit of doggerel which suggests that their ability to control their many legs might be easily disrupted. In 1889, E. Ray Lankester quoted these familiar lines in a letter to *Nature:*

> A centipede was happy—quite!
> Until a toad in fun
> Said, "Pray which leg moves after which?"
> This raised her doubts to such a pitch,
> She fell exhausted in the ditch,
> Not knowing how to run.

We think of scorpions as animals of tropical jungles and hot deserts, but they occur much farther north than that. There is even one species in southern Illinois. Most scorpions are fairly large, sand scorpions, among the largest, being almost 3 inches long. Scorpions subsist mainly on insects and to a lesser extent on other arthropods. One of the few studies of the feeding habits of a scorpion in nature was done by Gary Polis near Palm Springs, California; he used an ultraviolet lamp to follow the nocturnal activities of 750 desert sand scorpions, each marked with an individually distinctive pattern of paint that fluoresces under ultraviolet light.

The sand scorpions emerged from their burrow at dusk, moved on average about 3 feet away, and sat motionless until they detected prey. All returned to their burrows by dawn. These "sit- and-wait" predators captured their prey in three ways. About 80 percent of their victims were grabbed as they walked across the sand; about 10 percent were dug out of the sand; and an amazing 10 percent were snatched from the air or just after they landed on the sand. The scorpions probably sense air currents caused by flying insects. They detect movement in or on the sand from 1 or 2 feet away, and rush forth rapidly to grab the prey with their pincers. If the prey is small and does not resist, the scorpion immediately devours it, but if it is active, the scorpion first stings it. The scorpions' prey consists of 80 percent insects and 20 percent other arthropods. Like all other scorpions, sand scorpions are highly cannibalistic. Over 15 percent of their prey is scorpions, about 9 percent of them members of their own species. Over 42 percent of a sand scorpion's insect prey consists of darkling beetles, both burrowing larvae and adults found on the surface; about 11 percent consists of burrowing cockroaches; almost 8 percent of ants; and about 4 percent each of sand crickets and wasps.

The smallest of the insect-eating arthropods other than insects are the pseudoscorpions, tiny creatures usually less than a quarter of an inch long and vaguely resembling true scorpions because their pedipalps end in stout pincers. But there the resemblance ends. The abdomen of a "false scorpion" is short and stout, not elongated like a true scorpion's. Pseudoscorpions are seldom seen because most live hidden in the soil, in leaf litter, under rocks, or beneath flakes of loose bark. An unusual one lives in debris chambers in the nests of leafcutter ants in South America, where it eats scavenging insects. Most pseudoscorpions assiduously avoid light; as biologists say, they are photonegative. They are strictly carnivorous, preying mostly on springtails and tiny insects such as barklice and the wild relatives of the silverfish that we see in our homes. As Cloudsley-Thompson put it, "It is doubtful that the prey is actively sought after, but rather that false-scorpions lie in wait, with their claws open, until some suitable animal accidentally brushes

This desert scorpion, a relative of the sand scorpion, is about
to make a meal of a much smaller harvester ant.

against their sensory hairs, when it is seized with extreme rapidity." Venom
secreted by glands in the pincers is usually injected to paralyze the prey.

There are about 30,000 known species of spiders but thousands more
are still unknown to science, because few taxonomists work on these wonderfully interesting little animals. All spiders are predators, and most feed
chiefly on insects. Without insects there would be no spiders, and without
spiders there would probably be too many insects. Many spiders are trappers
that spin silken webs in which they snare their victims; others spin no webs
but are hunters that ambush or stalk their prey.

Different web-spinning spiders build different webs, but the most familiar
are flat orbicular webs with spirals of many sticky hooplike strands sup-

ported by long, straight, nonsticky strands that radiate from the center of the web like spokes of a wheel. Rainer Foelix, an expert on spider biology, described how the common garden spider of Europe, an orb weaver, handles prey caught in its web:

> When a fly becomes entangled in the sticky spiral thread of the orb web, it produces specific vibrations which immediately excite the spider. Even if the fly then remains quiet, the spider will pluck several radial threads, apparently to probe the load on each radius. In other words, it tries to find the exact position of the prey. Especially if the fly moves its wings again, the spider will rush out of the hub using exactly that radial thread which leads to the prey. The victim is briefly touched with the front legs and palps; then the hind legs wrap silk around it. Only thereafter follows a *brief* bite. Using its legs and chelicerae, the spider then cuts the neatly wrapped "package" from the web and carries it to the hub. There it is attached by a short thread before it is eaten. The feeding process always takes place in the hub, never at the actual capture site . . . if the prey is very small (such as a fruit fly), it is simply grasped with the chelicerae and carried to the hub. Large insects, which set up strong vibrations in the web, are also bitten immediately, but the bite lasts many seconds or even minutes. Such a *long* bite probably prevents a possible escape of strong prey animals. On the other hand, aggressive prey, such as wasps, are always wrapped first, and then bitten. Apparently in this situation it is safer for the spider to keep the dangerous prey at a distance.

The spiders that do not spin webs ambush or stalk their prey. Among the ambushers are the famous trapdoor spiders. I first saw these spiders on a tour of the magnificent botanical garden in Kingston, Jamaica. Our guide ended the tour with one of the garden's star attractions, spiders that live in burrows in the soil closed by hinged trapdoors camouflaged by a covering of debris from the ground. According to Foelix, they "hunt at night, just cracking open the trap door and stretching out their front legs. The more 'primitive' [species] jump out and chase passing prey, whereas more advanced species leave the [burrow] only if prey is within easy reach." The most sophisticated extend from their doorway an array of "tripping threads" that alert the spider when some creature stumbles into one of them.

Other hunting spiders have evolved their own ways of catching prey. Some crab spiders lurk quietly on a blossom until some nectar-seeking insect comes close enough to be grabbed by their long, powerful front legs, and some can, like chameleons, change their color to match the flower on which they lurk. Like crab spiders, jumping spiders spot their prey visually and hunt only in daylight. They are small, usually less than a quarter of an inch long, and have very large eyes that can spot small insects from a distance. They stealthily stalk their prey until they are close enough to leap upon it. Spitting spiders, Rod and Ken Preston-Mafham report, neither are fast enough nor see well enough to dash after prey as do other hunting spiders. Instead, when they are close to an insect, they immobilize it by squirting sticky gum on it. Only then do they approach it to administer the fatal bite. The remarkable bolas spider hunts at night, capturing moths by flicking at them its "bola," a strand of silk with a glob of very sticky glue at its end. These spiders capture only male moths, an amazing fact Mark Stowe and two colleagues were able to explain when they discovered that this spider uses "sex appeal" to lure its male prey. The spiders synthesize and release into the air compounds chemically identical to components of the sex-attractant pheromones female moths release to attract males from a distance.

🐜 🐜 🐜 Entomologists often say that insects are their own worst enemies. The late Robert L. Metcalf and his son Robert A. Metcalf expressed the same thought more explicitly: "There is no doubt that the greatest single factor in keeping plant-feeding insects from overwhelming the rest of the world is that they are fed on by other insects." At least 300,000 species of insects, some parasites and some predators, eat other insects, often plant eaters but sometimes other parasites or predators. Predaceous insects occur in virtually all habitats, from the driest of deserts to freshwater ponds and streams, and capture their prey by chasing, ambushing, trapping, or stealthy stalking. Parasites, like predators, occur in virtually all habitats and have evolved diverse ways to exploit other insects. With few exceptions, insects are parasites only during their larval stage, and the larva must somehow get into or—less often—onto its host. This, as you will see, is accomplished in several ways.

Dragonflies and their close relatives the damselflies are voracious predators throughout their lives. The adults are strong, agile fliers and, except for a few that pick insects from foliage, pursue flying insects in the air. The immature forms, nymphs, live in ponds and streams, ambushing many different kinds of aquatic insects. The familiar green darner, largest of the North American dragonflies, courses over ponds, swamps, and even fields as it searches for insects it will capture in flight. Like other dragonflies, it uses its spiny legs to form a "basket" in which it traps its prey, which is then moved to the jaws to be chewed up and swallowed. As Philip Corbet wrote in his book on these fascinating creatures, dragonflies may land to devour large insects, but they eat small ones, such as mosquitoes or tiny flies on the wing, ever eager to satisfy their voracious appetites.

Nymphal dragonflies are, to say the least, unusual animals. The ways in which they obtain dissolved oxygen from water and capture prey are unique. Like many other aquatic insects, the nymphs have gills that absorb dissolved oxygen from water, but their thin, overlapping gills line the inner wall of the muscular rectum and are continuously bathed by fresh water sucked in through the anus and expelled by the same route. To escape enemies, nymphs shoot through the water like a rocket by forcefully spurting water from the rectum. Unlike all other insects, immature dragonflies and damselflies have a long, prehensile labium (lower lip) that is jointed, has grabbing claws at its tip, and can be extended for a distance equaling at least a third of the body length. At rest, the labium is folded below the body, its broad outer end covering the face like a mask. When an insect comes within striking range, blood is forced into the labium, straightening it and swiftly thrusting it forward to seize the prey. Then the blood pressure is relaxed and the labium refolds itself to bring the unfortunate insect back to the jaws. Nymphal dragonflies ambush their prey from two different positions. Some lie on the bottom covered with mud or debris, their eyes poking up the way the eyes of a submerged alligator protrude above the water. When prey comes within about 2 inches, the nymph's labium strikes. Others, nymphs of the green darner among them, lurk on aquatic vegetation as they wait for a meal to come close.

Antlion larvae catch insects in a different way, trapping them in funnel-shaped pitfalls that may be as much as 2 inches deep and over 3 inches in diameter. The pits, in loose sandy or dusty soil, are usually protected from rain by an overhanging cliff or a leafy shrub. The hungry larva waits at the bottom of the pitfall, hidden in the sand with only its long, curved, sucking mandibles visible. If an ant or some other insect stumbles into the pit and begins to slide down its steeply sloped wall, the antlion is alerted by falling particles of soil. Walter Balduf described the process graphically:

When the larva has taken . . . the "readiness stance," it lies on the alert for prey with its jaws spread wide apart . . . This stance is assumed in particular by hungry larvae. If an animal, above all, an ant, now approaches the edge of the pit, it is in danger of dashing into the depth of the cone into the open jaws of the antlion. The unstable soil of the pit margin and the steep slope are conducive to the involuntary descent of the ant, or other prey animal. An ant which steps on the border of the pit tends to make very excited movements then very frequently finds itself going to the bottom in a single dash or slide. There the jaws of the antlion suddenly snap shut upon it.

If the prey tries to climb back up the slope, the antlion uses its broad, flat head like a shovel to bombard the escaping insect with sand grains, causing a tiny avalanche that sweeps the prey back down the slope. With its channeled jaws, the antlion injects its prey with digestive juices that liquefy all the soft tissues within its body wall. It then sucks up its predigested meal through the same channels.

In constructing its pitfall, the larva walks backward, using the tip of its abdomen like a plowshare to dig a circular furrow, all the while using its head to flip away loose soil. It digs downward in a spiral path that becomes tighter and tighter as the narrow bottom of the funnel is approached, continually throwing out loose soil with violent tosses of its head. As it descends, the column of soil that remains standing in the middle of the pit becomes ever narrower until its base is so narrow that it collapses into the bottom of the pit. Adult antlions resemble damselflies because of their elongate abdomens and narrow net-veined wings, which have a span of up to 2.5 inches. But they differ in having long, clubbed antennae and in being nocturnal. They are not predators, but drink nectar, nibble pollen, or do not eat at all.

As you walk down a path in the woods in the eastern half of the United States or southern Canada, you may see a shiny green beetle about half an inch long standing on the path. If you come too close, it will suddenly fly a few yards down the path, alight, and then turn to face you as if watching for another intrusion. If you walk on and again come too close, it will repeat this procedure and may even repeat it several times as you walk along the path. This conspicuous insect is the six-spotted tiger beetle, one of the many beetles that eat other insects. The spots, small and white and not always six in number, are on the rear ends of the two armor plates (actually modified front wings) that completely cover the abdomen of this and almost all other beetles. Like other adult tiger beetles, it is a fast-running, ferocious predator that chases small insects and other arthropods and seizes them in its sharply pointed, sickle-shaped jaws.

Female tiger beetles place their eggs in small holes, one for each egg, in bare ground, often along a path. The larva digs a vertical burrow that may be a foot deep and lurks motionless at the entrance, waiting to ambush a passing insect. Its broad, flattened head closes the entrance like a trapdoor, and its body is firmly anchored to the wall of the burrow by the claws on its legs and the hooks on the back of its abdomen. Short antennae and six eyes are on the upper surface of the head and the long, sickle-shaped jaws point upward almost vertically. As Walter Linsenmaier put it, "if a small insect happens to tread on this odd door—that is, on the larva's head—the trespasser is seized immediately by the larva's sharp, erect biting jaws."

Another group of insects that feeds on other insects is the wasps. With few exceptions, they are either predators or parasites of arthropods. Females of the approximately 15,000 predaceous species have a venom-injecting stinger, a modification of the ovipositor. The solitary species use their venom mainly to paralyze insects they feed to their larvae. The stings of most solitary wasps are not particularly painful to people, but there are notable exceptions, such as the wingless, ground-dwelling wasps known as velvet ants, which deliver a potent sting. The venoms of the social wasp species are generally very painful to people and are used mainly in self-defense or to defend the colony, a tempting source of abundant food, against mammals such as skunks and bears. The sting is seldom used to subdue prey. Most of the 800 social species build communal nests of paper they make from wood or other plant fibers. The larvae, fed insects by the adult workers, are raised in individual hexagonal cells in paper combs—a few dozen cells in the nests of some species and thousands in others. The solitary species raise their larvae in nests in the soil, tunnels in wood, hollowed-out plant stems, or chambers built of mud. A female may have several nests, each containing one larva and its food supply of paralyzed insects, or one or just a few nests, each containing several larvae isolated with food in separate cells.

Yellowjackets, the yellow and black wasps that annoy us at picnics in late summer and are often mistaken for bees, live in colonies that may ultimately contain thousands of cells and include thousands of workers. Their nests consist of several horizontal paper combs, one above the other, covered by a paper envelope with an entrance at its bottom. They look much like the large football-shaped nests of hornets that hang from the branches of trees.

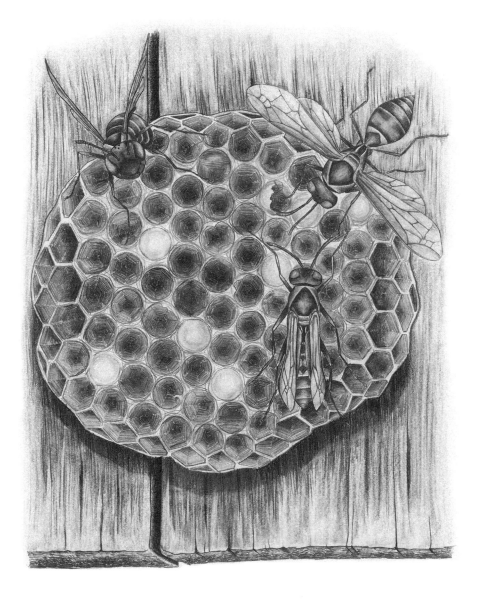

A view from underneath of *Polistes* wasps on a paper comb
hanging from the ceiling of a shed.

But they are usually in subterranean chambers close to the surface of the ground. Yellowjacket nests are seldom noticed by people unless they happen to step in one and provoke the wasps to attack. In their book on yellowjackets, Roger Akre and his coauthors show that these wasps are generalist predators which feed their larvae a wide variety of insects.

The nests of *Polistes* wasps, unlike those of yellowjackets, are a familiar sight because they often hang under the eaves of a house or from the ceiling of an open porch or shed. Each nest consists of a single uncovered paper comb, and consequently *Polistes* colonies are much smaller than those of yellowjackets. According to Hudson Reeve, an authority on these wasps, the most common species of the northeastern United States, *Polistes fuscatus*, averages only 180 cells per nest, with a maximum of 306. Nevertheless, *Polistes* wasps are abundant. As Edward O. Wilson pointed out, in Europe and North America their colonies outnumber those of all the other social wasps combined. These wasps prey mainly on caterpillars, a fact that led Robert Rabb to consider using *Polistes fuscatus* in North Carolina to control destructive tobacco hornworm caterpillars. Taking prey from workers returning to nests showed that over 93 percent of *fuscatus*'s prey consisted of caterpillars, about 23 percent of them hornworms. Rabb and F. R. Lawson greatly increased the number of *fuscatus* nests around tobacco fields by providing artificial nesting sites, small wooden boxes nailed open side down to posts. In 1956, these wasps occupied 206 of the 250 shelters that Rabb put up; and in 1957, 306 of 320. The wasps killed about 60 percent of the large hornworms in an experimental field. My own research showed that these large, almost fully grown hornworms are especially destructive, consuming over 86 percent of all the food eaten during the entire larval stage.

Most ants, like their wasp ancestors, are predators, although a few feed on plants. A colony of the nomadic army ants of Central and South America may include from 150,000 to 700,000 workers, relentless predators whose hunting swarms sweep over the forest floor on a front that may be as much as 60 feet wide. As T. C. Schneirla wrote, "Their normal bag includes tarantulas, scorpions, beetles, roaches, grasshoppers, and the adults and broods of other ants and many forest insects; few evade the dragnet." The prey are carried back to a temporary bivouac in some sheltered place. There is no nest. The bivouac is a ball of ants, which may be as much as 3 feet in diameter, that surrounds the queen, eggs, larvae, and pupae. During the ants' nonmigratory phase the colony remains in the same place for 2 or 3 weeks. At that time it includes few larvae and many developing pupae. Then, in a few days, the queen lays as many as 300,000 eggs. When new adult workers emerge from the pupae and the eggs have hatched, the ants go into their migratory phase and, carrying the newly hatched larvae, move the bivouac to a new site every day, a site whose insect population has not yet been decimated by the ants' incessant hunting. Thus every day the hunters provide a large supply of fresh prey to satisfy the ravenous appetites of the growing larvae. When the larvae molt to the nonfeeding pupal stage, the colony reverts to the nonmigratory phase, which requires less food, until a new brood of larvae is produced.

According to Askew, over 100,000 insects live by parasitizing other insects. About 10 percent are flies and the great majority of the rest are wasps. Adults are almost always free-living and the larvae are parasitic, usually living within the body of the host insect. The adults serve their offspring by finding hosts and placing eggs within or occasionally on or near the host's body. Wasps' eggs are inserted into a host's body by the piercing ovipositor, which is not converted into a stinger as in the predaceous wasps. Most parasitic flies do not have a piercing ovipositor, and must glue their eggs to the outside of the host's body. Some tachinid flies, for example, put their eggs just behind the head of a caterpillar, where they are out of reach of its menacing jaws. When the maggots hatch from the eggs, they burrow into the caterpillar's body.

With some exceptions, the hosts of both parasitic wasps and flies are immature forms: eggs, larvae, or pupae. Although most predaceous insects will eat almost any insect they can catch, many parasites are highly host-specific, parasitizing only one or a few closely related insects. There are two reasons for this specificity: First, host-specific parasites can survive only in insects to whose particular internal anatomy and physiological "climate" they are specifically adapted. Second, they must synchronize their development and seasonal occurrence with the host's. A parasitic larva may not be able to mature if it develops more slowly than its host, and hosts must, of course, be available during the usually brief lifespan of the parasite's mother.

Larvae of the tiny wasps of the family Mymaridae, inappropriately called fairy flies, are parasites in the eggs of other insects. The smallest of them—probably the world's smallest insect—is one eight-thousandth of an inch long, much smaller than some protozoa, single-celled organisms such as amoebas. It is, in fact, less than half the length of one of the larger amoebas,

Pelomyxa carolinensis. The approximately 1,300 species of fairy flies parasitize the eggs of many different kinds of insects. Some species parasitize the underwater eggs of aquatic beetles such as the predaceous diving beetles, whose voracious larvae are known as water tigers. The beetles lay their eggs one at a time in punctures in the stems of aquatic plants. As Ian Gauld and Barry Bolton explained, the adult parasites swim under water, using their wings as paddles. "Females need not exit from the water immediately after emergence [from the pupa], and mating may . . . take place underwater. Individuals are capable of remaining under water for 15 consecutive days." They can disperse to other bodies of water by flying away after climbing the stems of aquatic plants that extend above the surface.

At the other end of the size scale is a parasitic wasp of the genus *Megarhyssa.* The males are small and look something like little damselflies, but the much larger females have a body about an inch and a half long and appended to it a 3-inch ovipositor, not much thicker than a horse hair, that trails behind them in flight. They are strictly host-specific, parasitizing only the wood-boring larvae of the pigeon tremex horntail, a primitive member of the Hymenoptera, the same order to which *Megarhyssa,* the other parasitic wasps, predaceous wasps, ants, and bees belong. Horntail larvae burrow in the wood of dead trees, a half inch or more beneath the surface. John Henry Comstock, the first great teacher of entomology in the United States, described how a female *Megarhyssa* lays her eggs:

When a female finds a tree infested by this borer she selects a place which she judges to be opposite a *Trêmex* burrow and making a derrick out of her body proceeds with great skill and precision to drill a hole in the tree. At the beginning of the process the excess length of the ovipositor is coiled into a sack formed by the very elastic membrane between the sixth and seventh abdominal segments. When most of the ovipositor is deep in the wood the ovipositor sheaths which do not enter with it form a loop over her back . . . After the *Trêmex* burrow is reached, a long, thin egg flows down the tube in the slender ovipositor and is left on or near the host. The larva which hatches from the egg feeds upon the *Trêmex* larva by sucking its blood and eventually destroys it entirely. When full grown it pupates in the *Trêmex* burrow. The adult gnaws a hole through the bark to emerge. If the adult is a female, there are males waiting just outside and she is mated just after emergence or even before she can extricate herself from the hole.

Parasitic insects are often themselves parasitized by other insects. Jonathan Swift, the great eighteenth-century satirist and author of *Gulliver's Travels,* wrote the frequently quoted lines:

> So, naturalists observe, a flea
> Hath smaller fleas that on him prey;
> And these have smaller still to bite'em;
> and so proceed *ad infinitum.*

Swift was not talking biology, although he happened to make an interesting ecological point. The next two lines, seldom quoted, tell us that he was satirically expressing his opinion of literary critics:

> Thus every poet, in his kind
> Is bit by him that comes behind.

Fleas are not attacked by smaller fleas, but a parasitic insect may itself be the host of a smaller parasite, which biologists call a hyperparasite. There may even be parasites of parasites of parasites, but such chains cannot be infinitely long, as Swift's doggerel suggests.

In the 1930s Frank Marsh studied the parasites of the large, leaf-eating caterpillar of the cecropia moth in what was then an undeveloped area of southern Chicago. Many of the caterpillars were parasitized by a large ichneumonid wasp that kills the caterpillars only after the caterpillars have spun their large silken cocoons. This wasp, the primary parasite, is sometimes parasitized by a second and smaller species of ichneumonid, which in turn may be parasitized by a small chalcid wasp that is sometimes parasitized by a second species of chalcid. This is the longest chain of primary parasite and hyperparasites that I know of. It is a six-link food chain: cecropia's host plant, often a wild cherry or a silver maple, the cecropia caterpillar, the caterpillar's primary parasite, and the chain of three hyperparasites.

As a group, parasites have many physical and behavioral adaptations that further their life style. Some even have the extraordinary ability to alter their host's behavior so as to favor themselves. They control the host's brain, but no one knows how they do it. In their studies of parasitic wasps, Jacques Brodeur and Louise Vet found that groups of from 15 to 35 tiny wasp larvae live and feed within caterpillars of the cabbage white butterfly. When fully grown, they force their way through the caterpillar's skin, settle next to it on a leaf, spin a cocoon, and then molt to the pupal stage. The caterpillar is doomed to die and no longer feeds but will live on for as long as 6 days, until about the time the adult parasites emerge from their cocoon and fly off. The caterpillar's behavior during the remaining days of its life is truly amazing. It spins a protective covering of silk over the parasite cocoons and stands guard to keep away predators and hyperparasites that might harm its charges.

Equally astonishing is the story of a parasitic wasp larva related by William Eberhard in *Nature*. "On the evening that it will kill its spider host," this wasp "induces the spider to build a . . . 'cocoon web' to serve as a durable support for [the wasp] cocoon." This web, formed by unusually frequent repetitions of certain misdirected aspects of normal spinning behavior, looks nothing like the spider's normal prey-trapping orb web.

Fish, the first animals with backbones, are the ancestors of all the other vertebrates: amphibians, reptiles, birds, and mammals. So few insects live in the oceans and seas that they must be rarely seen by saltwater fish. But in fresh water, aquatic insects are varied and abundant. Insects are staples in the diets of many fish found in ponds, lakes, and streams. Indeed, the importance of insects in the diet of trout is reflected in the dedication with which some people who fish for them strive to make their lures, artificial "flies" of feathers and hair, convincing imitations of the stream-dwelling insects that trout eat.

A few fish have evolved amazing ways of hunting for insects that do not live in water. In *The Biology of Fishes,* Harry Kyle wrote that "the taste for flies has become so great, that one fish has developed into an expert sharpshooter in stalking and smothering flies—with a drop of water and mucus . . . Others again have taken partly to a land life, even to climb bushes, in search of their favourite food."

The amazing little 9-inch archerfish, the sharpshooter, lives in inshore waters, estuaries, and the lower reaches of rivers of the Indo-Pacific region. Alwynne Wheeler reported in *The World Encyclopedia of Fishes* that its ability to capture terrestrial insects near the water's edge has been well known since 1764. "It can spit drops of water with great accuracy at 1 m (3 1/4 ft) distance, and after a few ranging shots can hit a small target at 3 m (10 ft)." Although this fish lives in the water, it is also a component of terrestrial ecosystems. It eats mostly terrestrial insects and other arthropods that it knocks into the water from overhanging foliage, knocks out of the air, or leaps out of the water to seize.

The bizarre goggle-eyed fish known as mudskippers have become amphibious, but they evolved this life style quite independently of the frogs, toads, and salamanders that we think of as *the* amphibians. Mudskippers live in the brackish water of mangrove swamps of the tropical oceans of the Eastern Hemisphere. They "walk" on mudflats and climb out of the water on mangrove roots, to hunt for terrestrial insects. In *A Naturalist in North Celebes* (the former name of Sulawesi, a large island just east of Borneo), Sydney Hickson reported that he frequently found flies and mosquitoes in the stomachs of these fish. Mudskippers walk on their muscular front (pectoral) fins with some assistance from the tail. They absorb oxygen from air and water trapped in their gill cavities. Their eyes are raised well above the top of the head—even more so than the eyes of a frog—close together, and can rotate in all directions. The closeness of the eyes gives this fish binocular vision, enhancing depth perception and the ability to judge distances. This and its ability to see in all directions are major assets in catching insects.

By the early Devonian period, about 410 to 360 million years ago, the continents had been colonized by flourishing populations of plants and invertebrate animals, including many insects. Most of these insects belonged to orders (groups of families) that are extinct today. But the only vertebrates of that day, the fish, were confined to the seas and bodies of fresh water. But by the late Devonian period, amphibians, descended from lobe-finned fish, had made their way onto land and, with no competition from other vertebrates, flourished as they took advantage of the many new opportunities offered by their new habitat. They gave rise to so many species, specialized to exploit these opportunities, that the Carboniferous period, which followed the Devonian, is called the age of amphibians. From the beginning, the amphibians were predators that ate insects and the other invertebrates that had preceded them onto land. This dependence upon insects persists in many of their descendants, the reptiles that arose directly from the amphibians and the birds and mammals that arose from the reptiles.

Today only about 4,000 species of amphibians remain on earth, very much outnumbered and overshadowed by their descendants. All amphibians are carnivores, and most of them feed mostly on insects, at least in the adult stage. Most must return to the water to breed, and most have an aquatic larval stage. In the spring, frogs and toads return to ponds or even puddles to breed. The mating songs of the males, the whistle of a spring peeper or the deep "jug-o-rum" of a bullfrog, are one of the surest harbingers of spring, anticipated by many people. One of my most poignant New England memories is of the deaf Lewis Hall Babbitt, author of the *Amphibia of Connecticut,* gently touching the sides of a glass jar containing a male frog to feel the vibrations of the singing he couldn't hear.

When a female responds to one of these mating calls, the male mounts her, hugs her tightly, and spews out his sperm as she releases her eggs in the water. The difference between larvae and adults is most pronounced in the frogs and toads. With few exceptions—none in North America—their larvae are legless, swimming tadpoles that feed mainly on algae and other plants. After they metamorphose to the four-legged adult stage, frogs and toads eat mainly living insects. Less familiar than frogs and toads are the secretive, inconspicuous, and mute group of amphibians called salamanders. The differences between larvae and adults are less pronounced in the salamanders and in most species both stages are carnivorous. In his book on amphibians and reptiles, James Oliver reported that the diets of two common salamanders include about 60 percent insects, and those of the green frog and American toad about 76 percent.

Frogs and toads have a sticky tongue for catching insects. Attached to the front of the mouth, it can be flicked out and forward for some distance with great accuracy. Jörg-Peter Ewert, a German neurophysiologist, explained that toads identify their prey by movement and size. They pay no attention to objects that do not move, and scrunch down or run away at the sight of a moving object large enough to be a predator that might eat them. Their response to almost any small, moving object is to turn toward it, approach it if necessary, and then flick out the sticky tongue to catch it. (Some frogs, bullfrogs for example, eat creatures as large as other frogs or even small ducklings. These they grasp with their jaws rather than the tongue.) One experimenter induced toads to "capture" and swallow ball bearings he rolled along the floor. Such indiscriminate responses seem inappropriate to us, but in nature they serve their purpose. After all, if no interfering scientist is around, a small moving object will almost always be an insect or some other prey.

By 315 million years ago, late in the Carboniferous period, the first group of undisputed reptiles had branched off from the amphibians. Although they have a long, jaw-breaking name, Captorhinomorpha, they were small, agile, and lizard-like. In Olivier Rieppel's words, "these predators no longer had the . . . snapping bite characteristic of [some] amphibians, but had a hard crushing bite suited to deal with the chitinous armor of their insect prey." Paleontologists believe that all other reptiles evolved from these insect-eating inhabitants of the Carboniferous forests. Insectivory probably remained widespread among their now extinct descendants and persists to this day among most of their surviving descendants.

The most famous and awesome of the extinct reptiles are the dinosaurs, which first appeared during the Triassic period, about 230 millions years ago. Eventually they became masters of the earth, but then they suddenly went extinct at the end of the Cretaceous period, about 65 million years ago. Many, such as the immense *Diplodocus* and the rhinoceros-like *Triceratops,* were plant eaters, but others, notably the bipedal theropods, from which the birds descended, were predators. They ranged in size from the huge *Tyrannosaurus rex* to much smaller creatures, such as *Compsognathus,* which was no larger than a chicken. It is a reasonable assumption that this small species and other tiny dinosaurs ate insects. But even "Sue," the huge *Tyrannosaurus rex* that my grandson Benjamin Yates and I saw at the Field Museum of Natural History in Chicago, probably ate insects when it was newly hatched and very small. What we know of the American alligator, which can grow to 19 feet in length, supports this notion. Newly hatched alligators are only 8 inches long and subsist almost exclusively on insects, but when they grow larger they also eat large vertebrates—sometimes pet dogs, as Floridians know.

The fossil record tells us that reptiles were once far more diverse than they are now. Only about 7,400 species are left and, although they are ecologically important, they are overshadowed by their now dominant descendants, the warm-blooded birds and mammals. The large lizards known as iguanas eat plants, but most of the other lizards, all of the alligators and crocodiles, and all of the snakes are carnivores, and some turtles are omnivores that eat both animals and plants. The East African lizard (*Agama agama*) is also an omnivore. During the dry season it eats mostly vegetation and does not reproduce. But during the wet months from May to August, when the veldt turns from brown to green, it greatly increases its intake of the protein-rich beetles, caterpillars, and other insects that then abound. A. J. Marshall and Raymond Hook found that these lizards produce eggs only after the "long rains" have brought them protein-rich insects. But unlike iguanas and *Agama,* the great majority of lizards eat only invertebrate animals, mainly insects. Grasshoppers and spiders are among their favorite foods, but almost all lizards are opportunists that eat whatever animals are most easily caught.

An exception is the thorny devil, or maloch, of Australia, an unusually fussy feeder that subsists almost exclusively on ants.

Among the most extraordinary of the insect-eating lizards are the more than 130 species of chameleons of sub-Saharan Africa, southern Europe, and Asia. Most of them are sit-and-wait predators that live in trees and prey mainly on insects. Their opposable toes and prehensile tails give them a secure grip. Chameleons are justly famous for their ability to change color rapidly and often strikingly. They adjust their camouflage as they move about, concealing themselves from both their prey and their predators, and they also communicate with each other by changing color. A territorial male may warn off an intruding male by changing from his usual camouflage, often a dull green, to a vivid display of bright, conspicuous colors. Chameleons' tongues are marvelously adapted for swiftly snatching insects that land near them. The sticky tip has muscles that grip the prey. The thin "projectile tongue" is usually equal in length to or even longer than the body. Once the chameleon has taken aim, the tongue is shot out rapidly—usually reaching the prey insect in less than one hundredth of a second.

When my family and I lived in Cali, Colombia, for several months, we were often entertained by the little house geckos, charming pastel-colored lizards, that spent the day hiding behind pictures on our walls and came out at night to eat insects attracted by our lights. We seldom saw one catch an insect, but their tiny, dry droppings on the floor, swept up daily by our maid, showed that they were often successful. "The ability of these lizards to scale walls and to run upside down across ceilings," wrote Angus Bellairs, "is familiar to most people who have lived in the tropics; some species can even adhere in this way to the under surface of a sheet of glass." Of course, geckos evolved this ability to facilitate living in trees, but a few of them now use it to exploit a new and as yet unoccupied ecological niche, the predator-free walls and ceilings of human dwellings.

How geckos manage to adhere to smooth surfaces has been the subject of research for decades. Several hypotheses have been proposed and disproved. Recently, Kellar Autumn and his coauthors proposed a new and plausible hypothesis that is consistent with the known facts, although not yet conclusively proved. The sole of the foot of a tokay gecko is covered with a dense growth of about 500,000 tiny setae, hairlike structures. The tip of each seta is split into an array of about 500 microscopic filaments so fine that their tips can interact with a surface at the molecular level. Autumn and his colleagues postulated that a short-range attraction between molecules,

An African chameleon shooting out its long tongue to snatch a fly from a leaf.

known to physicists as van der Waal's force, is responsible for making the feet of geckos so extraordinarily sticky. The *San Francisco Examiner* quoted Autumn: "The hairs themselves are so sticky that if we were to fit a million or two of them onto the surface of a dime, it would be enough to lift a small child."

Some snakes include insects in their diets, and a few feed only on insects. In their field book, Karl Schmidt and D. Dwight Davis summarized the known feeding habits of the snakes of the United States and Canada. According to them, the tiny ring-necked snakes, the black racer, and even the venomous copperhead eat insects in addition to other prey. The closely related arboreal rough green snake and terrestrial smooth green snake feed almost exclusively on insects, and ant larvae are apparently the only food of the tiny burrowing banded sand snake of our southwestern deserts.

Birds inherited the insectivorous habit from their reptilian ancestors. The earliest known bird is the crow-sized *Archaeopteryx,* represented by several fossils found in 150-million-year-old limestone from quarries in Bavaria. *Archaeopteryx* may not have been the first bird to evolve, but it is surely a very early bird; with its mix of reptilian and avian characteristics, it is a "missing link" between the dinosaurs and the birds. Among its reptilian features are teeth and a long bony tail. Its most apparent avian characteristics are feathers and wings. The recent discovery in China of dinosaur fossils with feathers, which probably served as an insulating coat, leaves little doubt that birds descended from dinosaurs. No one knows what *Archaeopteryx* ate, but its small size and many small teeth suggest that it may have eaten insects.

There are about 9,600 species of living birds. Although bird species are relatively few compared to the huge number of insect species, birds live on all of the continents, including Antarctica, and are ecologically important because many are abundant and because, owing to their diverse feeding habits, they play many different roles in ecosystems. Hawks and owls eat the flesh of vertebrates; vultures and caracaras are carrion feeders; finches, sparrows, and grosbeaks eat seeds; and a great many birds eat insects. Among the insectivores are several hundred cuckoos, swifts, woodpeckers, and whippoorwills and their relatives. But the great majority of the insect eaters are perching birds (passerines), a large order that, according to the ornithologist Frank Gill, includes over 5,700 species, over 60 percent of all living

A loggerhead shrike with grasshoppers it impaled on thorns.

birds. Some perching birds eat almost nothing but insects, and others include them as a more or less important part of a mixed diet. Even the nectar-sipping hummingbirds eat insects. Some perching birds, such as finches, cardinals, and grosbeaks, eat only seeds as adults, but even they, with few exceptions, feed their nestlings protein-rich insects.

The first birds underwent an adaptive radiation: they evolved to form many new species that exploited as yet unused ecological niches—an ecological niche is not just the place where an organism lives, but includes all of the resources that it requires, such as food, nesting sites, and even a dust bath to rid it of lice. As the radiation progressed, insect-eating birds became specialized to seek and catch their prey in many different ways. Among our North American species, vireos, warblers, kinglets, and other perching birds with tweezer-like bills are leaf gleaners that search the leaves of trees and shrubs for caterpillars and other small insects. Nuthatches, brown creepers, and some woodpeckers take insects from crevices in the bark of trees. Creeping downward head first as they search, nuthatches find insects that woodpeckers and creepers, which always move upward, may have missed. Most woodpeckers are mainly wood and bark probers that use their chisel-like bills to cut into the branches and trunks of trees and their long barbed tongues to remove the beetle larvae and other insects they expose. Some of the New World flycatchers, including most of the species that breed in North America, are air salliers that zip out from an exposed perch to snatch passing insects from the air. Swifts, nighthawks, and swallows are gleaners of "aerial plankton," the insects and spiders that fly or float in the air from just above the ground to hundreds of feet above it. These birds divide up the space occupied by the aerial plankton. Swifts and common nighthawks hunt high in the air, in cities and towns well above the roofs of buildings. Swallows fly closer to the ground, barn swallows sometimes just above the grass in a hay field to snatch insects flushed by a mower or a passing dog.

"Darwin's finches" of the Galápagos Islands are an example on a small scale of an adaptive radiation into unoccupied ecological niches. These volcanic islands, evidently always separated from the mainland, straddle the equator in the Pacific Ocean about 600 miles west of the coast of Ecuador, the closest part of the mainland. The Galápagos were colonized only by the few organisms that managed to cross those 600 miles of ocean. Among them were the ancestors of Darwin's finches, which arrived long ago, probably a small flock that flew or was blown from the mainland. Faced with a plethora of unoccupied niches, this seed-eating finch gave rise to 14 distinct species, comprehensively described by David Lack, that are all different from any finch on the mainland. Seven are seed eaters with stout bills adapted for cracking seeds, but the bills of the 7 insectivores are variously adapted for catching and eating insects. The warbler finch has a pointy, tweezer-like bill, and so closely resembles a New World warbler that it was originally classified with that group. The most remarkable of the Darwin's finches are the woodpecker finch and the mangrove finch. Both have strong, pointed bills with which they dig in wood to expose burrowing insects, but lacking the long, barbed tongue of true woodpeckers, they use tools to extract beetle grubs from their burrows. They select a cactus spine or a twig of the appropriate size and then, as Roger Tory Peterson so vividly put it, they poke it into the insect's hole and extract the grub "much as one would spear a cocktail snack with a toothpick."

Most insectivorous birds are not immutably committed to their predominant hunting stratagem; they are opportunists that will resort to other tactics to take advantage of serendipitous opportunities. Many, including among others warblers, kinglets, and even woodpeckers and nuthatches, will dart out like flycatchers to snatch a morsel from the air. When I followed the tracks of an American crow in a field of recently germinated corn, I saw that it had stopped at the base of most plants along a row to dig for cutworm caterpillars, which feed on the corn at night but hide in the soil during the day. In wintertime, downy and hairy woodpeckers extract the internal tissues from the pupae of cecropia moths, which are in large, strong, thick-walled silken cocoons firmly attached to a twig of a tree or shrub. James Sternburg and I found that they peck a small hole through the double wall of the cocoon and into the pupa, and then lap up the semi-liquid tissues of the pupa with their long, barbed tongues. Normally, they cannot peck into promethea cocoons, which dangle loosely from a twig by a long, narrow strap of silk. But they are amazingly alert to unexpected feeding opportunities. Jim and I found that woodpeckers could pierce several promethea cocoons that were partially exposed but firmly held in place on the face of a melting snowbank that partially covered saplings from which the cocoons hung.

Some birds have other, more unusual ways of hunting for insects. The dippers of western North America (known as water ouzels in Europe) live on fast rocky streams and swim and walk under water to pick aquatic insects from the bottom. In Africa, oxpeckers walk about on the bodies of cattle, rhinoceroses, and other large mammals as they pluck ticks and insects from the skin. The greater honeyguide of Africa, one of the few animals of any sort that can digest wax, finds a colony of bees, and then leads a honey-badger or a person to the colony by flitting from perch to perch as it displays. After the nest cavity has been torn open and raided by the mammal, the honeyguide helps itself to the waxen combs left behind.

There are only about 4,500 kinds of mammalian descendants of the reptiles, but they are varied and are numerous as individuals, and many—ranging from tiny shrews and mice to gigantic grizzly bears—include insects in their diets. The aardvark—"earth pig" in Afrikaans—of the African grasslands and savannas south of the Sahara feeds almost exclusively on ants and termites. It is a large mammal, usually about 6 feet long and weigh ing around 130 pounds. Its powerful front legs with their huge, blunt claws excavate ant nests, tear open termite mounds, and dig the burrow in which it spends the day sleeping. The tapered head forms a long, slim snout. A long, sticky tongue, useful for lapping up insects and probing into termite galleries, can be extended as much as a foot from the narrow mouth at the end of the snout. Aardvarks sometimes lap up columns of foraging termites from the ground. Interestingly, since the aardvark swallows its food whole and therefore has no need to chew, its teeth are reduced in size and number and are not coated with enamel.

Ants and termites are small, but they occur almost everywhere on earth and their colonies may contain hordes of individuals, a bountiful supply of food. Consequently, unrelated animals have evolved, quite independently of each other, similar anatomies and behaviors for feeding on ants and ter-mites, a phenomenon that evolutionary biologists call convergent evolu-tion. Among them is the echidna, the spiny anteater of Australia and New Guinea, one of the egg-laying mammals, the monotremes, the most ancient and primitive of the mammals. Much like an aardvark, the echidna digs into ant and termite nests with its powerful claws and laps up its prey with a long and sticky tongue that can be protruded from its long, narrow snout. In Australia there is also a marsupial anteater, the numbat, that is simi-larly equipped for lapping up the termites it exposes by ripping open their nests. The nine-banded armadillo, the only species of armadillo whose range extends north from Central and South America into the United States, is named for the bands of armor that cover its body—armadillo is the Spanish diminutive for "an armed man." Like the aardvark, echidna, and numbat, it

A giant anteater of South America standing in front of a termite mound that it will soon break open.

is a powerful digger, has a rather long snout, a long sticky tongue, and only a few vestigial teeth. It feeds mainly on ants and beetles, which it finds by furrowing through leaf litter and loose soil with its snout. The giant anteater of South America is almost a copy of the aardvark. It is much larger than the echidna, numbat, or armadillo, but at an average weight of 30 pounds, smaller than the aardvark. It has a long, narrow snout and a sticky tongue, which can be extended up to 2 feet, with which it licks up termites and ants—as many as 30,000 per day—that it exposes by tearing open their mounds with its large, powerful front claws.

Black bears use just their wet tongues to lick up ants and other insects, but the snout and mouth of the sloth bear of Asia are specially modified to suck up the myriad ants, termites, and other insects that it exposes by digging with its strong 3-inch claws. As Adele Conover wrote in the January 2000 issue of *Smithsonian*, the holes this bear digs are sometimes huge, "deep enough for a man to stand in up to his chest." The sloth bear's extended snout, long, flexible, and "almost prehensile" lips, concave palate, and the gap left by missing front teeth form a functional vacuum device for sucking up the tiny insects that are a significant part of its omnivorous diet. The bear first blows into a nest of ants or termites and then vacuums up its prey with its nostrils closed to keep from inhaling the insects. The population of these wonderful bears is in steep decline. They are killed for their gallbladders, which Asians value for their supposed medicinal powers, and the cubs are kidnapped to be trained as the dancing bears that are popular in India.

Bats are the only living mammals that can fly like birds—not just glide like flying fish or flying squirrels. Most birds are active during the day but, generally speaking, bats are not. They are the masters of the night sky. Some bats eat fruit, fish, or even blood, but the 45 species in America north of Mexico feed almost exclusively on insects, most of them on flying insects they catch on the wing. Bats have functional eyes but in the dark can locate insects and avoid obstacles by means of echolocation, or sonar. As they fly, they emit sounds that echo back to their sensitive ears from flying insects or obstacles such as twigs. The bat "sees" insects by means of these echoes. Unimpaired bats, M. Brock Fenton has shown, can catch insects and avoid obstacles in total darkness, but bats that have their ears plugged cannot do so. Bat cries are too high-pitched to be heard by humans, a good thing, because close to a bat the sound pressure level of its cries may be more than 110 decibels, 14 times as noisy as the traffic on a busy highway.

In laboratory experiments, Shelley Kick found that the effective range of echolocation by the common big brown bats of North America depends upon the size of the "target." They detected nylon spheres about two-tenths of an inch in diameter at a distance of about 9.5 feet and similar spheres about three-quarters of an inch in diameter from 16.7 feet away. Their ability to detect insects is presumably similar. A bat's relatively short detection range obviously limits its success as a hunter, especially because many moths, the principal prey of many bats, can detect bat cries from about 130 feet away and take evasive action before a bat comes close enough to detect them.

Mexican free-tailed bats are famous for their immense colonies in caves in the southwestern United States. Every year hundreds of thousands of tourists come to Carlsbad Caverns National Park in New Mexico to view the mass exodus of as many as a half million of these bats from their cave at dusk. On a day in 1936, an observer estimated—possibly overestimated—that a column of about 8.7 million Mexican free-tailed bats poured out of an opening from Carlsbad Caverns during a period of 15 to 20 minutes. By 1973 the population had shrunk to 200,000. Poisoning by DDT caused much of this decline. There was only a slight recovery after DDT was banned in the United States in 1972, probably because this nonbiodegradable insecticide is still being used in Mexico, where most of these bats spend the winter. Nevertheless, there has been some increase in this bat's numbers; David Roemer, a biologist at Carlsbad Caverns National Park, told me that in 1999 the population was about 500,000.

Roemer said that the first bats return to Carlsbad in late March, and that most of them have left by the end of October. They are essentially a maternity colony that includes mainly females, most of them nursing a pup. The mothers require a great deal of energy to secrete milk and to fly throughout the night, making round trips of as much as 60 miles from the cavern to a feeding site. They may eat from a half up to their full body weight of insects in a night—mostly small moths and beetles. The 8.7 million bats that left Carlsbad Caverns on that night in 1936 weighed about 12 grams each, a total of about 230,000 pounds, or 115 tons. If we assume that the average bat ate only half its own body weight in insects, on that one night the bats of the Carlsbad Caverns ate at least 58 tons of insects.

Among the most remarkable insect eaters are shrews, tiny mouselike creatures. The shorttail shrew of the eastern half of southern Canada and the United States weighs little more than half an ounce and, including the tail, is only about 4.5 inches long. The newborn babies are naked and about the size of a honey bee. But ounce for ounce this shrew is among the most insatiable predators on earth, for its size far fiercer than a tiger or lion. It hunts both by day and by night and because of its high metabolic rate must each day eat a quantity of food, mostly insects, equivalent to its own body weight. Shrews and moles are the North American representatives of the mammalian order Insectivora (insect eaters in Latin). It has been said that it's a good thing that shrews are not as big as dogs, for no large animal would be safe from them. But the truth of the matter is that if shrews were that big, their metabolic rate would be much lower, and they would not have to be nearly so voracious.

The salivary glands of the shorttail shrew secrete a powerful venom, very toxic to mice. Irwin Martin demonstrated experimentally that it also immobilizes but does not immediately kill insects, the usual prey of these shrews. Martin found that crickets and cockroaches envenomated by shorttail shrews lived for as long as 5 days, but were immobilized and could be stored with no possibility they would escape. They did not decompose and thereby lose nutritional value before the shrew ate them a day or two later.

The ancestor of the primates, the order to which monkeys, apes, and people belong, was a shrew or shrewlike insectivore that lived more than 65 million years ago. The most likely candidate is a fossil tree shrew that resembles those that still survive in Asia and the East Indies. Tree shrews and primates have similar life styles. Many tree shrews are arboreal, as are most primates, and almost all eat a diet of insects and fruits, as do all but a few primates.

The more primitive primates, the prosimians, include the lemurs and aye-ayes, found only on Madagascar, and the pottos, lorises, and galagos of tropical Africa and southern Asia. All are arboreal. With the exception of a few lemurs that seem to be strictly vegetarian, all include insects in their diets. Some feed mainly on insects and others make insects an important part of an omnivorous diet that may also include fruits, leaves, and in a few species bird eggs and small vertebrates. The dwarf lemur is remarkable because in the wet season it eats flowers, fruits, leaves, and a few insects, but in the dry season switches to a diet composed mainly of planthoppers and the honeydew they excrete.

All of the prosimians, like their more advanced anthropoid relatives, monkeys and apes, have major adaptations that facilitate living in trees and capturing insects to supplement the basic primate plant diet. Their vision is acute and their eyes are close together on the front of the face rather than on opposite sides of the head as in other mammals. This arrangement gives them the binocular vision and acute depth perception that make possible the precise coordination of hand and eye necessary to grab a swiftly moving insect. They also have opposable thumbs on both hands and feet for deftly holding insects, picking fruit, and gripping twigs and branches as they move through trees. These traits are the essence of being a primate, and most of them persist in humans, who use them to practice such skills as tool making, planting seeds, hunting with weapons, writing, and painting.

These are all broadly applicable adaptations rather than highly specialized ones that can lead to evolutionary cul-de-sacs. The exceptional aye-ayes of Madagascar do, however, have a highly specialized adaptation for capturing insect larvae that live in decaying wood, an extremely long and thin third finger with a long, sharp, talon-like nail. An aye-aye taps the surface of the wood with its long third finger and listens for the sound of burrowing larvae. When it locates a larva, it bites through the wood with its sharp and powerful front teeth and inserts its long third finger to crush and extract its prey.

Insect eating is also widespread among the anthropoid primates. Of the 182 species of monkeys, only 31 species—the howler monkeys of the New World and the Old World langurs, proboscis monkey, and colobus monkeys—are strict vegetarians. The other 151 species all eat insects. Tarsiers eat mainly insects, and the others include them to varying degrees in omnivorous diets that consist mainly of fruit and other plant parts, and in some species also birds, their eggs, frogs, or other small vertebrates. The baboons are omnivores, but ignore other foods when there are massive outbreaks of grasshoppers and other insects. The drill and the mandrill forage on the ground, as do baboons, and have been seen turning over stones and debris as they search for insects.

All apes eat at least some insects. The gibbons, the orangutan, and the chimpanzee are omnivores that regularly eat insects. Even the largely vegetarian gorilla and the pygmy chimpanzee eat the occasional insect. Chimpanzees have been famous as eaters of insects since the early 1960s, when Jane Goodall discovered that they use tools to capture termites and ants. In Tanzania she found that chimpanzees feed on mound-building termites during the early part of the rainy season. At that time, the colonies produce thousands of winged males and females of the reproductive caste that swarm forth to found new colonies. Workers extend exit passages to the surface of the mound and lightly seal them until conditions favor the release of the reproductives. As Goodall described the feeding process, a chimpanzee stands or squats on a termite "hill" as it "peers at the surface and, where it perceives one of these sealed-off holes, scratches the layer of soil from the opening with its index finger." It then pokes a tool, often a thin twig or a length of grass stem, into the hole. After a moment it pulls out the tool and draws it sideways through its mouth to pick off the termites that cling to it. The tools are selected and altered to fit their purpose. Goodall watched a male chimpanzee that found no tools near a termite mound go to a clump of dried grass some yards away and carefully select four stems and carry them back to the mound. Before using a stalk to "fish" for termites, he trimmed it to a length of about 9 inches. He tucked spare stems in his groin or laid them on the ground. She saw chimpanzees using a similar method to fish for ants that build "a round nest, about the size of a football, up in a tree." They broke a stick from a dead branch, poked it into an ant nest, and then withdrew it to lick off clinging ants.

Insects are eaten by people of most cultures, sometimes just as a delicacy served in gourmet restaurants but often as a nutritionally significant component of the regular diet. This is to be expected. After all, most other primates eat insects, and insects are rich in protein and all or most of the other nutrients required by humans and other animals. In an article in the *Bulletin of the Entomological Society of America,* Gene DeFoliart wrote that caterpillars may contain 28 percent protein—not bad when compared to beef at 26 percent, kidney beans at 23 percent, and fish at 16 percent. People often eat insects because they like them, not because they have to. Of the 200 species of insects eaten in Mexico, many are served as delicacies in fine restaurants. Pacific Islanders who eat an abundant and balanced diet of fruit, vegetables, fish, and meat nevertheless eat a wide variety of insects by choice.

But people of Western cultures are strongly prejudiced against eating insects and are usually not aware that insects are nourishing and regularly eaten by most other peoples. We make an exception of honey, an insect product, and the occasional brave soul may nibble a chocolate-covered ant on a dare, but most members of our culture abhor the very idea of eating insects. This attitude seems incongruous. We do, after all, relish crustaceans such as shrimps and lobsters, insect relatives that can—not unreasonably—be thought of as big sea-going bugs. Furthermore, we constantly but unknowingly eat insects without ill affect. Recognizing that it is economically impracticable to keep all foods "pure," the U.S. Food and Drug Administration tolerates some insect contamination in foods that are to be sold. For example, peanut butter is allowed to contain up to 30 insect fragments per 100 grams (3.5 ounces) and frozen broccoli up to 60 aphids per 100 grams.

The sub-Saharan Africans' fondness for termites was picturesquely conveyed by Herbert Noyes: "As a Bayere chief, who, calling on Dr. Livingstone, was offered apricot jam, remarked: 'Ah, you should try roasted Termites.' So, in Central Africa, natives welcome the rainy season in very much the same way as obese British gourmands hail the advent of the oyster season and journey from afar to gorge themselves at Colchester on living food." (I think that anyone who would eat a raw oyster should be willing to try insects.) In *Insects as Human Food,* Friedrich Bodenheimer, an Israeli entomologist, described some of the ways in which Africans catch termites. Children in Zaire catch them the way chimpanzees do. They poke a palm leaf into a hole in a mound that has been broken open and withdraw it to collect the termites that cling to it. Bodenheimer described the making of an ingenious trap for collecting large numbers of winged reproductives as they leave the mound on their nuptial flight: "They [the trappers] tightly enfold the termite mound in several layers of . . . broad leaves . . . the interstices soon being closed by the termites, which usually join the inner leaves to the nest. A projecting pocket, built on one side of the leaf cover, serves as a trap; for when the winged termites begin to swarm, they find no egress and finally drop in masses into the pocket from which they are scooped out." Termites are eaten raw, boiled, broiled, or roasted in an iron pot. According to DeFoliart, in Zimbabwe many people of European background eat termites but not in the quantities that the local people do.

Some Native Americans used to stage cooperative hunts to round up large quantities of grasshoppers, one of their favorite foods and a staple for some. In an article on the food products of Native North Americans in the 1870 annual report of the U.S. Commissioner of Agriculture, there is an account of such a roundup:

> By the Diggers of California and the Plains grasshoppers are caught in great numbers. When the insect attains its best condition, the Indians select some favorable locality and dig several little pits; in shape somewhat like inverted funnels, the aperture being narrower at the surface than at the base, the object being to prevent the insect which chances to tumble in from hopping out again. The pits being ready, an immense circle is formed, the surrounding grass is set on fire, and the Indians, men, women, and children, station themselves at proper intervals around the fiery belt, keeping up a continual ring of flame, until the luckless grasshoppers are corralled in the pits or roasted at the brink. They are eaten after being mixed with pounded acorns, and constitute one of the national dishes. Grasshoppers are sometimes gathered into sacks saturated with salt, and placed in a heated trench, covered with hot stones, for fifteen minutes, and are then eaten as shrimps, or they are ground and put into soup or mash . . . Grasshoppers are also pounded up with service, hawthorn, or other berries. The mixture is made into small cakes, pressed hard, and dried in the sun for future use.

This author expressed the unreasoned European prejudice against insects as food: "The review of the articles of food consumed by the Indians will show that many of the substances are not only distasteful but disgusting to civilized persons, and many, also, are not of a nutritious character. It is barely possible that there is a flavor in some of these undetected by the whites because untried."

Insects are definitely not disgusting to all "civilized persons," unless you consider only members of European culture and its offshoots in places such as the United States, Australia, and French Canada to be civilized. In Japan, insects ranging from grasshoppers to honey bees are widely consumed. An American who ate in one of the best restaurants in Tokyo very much enjoyed the appetizer of fried bees, whose flavor he described as being halfway between that of pork cracklings and wild honey. Pupae of the silkworm are a byproduct of silk production. They are what is left after the silk of the cocoon has been unwound. In Asia they are not wasted. According to Bodenheimer, in China the pupae are dried to preserve them for future use. The dried pupae are "softened in water and then fried either with chicken eggs in the form of an omelet or simply fried with onions and sauce." Pupae are also defatted and ground into a meal that is eaten by humans or fed to poultry. According to DeFoliart, in India alone over 20,000 tons of this meal is produced annually.

The eating of insects can be traced back to biblical times. And the Bible is, of course, an essential part of the foundation of Western culture. Leviticus 11:20 admonishes the Israelites that insects are unclean: "All winged creeping things that go upon all fours are an abomination unto you." Deuteronomy 14:19 repeats the warning: "And all winged creeping things are unclean unto you: they shall not be eaten." No mention is made of what makes creeping creatures unclean. But it seems that the great swarms of locusts that often descended upon biblical lands were just too much food to resist. Leviticus 11:21–22 specifically exempts locusts, grasshoppers, and their relatives: "Yet these may ye eat of all winged creeping things that go upon all fours, which have legs above their feet, wherewith to leap upon the earth . . . the locust after its kind, and the bald locust after its kind, and the cricket after its kind, and the grasshopper after its kind."

6

GIVING PROTECTION

Many ants and some wasps and bees intentionally, inadvertently, or even unwillingly and with fatal outcome protect numerous birds, some caterpillars, and many aphids and related sap-sucking insects against predators and parasites. Some relationships between protector and protected are beneficial to both participants. For example, ants are rewarded with food by the caterpillars, aphids, scale insects, and treehoppers they protect from predators and parasites. Other relationships are benignly one-sided. Colonies of stinging wasps are an umbrella of protection for birds that nest close to them, but the wasps are neither benefited nor harmed by the birds. Some interactions, such as those that involve birds crushing ants to make use of their insecticidal secretions, are blatantly and destructively one-sided.

 The amazing behaviors by which many birds, including some quail and a multitude of perching birds, use or solicit ants to free their bodies of lice and parasitic mites are known as anting. Birds ant in two quite different ways. Many use the active method of anting, picking up ants with their bills—usually crushing them—and wiping them against their plumage to anoint it with ant secretions that are toxic to mites and lice. But a few birds, mainly some crows, thrushes, and finches, use the passive method of anting. They squat or lie on an ant nest and allow the insects to board their bodies and roam, unharmed, through their plumage to search for and destroy external parasites.

In a brief 1947 note in the journal *British Birds,* W. Condry described passive anting by an inexperienced, hand-reared carrion crow. He placed it on the ground near a large slab of stone under which there was a horde of ants. After he turned over the stone, the crow immediately became obviously excited. It hesitated for a few seconds but then "stepped into the middle of the swarming ants . . . When some of the ants found their way via his legs to his feathers, the bird showed apparent pleasure and slowly settled down among the ants like a brooding hen, with wings outspread and tail fanned." Then, in Condry's words, it acted as if "swooning," slumping down and lowering its head until it was flat against the ground. Condry concluded that his crow's behavior was innate, programmed in the genes. After all, it had never before seen an ant or another bird anting.

Active anting coats the plumage with noxious secretions. In fact, birds use only ants that secrete formic acid or other toxic fluids. In an article in the *Wilson Bulletin,* Leon Kelso and Margaret Nice wrote that, in a little-known article in Russian, Vsevolod Tmbhili reported that he found drops of liquid that smelled of formic acid on the feathers of steppe pipits that had been actively anting to rid themselves of mites. The pungency of all substitute substances that birds sometimes smear on their plumage when ants are unavailable leaves little doubt that their purpose is to deter parasites. Among those listed by Lovie Whitaker and other researchers are toxin-oozing millipedes; grasshoppers, which regurgitate a noxious brown liquid; wasps; raw onion; lime fruits; burning matches or tobacco; prepared mustard; vinegar; and mothballs.

A. H. Chisolm vividly described anting by a flock of starlings in Australia:

Each bird snatched up an ant from a gravel path and dabbed it quickly first under one wing and then under the other, after which the insect usually was dropped . . . All the actions of the Starlings were very rapid. Two birds in particular nearly fell over backward while rearing up smartly and applying ants beneath their tails. I saw no evidence of the insects being eaten. When the birds departed, the path was bespattered with dead and maimed ants, some 50 percent of which had their abdomens burst, while the others were more or less intact. The species was . . . a type that bites and sprays quickly, which possibly helps to explain the rapidity of the Starlings' actions.

Some birds, according to K. E. L. Simmons, of the Department of Psychology of the University of Bristol in England, are more sophisticated and use ants with considerable finesse. The Pekin robin holds an ant by its thorax, leaving free the abdomen, which contains the formic acid glands and the mechanism that sprays the acid. Depending upon from which side of the bill the abdomen protrudes, the bird turns its head and applies the ant to the appropriate wing. At first it uses the ant as a "bug bomb," holding the struggling creature so that it sprays formic acid onto the undersurface of the flight feathers. But eventually the ant dies. Then the bird daubs the fluids that ooze from the ant's crushed body onto its feathers.

Active anting really does kill parasites. Lice sprayed with formic acid by a German biologist died after a few minutes, while control lice sprayed with water were not harmed. Dubinin's examination of birds that had anted in nature provides even more convincing proof. In the summer of 1943, Dubinin watched four steppe pipits standing on ant hills as they picked up wood ants and rubbed them on their flight feathers. After the birds had been anting for 20 to 40 minutes, he collected them and four other steppe pipits that had not been anting. Only the feathers of pipits that had anted smelled of formic acid and had dead mites sticking to them.

> The rest of the mites on the four pipits which had anted were crawling over the feather surfaces at random. On the four non-anting pipits, the mites remained undisturbed . . . Of 642 live feather mites taken from the four anting pipits, 163 died within 12 hours, and 8 more within 24 hours. Simultaneously from the four control pipits taken at the same time from near their nests, 758 live [mites] were collected, of which only five died within 12 hours and two more within 24 hours.

Dale Clayton and J. G. Vernon showed that a substitute substance used by "anting" birds kills bird lice. After watching a common grackle "ant" with a piece of lime fruit, they tested its effect on bird lice. They put a feather occupied by 52 lice in a covered glass petri dish with a slice of lime. Nine hours later, 35 of the lice had died and the rest were obviously moribund. But after 9 hours only one of 31 lice held in a control dish lined with moist paper had died, and all but one of the others looked healthy. Testing lime juice and lime peel separately showed that only the peel is toxic to lice.

🐜 🐜 🐜 In both the Old and New World tropics, some birds nest near or even within the nests of colonial insects such as termites, ants, or bees. Some birds excavate nest holes in the large, arboreal nests of termites, often called termitaria. They provide convenient shelter for a bird's nest, but it is not readily obvious that they offer any other protection to eggs or nestlings. Termites do not pour out of their nests to attack approaching animals, but they do defend their nests against animals that try to break in. J. G. Myers reported that when he broke through the wall of a termitarium, "the workers became aggressive, and bit sharply, often leaving the head attached to my skin." Thus defense of the termitarium is also an inadvertent defense of any bird nest that it may contain. A surprising variety of birds nest in arboreal termitaria, among them many but not all species of parrots, kingfishers, and woodpeckers and a few trogons, puff birds, jacamars, and cotingas. Excavating in termitaria is relatively easy because they are constructed of "carton," a relatively soft material that termites make by gluing together feces and bits of wood.

The orange-fronted parakeets of Central America sometimes nest in natural hollows in trees but they also excavate cavities in termitaria. At first, Wolfgang von Hagen reported, the termites attack the intruding birds, but the insects become amazingly tolerant of them once they have become established. The parakeets bring in no nesting materials, but lay their eggs on the bare floor of the cavity.

In Australia, Keith Hindwood noted, kingfishers frequently nest in arboreal termitaria, and near Sydney "most termites' nests of suitable sizes are usually occupied during the breeding seasons of the various kingfisher species." Near Brisbane, 26 pairs of kingfishers of four different species, including the famous kookaburra, nested in termitaria in an area of no more than 50 acres of open forest. According to Hindwood, kingfishers begin excavating by flying at the termite nest and smashing into its outer wall with their strong beaks. The male and female work alternately, each calling loudly as it awaits its turn. After they have punched a large enough opening, they cling to its side and continue to excavate, jabbing away with their beaks as they dig a tunnel and then hollow out a nesting chamber, pushing out loose material, which falls to the ground. The birds complete the nest in about 2 weeks and lay eggs a few days later. Hindwood went on to say:

> As the excavation progresses so the termites instinctively seal the exposed portions of their nest. The galleries facing both the tunnel and the nesting chamber are thus plugged so that there is no actual contact between the birds and the insects. Both are able to carry out their respective activities without interference, one from the other. I have cut open and examined several arboreal nests that have been used by kingfishers (Kookaburras and Sacred Kingfishers) and, for the most part, the ends of the galleries have been closed . . .

Thus it will be seen that the "problem" of how nesting birds can live

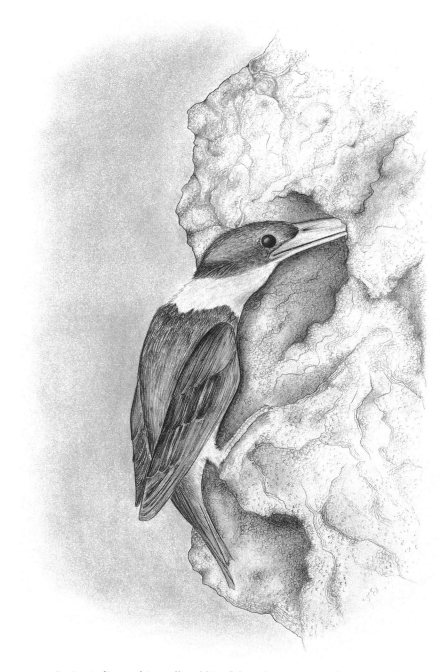

In Australia, a white-collared kingfisher about to enter the nest cavity
it and its mate excavated in an arboreal termite nest.

within a termitarium containing hundreds of thousands, perhaps millions, of insects, is really no problem at all.

On previous pages you met stinging ants that protect the trees on which they live by driving away plant-eating animals such as insects and even browsing deer and cattle. The ants, unable to distinguish between herbivorous and predaceous mammals, attack and drive away both with equal vigor. Many predators, such as ringtails, and omnivores, such as certain monkeys, eat eggs and nestlings. Consequently, birds that nest in ant-protected trees are less likely to be victims of vertebrate marauders than are birds that nest in unprotected trees. Through natural selection, birds of several species have "learned" to take advantage of this opportunity by more or less consistently nesting in association with stinging ants.

Daniel Janzen found that birds prefer to nest in bull's-horn acacias rather than in other shrubs that are not protected by ants. With the help of another observer, he counted nests in the shrubs on both sides of 22 miles of road in southern Mexico. The observers were in a slowly moving car, but nests were easily seen because it was the dry season and shrubs and trees were almost bare of leaves. They spotted 253 nests of several different species of birds. One hundred and fifty-nine of these nests, about 62 percent, were in bull's-horn acacias. Janzen estimated that bull's-horn acacias constituted only about a tenth of 1 percent of the volume of vegetation on the roadsides. Thus about a thousandth part of the vegetation available held over 60 percent of the nests found. But why don't the ants drive off the nesting birds? The most likely explanation is that they soon habituate to the frequent and gentle disturbances caused by the birds, although they give a vicious response to an unfamiliar disturbance if the foliage and twigs of their acacia are disturbed by people or other animals.

There is no direct proof that acacia ants protect nesting birds, but it is very likely that they do, and as you will read, there is good evidence that birds benefit from nesting near colonies of stinging wasps, which are painfully effective evictors of intruding vertebrates. According to Frank Joyce, an ecologist, at least 100 species of perching birds of the Old and New World tropics often nest within 5 feet or less of a colony of wasps or meliponine bees. Among them are some species of cuckoos, tyrant flycatchers, Old World flycatchers, South American ovenbirds, sunbirds, wrens, flowerpeckers, bananaquits, orioles, caciques, oropendolas, finches, and weavers. There is convincing evidence that this choice of nest site gives at least some of them significant protection against nest predators. The protectors with which these birds associate are all social, form large colonies, and with painful stings or bites drive away intruding mammals or other vertebrates, some of which are likely to loot birds' nests. The toxicity of the venom that wasps inject with their stingers varies with species, but some venoms are so potent that just a few stings can cause illness in humans. The effect on a small mammal such as a monkey is surely much greater.

Meliponine bees, also known as stingless bees, inflict painful bites with their jaws. Although they cannot sting, they very effectively defend their nests against human intruders. "They swarm over the body," wrote Edward O. Wilson, "pinching the skin and pulling hair, occasionally locking their mandibles in catatonic spasms so that before the grip can be broken, their heads tear loose from the body." Some tropical American species "also eject a burning liquid . . . which in Brazil has earned them the name *capafogas,* meaning 'fire defecators.'"

When my family and I vacationed near Port Maria on the north coast of Jamaica, little nectar-eating bananaquits came to our outdoor table and perched on the rim of the sugar bowl as they helped themselves. These birds, I learned later, nest at almost any height in many different kinds of shrubs and trees. Many of their nests, Joseph Wunderle and Kenneth Pollock report, are within a yard of a nest of polybiine wasps, fierce stingers that aggressively defend their colony against vertebrates that shake the branch from which their nest hangs. Since not all bananaquits nest near wasp nests, Wunderle and Pollock were able to compare the nesting success of pairs nesting near or away from polybiine colonies. Over 43 percent of nests not near a wasp colony were destroyed by predators, probably snakes, grackles, or rats. But in the same area only 13 percent of the nests on the same branch with a wasp nest and within a yard of it had been attacked by predators.

Other experiments done by Frank Joyce in the dry Guanacaste Conservation Area of Costa Rica leave no doubt that birds can benefit from the presence of a nearby colony of a notoriously aggressive, stinging polybiine wasp that, according to Joyce, viciously attacks people and other vertebrates that come within a few yards of its nest. These wasps characteristically build their nests near colonies of acacia ants. The wasps probably choose to nest in aca-

cias because the aggressive acacia ants protect them from their chief enemies, army ants—against which they are defenseless.

The common and conspicuous rufous-backed wrens almost always nest in
ant acacias, 93 percent of 338 nests surveyed by Joyce in Costa Rica. Fourteen percent were within 5 feet of a polybiine wasp's nest. Joyce's experiments showed that wren nests protected by a wasp colony were more likely
to produce offspring that survived to leave the nest than were nests not protected by wasps. He first located nests in ant acacias that did not also harbor a
wasp colony. Then, working at night and wearing a beekeeper's suit, he
taped an active wasp nest near—usually within 2 feet of—each of a randomly selected group of wrens' nests. No wasp nests were placed near a randomly selected control group of the same number of nests in the same area.
The results are unequivocal. Nestlings survived in about 54 percent of the
nests protected by wasps, but survived in less than 10 percent of the unprotected nests. How much protection the wrens receive from the acacia ants
themselves cannot be ascertained from these experiments, but the wasps obviously give protection over and above that afforded by the ants alone.

Many vertebrates, among them snakes, birds, and mammals, eat the eggs
and nestlings of birds. But according to Joyce, white-faced monkeys were
the most important predators of rufous-naped wrens' nests at Guanacaste.
He noted that in 59 percent of 184 failed nesting attempts, the nests had
been damaged in a manner typical of the way these monkeys raid nests.
Scott Robinson, an ornithologist at the University of Illinois, observed that in
Peru three different species of monkeys try to raid the nests of yellow-
rumped caciques, but are repelled by stinging wasps from nearby colonies.

In *Our Search for a Wilderness,* Mary and C. William Beebe wrote that in
Venezuela yellow-backed caciques show "a real intelligence in the selection of a site for their nests." They nest in colonies of as many as 150 pairs
per tree, and their long, pendulous nests, which swing from the tip of a
branch, are impossible to conceal. Consequently, they need outside protection against nest predators. Over 90 years ago the Beebes observed that in
northeastern Venezuela yellow-backed caciques nest in trees where they are
protected by either wasps *or* humans. If they nest in the forest, they choose
trees with wasps' nests, sometimes placing their nests so close to a wasp nest
that wasp and bird nests bang together in the wind. "These insects," the
Beebes wrote, "are usually large and venomous, and one sting would be
enough to kill a bird; indeed a severe fever often ensues when a man has

been stung by a half dozen." The wasps do not sting the caciques. When the birds "returned to their nests with a rush and a headlong plunge into the entrance, the whole branch shook violently. Yet the wasps showed no excitement or alarm . . . but when I reached up and moved the branch gently downward, the angry hum [of the wasps] sent me into the underbrush in haste."

The cacique-wasp association has been going on for hundreds of thousands if not millions of years. But people occupied South America only after they crossed the land bridge that once connected Siberia and Alaska, probably about 20,000 years ago. During this relatively short time, the caciques have "learned" that humans can substitute for wasps as protectors. They nest in villages and they "choose a solitary tree which fairly overhangs some thatched hut . . . and weave their nests and care for their young almost within arm's reach of the thatched roofs." The villagers never trouble the birds, and "no monkey dares venture here, and mongrel dogs keep off all the small nocturnal carnivores."

🐜 🐜 🐜 You will recall that elaiosomes, nutritious appendages of seeds, induce ants to collect these seeds, carry them to their nests, eat the elaiosomes, and then discard the otherwise intact seeds on their trash piles. As amazing as it may seem, a few insects have discovered that it is useful to have elaiosome-like structures, known as capitula, on their eggs. According to J. T. Clark, certain species of walkingsticks, so called because of their deceptive resemblance to twigs, make no attempt to guard or hide their eggs, as do some other walkingsticks. As they crawl along feeding on the foliage of trees, they simply let their eggs fall to the forest floor. Lying plainly visible on the ground, the eggs are subject to destruction by certain parasitic cuckoo wasps that insert their eggs into walkingstick eggs. Some of the stick insect eggs are rescued by ants that bring them to the safety of their nests to eat the capitula. They do not harm the eggs, and when they hatch, the little walkingsticks make their way to the surface and crawl up into the trees. In Australia, L. Hughes and M. Westoby buried capitulum-bearing walkingstick eggs at various depths in potting soil and discovered that 98 percent of the newly hatched nymphs from eggs buried at a depth of 2.4 inches or less made it to the surface. About 11 percent reached the surface even when buried almost 5 inches deep.

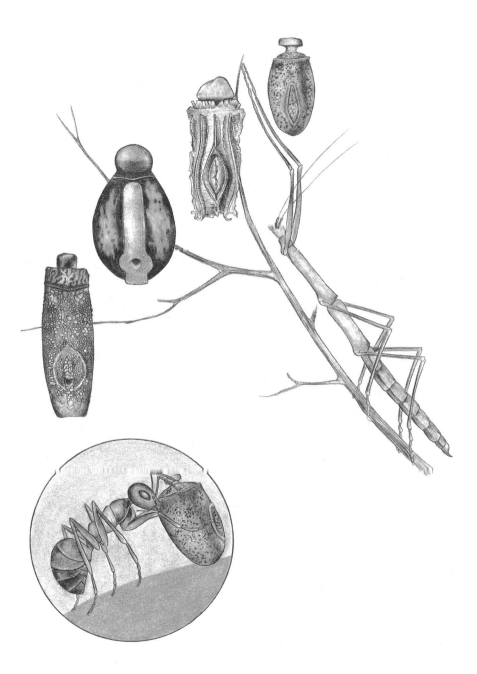

Capitulum-bearing eggs of four species of walkingsticks and an ant
about to carry a walkingstick egg to its nest.

🐜 🐜 🐜   All insects of the order Homoptera suck sap from green plants, and most green plants are host to one or more of them. Some, among them aphids, certain treehoppers, and some other gregarious species, cluster in colonies that usually include only the offspring of one mother. The majority of Homoptera take sap only from a plant's phloem tubes. Among these phloem feeders are treehoppers, leafhoppers, planthoppers, aphids, mealybugs, and scale insects. Phloem sap carries nutrients produced by the leaves down to the stems and roots. It is under positive pressure and can be sipped with little or no suction. Indeed, as J. S. Kennedy and T. E. Mittler showed, if you snip an aphid's body from its beak, leaving the latter embedded in a phloem tube, sap continues to flow from the beak although the sucking apparatus was removed with the body. Phloem sap is a dilute solution of nutrients in water, rich in sugars but also containing low concentrations of other nutrients. Homopterans excrete most of the water and some of the sugar. Their excrement is thus a clear, sweet-tasting, liquid—honeydew.

A colony of Homoptera excretes copious quantities of honeydew, often visible as a glaze of droplets on the leaves of trees and other plants. The amount produced is prodigious. Jacques Auclair reported that the sap intake of several aphids ranged from 10 to 133 percent of the individual's body weight per hour, and that most of it is passed out as honeydew. Many insects and even some people include this abundant and virtually ubiquitous substance in their diets. The manna from heaven that saved the Israelites from starvation was probably honeydew, and to this day, some peoples of the Near East use honeydew, which is known as *man* in Arabic, to make candy. Among the insects that eat honeydew are many species of wasps and bees. Honey bees, for instance, collect large quantities and convert it to honey. Many moths and butterflies suck honeydew from leaves with their long, thin, soda-straw mouthparts, its sugar providing a ready source of energy to fuel their flight "motors." In their own unique way, many flies eat honeydew. Their stubby mouthparts end in two spongelike pads that soak up fresh, liquid honeydew or even a crust of dry honeydew after it has been liquefied by the fly's regurgitant or saliva.

But ants are by far the outstanding users of honeydew. Bert Hölldobler and Edward O. Wilson noted that it is eaten by the great majority of species in the three largest and most evolutionarily advanced groups of ants, which probably include about 80 percent of the almost 9,000 known species of ants. Ants differ in how much they depend upon honeydew and in the closeness and complexity of their interactions with honeydew producers.

Many, perhaps most, are mere opportunists that collect fallen honeydew where, when, and if they can find it. But others go a step further and solicit it directly from its producers. When they "milk" an aphid by stroking its hind end with their antennae, the aphid responds by oozing out a droplet of honeydew. Aphids of some species accommodate temporarily absent ants by retaining expelled droplets in a "basket" of bristles surrounding their anus. Some ants guard "wild" colonies of Homoptera against predators, sometimes even building shelters for them. A few "domesticate" aphids or other Homoptera and care for them just as people care for cows. Generally speaking, ants and honeydew-producing homopterans have mutually beneficial associations. The Homoptera provide food, while the ants provide protection. As ecologists put it, ants provide the homopterans with "enemy-free space."

While some ants collect little or no honeydew, others collect large quantities and are to varying degrees dependent upon it. Gerhard Zoebelein found that the common wood ants of Europe cannot survive without the honeydew they collect from colonies of aphids and other homopterans that live in trees near their nest, a huge mound of twigs, pine needles, and bark that may be more than 6 feet high and house a million workers. The ants capture and eat many insects, but they collect even greater amounts of honeydew, sometimes well over a thousand pounds per nest per year, which contains a dry (water-free) weight of well over 200 pounds of sugar.

But do we really know that Homoptera benefit from being tended by ants? One benefit is that ants remove honeydew that might otherwise gum up an aphid's body. When not tended by ants, most aphids keep themselves clean by flicking droplets of honeydew away with their hind legs or "shooting" them away with forceful contractions of the rectum. But some aphids have become so completely dependent upon ants that they have lost the ability to keep themselves clean and must rely upon their attendant ants to do the job for them. Ants also protect aphids and other homopterans from parasites and predators. Some ants provide shelter for their homopteran "cattle," thereby giving them at least some protection from the weather and even more protection from their enemies. Others do not shelter their "cattle" but leave them on the "open range," the leaves of whatever species of plant hosts the aphid. So it is with wood ants. They guard free-ranging Homoptera, protecting them by killing or driving off other insects that come to the tree on which the aphids live.

Catherine Bristow measured the effect of protective ants on colonies of treehoppers that lived on New York ironweed, a tall wildflower of the eastern United States. Over 95 percent of these colonies had attracted attending ants within a week of their initiation. Bristow compared the fate of colonies tended by ants with the fate of colonies from which ants had been excluded by a barrier of Tanglefoot. The ants were very effective guards. She found very few predators on ant-guarded plants, but over 15 times as many on plants without ants. When she placed predators on plants with ants, they were soon discovered and attacked. On average, predators survived for only 16 minutes. Because the members of a treehopper colony are all offspring of the same mother, these guarding ants greatly increased her evolutionary fitness by more than quadrupling the number of survivors that could pass her genes on to the next generation. Guarding may further increase the number of offspring that a female produces in an indirect way. Mothers usually care for their young until they are grown. But if ants are present, they may leave partly grown young, relying on the ants to "babysit," and go off to produce another batch of young sooner than would otherwise be possible.

Many marvelous stories of shelter-building ants have been told by naturalists. In 1888, Thomas Belt wrote in *The Naturalist in Nicaragua* of a scale insect that was a pest in his tropical garden:

> My pineapples were greatly subject to the attacks of a small soft-bodied brown [scale insect] that was always guarded by a little, black, stinging ant *(Solenopsis)*. This ant took great care of the scale-insects, and attacked savagely any one interfering with them, as I often found to my cost, when trying to clear my pines, by being stung severely by them. Not content with watching over their cattle, the ants brought up grains of damp earth, and built domed galleries over them, in which, under the vigilant guard of their savage little attendants, the scale-insects must, I think, have been secure from the attacks of all enemies.

Among the weaver ants that protect homopterans is a Malaysian species of *Camponotus*, the same genus that includes the big, black carpenter ants of North America. While the carpenter ants nest in tunnels they excavate in dead wood, the Malaysian species occupies many small silken nests that the workers, using their larvae as the source of silk, spin on the undersides of the leaves of their host tree. Construction is a cooperative effort of four teams of workers that simultaneously perform their specific tasks: weavers hold and manipulate the silk-spinning larvae; fetchers bring in sand particles and bits of plant material that they incorporate in the wall of the nest; a third crew manipulates the silk to tighten the weave; and a fourth crew gathers and places scale insects in the new nest. Ulrich Maschwitz and his coauthors found that one colony of these ants had built 207 such nests, but only 50 of them were occupied. The occupied nests contained both ants and leaf-sucking, honeydew-producing scale insects, an average of 13 ants and 34 scale insects. Workers move scale insects from old to newly built nests. They protect the scales in their nests, but after fully populating a nest with scales, they behave aggressively toward scales outside the nest, hurling them from the tree and thereby preventing the establishment of "wild" colonies that would compete with their "domesticated" scales.

The amazing association of the cornfield ant with the corn root aphid is a close parallel to that of a dairy farmer with his cows. Cornfield ants keep corn root aphids as domestic animals, pasturing them on the roots of corn and other plants and "milking" them for their honeydew. In 1919, W. P. Flint, then chief field entomologist of the Illinois State Natural History Survey, wrote of the detrimental effect of this ant-aphid partnership on corn (maize): "If the old adage holds true that possession is nine points of ownership, the corn root-aphis has an owner, and it is this owner, the little brown ants of the cornfield, that we must combat if we are to keep our corn roots safe from the aphis."

In 1906, three years before he became the first head of the University of Illinois's Department of Entomology, Stephen A. Forbes reported his studies of this complex ant-aphid association. He found that the seasonal history of the corn root aphid is similar to that of most other aphids. They survive the winter as eggs in diapause, a quiet resting state that is the insects' version of hibernation. Early in the spring the nymphs hatch, every one of them a wingless female that, when mature, will reproduce parthenogenetically and give live birth to an average of four young per day, a second generation that also consists of only live-bearing females. And so it goes for an average of 12 generations per year except for the occasional production of winged females that may leave the nest and fly away. Forbes calculated that, given optimal conditions, after 3 generations a single female will have 66,000 descendants. In the fall, the eleventh generation gives birth to egg-laying females and the only males of the year. These last females are inseminated by the males and lay eggs that will not hatch until the following spring.

Throughout the winter, the ants store the aphid eggs in piles in their underground nest, when necessary moving them to places with favorable moisture and temperature levels. "In cold weather," wrote H. T. Fernald and

Harold Shepard in 1942, "the ants carry the eggs down below the frost and on warm days bring them up to warmer levels." Beginning early in April, the eggs hatch. The ants then tunnel to the roots of smartweed and other plants and put the young aphids to pasture on them. But if and when corn germinates nearby later in spring, the ants—apparently knowing that the sap of its roots is their aphids' favorite food—tunnel among corn roots and transfer the aphids to them. (Robert L. Metcalf and Robert A. Metcalf reported in their textbook on insects that one colony of cornfield ants moved aphids and their young 156 feet from a field of grass into a corn field.) The ants' burrowing can be extensive. A colony, noted Forbes, "may extend its burrows in the corn field under an area 3 or 4 feet in diameter and to depths usually varying from 1 to 4 or 5 inches."

Ants are not the only insects that tend Homoptera, although they greatly outnumber all the others. Belt described a wasp that tended colonies of treehoppers and chased off competing ants, and although he does not say so, it is a fair surmise that they also chased away insects that might parasitize or prey on the treehoppers.

The wasp stroked the young hoppers, and sipped up the honey when it was exuded, just like the ants. When an ant came up to a cluster of leaf-hoppers attended by a wasp, the latter would not attempt to grapple with its rival on the leaf, but would fly off and hover over the ant; then when its little foe was well exposed, it would dart at it and strike it to the ground . . . I often saw a wasp trying to clear a leaf from ants that were already in full possession of a cluster of leaf-hoppers. It would sometimes have to strike three or four times at an ant before it made it quit its hold and fall. At other times one ant after the other would be struck off with great celerity and ease, and I fancied that some wasps were much cleverer than others. In these cases where it succeeded in clearing the leaf, it was never left long in peace. Fresh relays of ants were continually arriving, and generally tired the wasp out. It would never wait for an ant to get near it, doubtless knowing well that if its little rival once fastened on its leg, it would be a difficult matter to get rid of it again. If a wasp first obtained possession, it was able to keep it; for the first ants that came up were only pioneers, and by knocking these off it prevented them from returning and scenting the trail to communicate the intelligence to others.

🐜 🐜 🐜   Caterpillars of at least 10 families have, according to Matthew Baylis and Naomi Pierce, evolved associations with ants. The great majority of them belong to the family of gossamer-winged butterflies (Lycaenidae), which includes the coppers, hairstreaks, and the blues that are familiar to many. Almost all of the others are in a closely related family, the metalmarks (Riodinidae).

Of the over 4,000 species of lycaenids, almost a quarter of the known butterflies, at least half serve ants as "cows" and interact closely with them. Such interactions—often complex and close—seem to be especially frequent among the over 1,000 species of blues. Many lycaenids invite the attention of ants by secreting (*not* excreting) from their Newcomer's gland, on the upper side of the abdomen near the tail, a sugar-rich liquid ants collect as avidly as honeydew. It was for years assumed that this secretion, often referred to as "honey," is rich in sugars, but that was not proved until Maschwitz and two of his colleagues analyzed the substance. They described it as being viscous, odorless, as clear as water, and containing from 13 to 19 percent sugars and mere traces of protein and one amino acid. But the composition of the "honey" varies with the species. That of an Australian lycaenid contains concentrated amino acids as well as sugars. Amino acids are also secreted by single-celled glands on the skin of lycaenid caterpillars. Ant-associated caterpillars also secrete various other substances whose function it is to appease ants, probably by mimicking their pheromones and thus making the ants think the caterpillars are fellow ants.

The honeydew excreted by aphids and other Homoptera costs them little or nothing in energy or nutrients. It is, after all, a waste product that passed through the intestines without being assimilated. But the sweet secretion of a lycaenid caterpillar is not a waste product. Producing it requires the expenditure of energy and nutrients that could have been used for growth if they had not been diverted to make honey. The metabolic cost of secreting these bribes for ants is reflected by the fact that caterpillars raised in the laboratory with attendant ants ultimately gained less weight than caterpillars raised without ants. As we would expect, most caterpillars dole out this expensive secretion sparingly, releasing it only when solicited by an attending ant.

The secretions of lycaenid caterpillars are a welcome and perhaps sometimes necessary supplement to the ants' basic diet of insects. "Lycaenid [caterpillars]," wrote Baylis and Pierce, "benefit from ant associations in at least two ways. First, by secreting chemicals that appease ants, they are protected

against the ants themselves that might otherwise be threatening predators
. . . Second, experiments with several species have shown that attendant
ants protect lycaenid larvae from predators and parasites."

Ants protect their caterpillar "cows" in two ways: by killing or chasing
away predators and parasites or, less often, by building shelters that shield
them from these enemies. In 1878, W. H. Edwards described an ant chasing
a parasite away from an azure blue caterpillar:

> On 20th June, in the woods, I saw a mature larva on its food-plant, and
> on its back, facing towards the tail of the larva, stood motionless one of
> the larger ants . . . At less than two inches behind the larva, on the stem,
> was a large ichneumon fly [a parasitic wasp], watching its chance to
> thrust its ovipositor into the larva . . . The [wasp] crawled a little nearer
> and rested, and again nearer the ant—making no sign. At length, after
> several advances, the [wasp] turned its abdomen under and forward,
> thrust out its ovipositor, and strained itself to the utmost to reach its
> prey. The sting was just about to touch the extreme end of the larva,
> when the ant made a dash at the [wasp], which flew away, and so long
> as I watched—at least five minutes—did not return.

One of the most remarkable instances of ants building shelters for caterpillars was described by Gary Ross. When a worker ant of a species of *Camponotus* finds a large untended metalmark caterpillar on a croton plant, it becomes extremely excited (it ignores small caterpillars because they have no
honey glands). The ant runs back and forth on the caterpillar's back, sips
some of its honey, and soon thereafter leaves the plant. When it finds one of
its nestmates, the two return to the caterpillar, but they soon descend to the
ground, and as Ross wrote:

> After approximately 1/2–1 hr, the ants begin excavating a small depres-
> sion surrounding the base of the plant. During this excavation, which
> takes several hours to complete, the ants frequently return to the larva
> to imbibe some of the honey. When complete, the depression is approx-
> imately 0.5–0.75 in. deep. The ants then return to the larva and by
> some means cause the larva to crawl down the stem and into the hol-
> low. Once the larva is in the depression, the ants begin the task of en-
> closing the depression with small pellets of soil. They continue until
> they have the chamber completely sealed off from the external envi-

ronment. Thus, the pen is complete. During the excavation process, several additional ants may join the original group, so there may be from 4 to 6 ants present before the construction of the pen is completed.

The ants guard the pen throughout the daylight hours, and then at dusk remove some of the soil pellets to make an opening. Before the caterpillar leaves the pen, the ants crawl over every leaf and stem of the croton plant. Ross thought that they might be "policing" it for creatures that might do harm to their caterpillar. He tested this idea by placing a few predaceous insects and spiders on a plant that would soon be policed by ants. As soon as the ants discovered them, they "immediately seized them with their large mandibles and carried them down the stem to the ground and on a distance of 6–12 in. from the plant. There they deposited the almost lifeless bodies of their victims" and then returned to the plant to continue their policing. After the ants had inspected the plant for 10 to 15 minutes, the caterpillar climbed the plant and fed on its leaves intermittently throughout the night, constantly guarded by ants. When dawn approached, the ants herded the caterpillar back to the safety of its pen on the ground. The ants literally compel the caterpillars to leave the plant at daybreak. If Ross kept ants away from caterpillars by covering plants with mesh cages, the caterpillars remained on the plant throughout the day and never returned to their subterranean pens.

Ross showed that the caterpillars did not survive without the protection of the ants; they were destroyed by predators if they were not in pens during the daylight hours. To prove the point, he removed several caterpillars from their pens and placed them on distant croton plants not patrolled by the ants that usually protected them. The following day, all these caterpillars were gone. He repeated the experiment, this time watching the caterpillars continuously, and found that they were killed and removed by predaceous ants of two species other than their *Camponotus* guards.

As cooler weather approaches, the ants deepen and widen the pens. By early February, the pens have been enlarged to long tunnels that may be 4 to 6 inches deep and have an entrance about an inch wide, which may or may not be sealed with debris. During the relatively cool nights of February and March, the caterpillars seldom leave the tunnel to feed—and then only briefly. They molt to the pupal stage in the deep tunnels, still tended by ants, and become adult butterflies in late April. Ants tend the pupae although they secrete no honey, apparently enticed by an attractive but nonnutritive substance and by sounds that the pupae make by stridulating, rubbing two body parts against each other.

A Malaysian lycaenid called *Athene emolus* is completely dependent upon ants of the species *Oecophylla smaragdina,* weavers that live in nests composed of living leaves they bind together with silk. If not guarded by its ant associates, the lycaenid is killed by predators and parasites. Furthermore, without the help of the ants it could not—except by pure chance— even find its food, young and tender leaves of its host plant, a tree called *Saraca thaipingensis.* Female butterflies of this species lay their eggs in masses of from 20 to 150, but only where their ant associates are present. Of 78 egg masses found by Konrad Fiedler and Ulrich Maschwitz in 1989, 24 were on leaf nests of the ants, 26 were on ant trails near nests, and 27 were in places where ants were tending honeydew-producing homopterans or *Athene emolus* caterpillars. But only one egg had been laid on a young leaf of the host plant, which the caterpillar will prefer to old leaves. The ants attack the butterflies unless they have already begun to lay eggs, which presumably give off a chemical signal that mollifies the ants. Within a few minutes, butterflies that have just emerged from their pupal skin and are not yet ready to lay eggs are killed, torn to pieces, and carried to the ants' nests to be eaten.

The caterpillars molt three times before the final molt to the pupal stage. Only after the second molt do they secrete honey. Nevertheless, the ants pick up newly hatched caterpillars and place them inside their nests. (Fiedler and Maschwitz think that the young caterpillars attract the ants with a chemical that resembles a pheromone given off by ant larvae.) The newly hatched caterpillars feed by scraping at the inner surface of the leaves that form the nest. After their first molt, they leave the nest. The ants pick them up and carry them, sometimes as far as 16 feet, to young shoots with tender leaves. There the caterpillars feed until their final molt. The older ones eat ravenously and are closely tended by their ants, which frequently milk them. The full-grown ones molt to the pupal stage near where they have been feeding. After 6 or 7 days, the adult butterflies emerge from the pupal skin.

The ants are well fed by the caterpillars. Every 10 minutes, full-grown ones secrete about 50 drops of honey with a total energy content of almost 50 calories. The caterpillars on one shoot of a plant, often more than 50 individuals, produce each day a quantity of honey with an energy content of almost 2,400 calories, enough to keep alive about 8,250 inactive ants for one day.

The caterpillars are very well protected against predators and parasites. The newly hatched ones within the ant nest are hidden from nearly all their enemies. Even after the caterpillars leave the nest, they are closely guarded by the ants. The caterpillars themselves warn the ants of approaching danger. If its twig is slightly shaken or at the approach of ants that might be enemies rather than friends, a caterpillar everts two bristly tubes, which probably emit a warning pheromone, from near the end of its abdomen. Nearby "guard" ants immediately go on the alert, ready to attack intruders. They spread their jaws wide, raise their front legs, and walk forward with jerky movements. They never attack the caterpillar but would presumably attack an intruding predator or parasite.

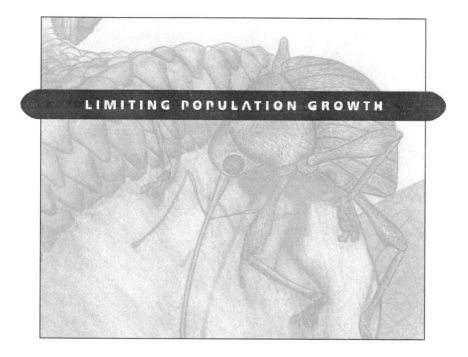

LIMITING POPULATION GROWTH

7

CONTROLLING PLANT POPULATIONS

On a bright summer day in 1944, a rancher in Humboldt County, California, rested in his saddle as he surveyed the broad expanse of rangeland before him. He saw a sea of bright yellow covering the land—millions upon millions of the large, yellow blossoms of St. John's wort. It was undeniably a beautiful sight, but bad news for the rancher. The dense infestation of St. John's wort—a pernicious weed—had depreciated his land to about a third of its former value. And for good economic reasons! Not only did this prolific alien weed from Europe crowd out nutritious forage plants, but sheep and cattle that fed on it actually lost weight because it is somewhat toxic. Practically speaking, it was impossible to destroy these plants. Weed killers were prohibitively expensive and there was no other affordable method.

In North America, St. John's wort was first seen in 1793, in Pennsylvania. By 1900, reported Curtis Clausen, an expert on using natural enemies to control weeds, "it had spread westward into California, where it became known as the Klamath weed because it was first detected along the Klamath River in the northern part of the state." By 1929, 156 square miles of open-range grazing land were infested in northern California; by 1940, about 390 square miles. By 1945 this aggressive weed had spread to nearby states and to Canada, and over 7,700 square miles—an area about the size of New Jersey—was heavily infested and almost useless for grazing. And there was every reason to believe that St. John's wort would continue to spread like wildfire.

The rapid spread of St. John's wort was possible because of favorable growing conditions, its rapid reproduction, and, as is now known to be most important, freedom from its natural enemies, which had been left behind in Europe. This plant is a very prolific breeder. A large one may produce more than 30,000 seeds, light and small enough—only a sixteenth of an inch long—to be dispersed for long distances, often adhering to the oily hair and hides of animals such as sheep. If conditions are not favorable for germination, the seeds can survive in the soil for years. Moreover, established plants also propagate by cloning, sprouting new shoots from their shallow, spreading rootstocks. In Europe certain small leaf beetles greatly curb the increase in the plant's population. Released from this restraint in the New World, populations of St. John's wort burgeoned explosively wherever growing conditions were favorable. Our California rancher would probably be surprised to learn that this plant, used in herbal medicine, is now cultivated in some places.

Previous successful biological controls (introduction of natural enemies) of a number of aggressive alien weeds in various parts of the world—including St. John's wort in Australia—suggested that the importation and establishment of insects that feed on this plant would reduce its population in California to a nonthreatening level. In the early 1940s the U.S. Department of Agriculture agreed to the introduction into California of three European beetles that feed on St. John's wort, with the proviso that tests would first show that they did not feed on important crop plants. Because World War II was raging in Europe at the time, it was impossible to obtain these beetles from that continent, but they were provided by the Australians, who had imported them from Europe before the war. In 1945, entomologists at the University of California began to release these three beetles in the northern part of the state.

One of them, a leaf beetle called *Chrysolina quadrigemina,* overwhelmed the other two and proved to be a superbly efficient destroyer of St. John's wort. "Within a decade after release of the insects," wrote Paul DeBach in 1964, "Klamath weed was reduced from an extremely important pest of range lands to a road-side weed . . . It is now less than one percent of its former abundance and has been removed from the list of California's noxious weeds."

This and many other successful biological controls using insects to combat unwelcome nonnative plant pests are essentially ecological experiments that demonstrate the importance of insects as regulators of plant populations. The results of these experiments tell us that all ecosystems must contain as yet unsung heroes—unsung because we are not aware of their essential roles in controlling plant populations, unsung because we have not seen what would happen if they were not there. But certain ecological observa-

tions and intentional experiments give us glimpses of how native herbivorous insects control the plants in their own ecosystems.

This story has an interesting ecological corollary. An ecological survey of
Humboldt County done today would find St. John's wort growing mainly in
the shade and supporting only sparse populations of *Chrysolina* beetles. If we
didn't know about the biological control of this plant, we might falsely con-
clude that it is shade-loving and doesn't grow well in the sun, and that
Chrysolina is naturally rare. But we know St. John's wort grows all too well
on the sunny open range, and that the beetle was obviously far from rare
when it was destroying these plants on open grazing lands. The truth is that
today this plant rarely survives in the sun because the beetle, which thrives
in the sun, soon kills it. Shady places are refuges for St. John's wort, because
the beetles do not thrive there and consequently do not become numerous
enough to kill their host plants. These facts led Peter Price, in his book on insect ecology, to ask two penetrating questions: "Could the ecology of these
species be properly understood if this interaction had not been witnessed?"
The second question, highly pertinent to this chapter, follows from the first.
"How many other plant species have distributions dictated by herbivore
pressure?" The point is that, unknown to us, plant-feeding insects are probably limiting the population expansion of many plants by driving them out of
the habitats where they grow best.

🐜 🐜 🐜 The most famous biological control of a plant is the vanquish-
ing of the prickly pear cactus *(Opuntia)* in Australia. Prickly pear, like all
other cactuses, occurs naturally only in the New World. Imported to be
grown in Australian gardens, prickly pear plants escaped from cultivation,
and, unchecked by natural enemies, they thrived. By 1925, dense and often
completely impenetrable tangles of them covered 44,000 square miles of
rangeland, an area almost twelve times the size of New Jersey. In the New
World, where they are held in check by many natural enemies, prickly pears
generally occur only as scattered clumps. In 1925, Australia's prickly pear
problem was alleviated by the introduction of many species of cactus-eating
insects from the Americas. By far the most effective of them was a caterpillar
aptly named *Cactoblastis*. By 1937, the last dense stand of these plants was
gone, and to this day prickly pears, controlled by *Cactoblastis*, are no longer
pests in Australia, where they now grow only in scattered clumps, never in
dense tangles.

🐜 🐜 🐜 Plants, like animals, have two quite different strategies—really opposite ends of a continuum—for improving the odds that parents will be survived by their offspring. At one end of the continuum are prolific species that invest a minimum of resources in each seed but produce a great many small seeds in the hope that a few lucky ones will germinate and grow. At the other end of the continuum are restrained species that produce a few large seeds that contain abundant nutrients to get seedlings off to a good start. The champion prolific species are orchids. A single blossom of some species, L. A. Nilsson tells us, can produce 2 million dustlike seeds. Cottonwood trees are not nearly that prolific, but they do produce very large crops of tiny seeds, each with a tuft of fine white down on which it floats with the wind. To judge by the clouds of cottonwood seeds in the air on breezy days in spring, a large tree can produce hundreds of thousands of seeds in a single year and many millions during its lifetime. Oaks are more restrained, producing relatively few large acorns, with a large store of nutrients that hastens the growth of seedlings. Victoria Sork and Judy Bramble found that some species of oaks produce only about 3,000 acorns each in a mast year, a time of high fecundity that occurs in oaks every 2 to 4 years. Thus during its lifetime of a hundred years or more, an oak may produce fewer than 100,000 seeds. These two reproductive strategies are, Daniel Janzen pointed out, even more strikingly illustrated by two North American trees of the Leguminosae, the family that also includes peas, beans, and clovers. In one year a small honey locust bore about 10,000 small seeds that weighed only about four-thousandths of an ounce each. In contrast, a Kentucky coffee tree of about the same size produced only about 400 seeds; each weighed about six-tenths of an ounce, over 15 times as much as a honey locust seed. Most of the extra weight was a large store of nutrients.

🐜 🐜 🐜 All plants produce more than enough offspring to replace themselves. If extra offspring consistently survive, more than one in self-pollinating species and more than two in species that require cross-pollination, the population will sooner or later increase to a level that threatens the stability or even the survival of that species' ecosystem. This rarely happens in undisturbed ecosystems; if too many seeds survive the gauntlet of competition and physical hazards, the excess will ultimately be eliminated by biological factors such as seed- or leaf-eating insects.

Mortality cancels out excess reproduction over the long run, but popu-

lations of the hundreds or thousands of species in an ecosystem are constantly fluctuating in size because of changes in the severity of mortality factors. Nevertheless, in the absence of disturbances from outside the system—
often caused by the activities of people—these fluctuations generally remain
within limits that an ecosystem can tolerate, because they do not upset its
long-term stability. We sometimes speak of the "balance of nature," a phrase
that ecologists avoid because it may suggest a perfect and permanent balance by calling to mind an apothecary's balance with its two pans in perfect
equilibrium, because each holds the same weight. But "balance" is useful
shorthand in speaking of ecosystems as long as we realize that the balance is
approximate, actually a dynamic equilibrium between populations that are
constantly fluctuating in size. For example, Peter Turchin and his coworkers
showed that the cyclical ups and downs of southern pine beetle populations
were correlated with the population fluctuations of their predators. When
predator populations are up, beetle populations are down, and vice versa.

The enemies of flowering plants are numerous and varied. Viruses, bacteria, and fungi cause plant diseases, many of them transmitted by
insects. And a few plants are parasites of other plants. Among them is dod-
der, a leafless relative of the morning glories, whose long, bare, stringlike
stems, orange in color because they lack chlorophyll, twine in a tangle over
and around their host plants. Thousands of different kinds of mites, not insects but eight-legged relatives of spiders and scorpions, suck nutrients from
the leaves of green plants. Geese, the hoatsin of South America, a strange
bird of the Amazon forests, and a few other birds feed on grass or the foliage
of broad-leaved plants. Sparrows and other finches eat small seeds. Turkeys
feed heavily on acorns, swallowing them whole. Most mammals from rodents to pandas are at least part-time vegetarians. Chipmunks and other rodents store seeds in their burrows; deer and moose graze and, especially in
winter, browse on the twigs and buds of shrubs and trees; the giant panda of
China eats virtually nothing but bamboo. But by far the most important
plant-eating animals are insects, because there are so many species and because the number of individuals and their biomass are so immense. About
450,000 species of insects are plant feeders—about half of the insects known
to science and approximately 38 percent of the currently known animals.

The association between insects and plants is ancient. In the fossil record,
the first primitive, wingless insects appear about 400 million years ago—

about 20 million years after their arthropod ancestors left the oceans and about 50 million years before the first primitive reptiles appeared. Some of the earliest insects probably fed on the primitive land plants that arose about 430 million years ago. But from about 345 to 280 million years ago there was a great radiation of plant-feeding insects, a major increase in their diversity, coincident with the growth of the great forests of trees whose wood became the coal we burn today, trees related to the still extant club mosses, horsetails, ferns, and conifers. Most of these insects became extinct as the coal forests declined, but from about 135 to 65 million years ago they were more than replaced by a second great radiation of insects that fed and still feed on the angiosperms that came to dominate the land. Many of these fossil insects are very similar or even identical to the plant-feeding insects of today.

Herbivorous insects feed in several ways, and among them they attack every part of virtually every species of flowering plant. Almost all of the herbivorous insects are contained in 9 of the 30 orders of insects: the grasshoppers, crickets, and katydids (order Orthoptera); the walkingsticks (order Phasmatodea); the thrips (order Thysanoptera); the true bugs (order Hemiptera); leafhoppers, aphids, scale insects, and their relatives (order Homoptera); the beetles (order Coleoptera); the flies (order Diptera); the moths and butterflies (order Lepidoptera); and the sawflies, wasps, ants, and bees (order Hymenoptera).* Without exception, all of the walkingsticks and all members of the order Homoptera are plant-feeding insects. All the other orders mentioned above also include anywhere from many to a few species of parasites, predators, or scavengers that do not feed on living plants.

How could anyone understand the ways of this immense assemblage of 450,000 herbivorous insects? They must obviously be considered as groups united by some common characteristic. The list that you just read tells what *kinds* of insects eat plants. But several of these taxonomically defined groups—the orders—include different species that feed on different parts of a plant and sometimes by quite different methods. One way to come to grips with this confusion is to group insects according to their "feeding strategies." We could group them in guilds, not unlike the medieval European trade

---

* Do you wonder why almost all of these ordinal names end in *ptera?* This suffix comes from the Greek word for wing. Hence: Coleoptera (*koleon* is sheath in Greek), referring to the hard front wings of beetles; Diptera (*di* means two in Greek), a reference to the absence of a second pair of wings in flies; and Lepidoptera (*lepis* is scale in Greek), an allusion to the scale-covered wings of moths and butterflies.

guilds, that include species that may be taxonomically unrelated but feed on the same plant part in similar ways. There are guilds of leaf chewers, sap suckers, stem borers, seed eaters, and others.

The leaf chewers are grasshoppers, katydids, stick insects, many of the beetles, almost all caterpillars, and most of the sawflies. Eating a leaf by chewing off bites and swallowing seems simple and straightforward. Not so. The behavior of leaf chewers is variously modified for rapid and ef- ficient feeding, for making a shelter out of the leaf being eaten, for blocking the ingestion of toxic sap, and for eliminating leaves damaged by feeding, which are visual clues that can lead a hungry bird to the leaf-chewing insect.

Hornworm caterpillars, as Bernd Heinrich reported, are especially effi- cient at consuming a leaf. Because of its large size, a hornworm must keep a secure grip on the sturdy stem (petiole) or adjoining midrib of the leaf with the tiny hooks on the five pairs of fleshy prolegs near the end of its long ab- domen. As it chews, it clutches the edge of the flexible leaf blade with its short thoracic legs, which are just behind the head. Since it takes more than one leaf to make a meal, the most efficient way for the caterpillar to proceed is to eat all of a leaf before moving on to another. But how can it do this without releasing its safe grip on the petiole if much of the leaf is out of reach because it is longer than the caterpillar's body? The answer is that the caterpillar bends the leaf so as to bring all of it within reach. It does so by keeping its abdominal prolegs in place on the petiole as it "walks" forward on the edge of the leaf blade with its thoracic legs—just as a person can bunch up a rug by kneeling on one end and pulling back on the other part with the hands.

Some caterpillars protect themselves by constructing a shelter. Evergreen bagworms, commonly seen feeding on juniper or arborvitae in the eastern United States, weave a silken bag, usually festooned with bits of leaves or other plant parts, and live within it much as a hermit crab lives in an empty seashell. The bagworm caterpillar's head and thorax can be protruded through an opening at the front of the bag, and fecal pellets are expelled from an opening at its other end. Before a fully grown caterpillar pupates, it closes the front opening and fastens that end of the bag to a twig with a strong band of silk. It then turns around to face the opening at the rear end of the bag. An adult male sheds his pupal skin and exits his bag to look for a mate, finding her by following the odor trail of a sex-attractant pheromone

A bagworm caterpillar (left) about to nibble on a juniper leaf and, hanging from
a twig, a closed bag in which a bagworm will molt to the pupal stage.

she released into the air. The degenerate adult female, lacking eyes, antennae, and all other appendages, separates from her pupal skin but remains within it. Copulation is not easy for bagworms. A male makes genital contact with his mate by thrusting his abdomen, extended to three times its usual length, into his mate's bag and up through a narrow slit in her pupal skin. The female deposits her eggs, up to 2,000 of them, in the back end of the pupal skin, plugs its front end with hairs from her body, and then dies and falls to the ground. A female's bag does double duty. It protects her in the summer and her eggs throughout the long winter.

Among the insects that make a shelter out of the leaf on which they are feeding is the rice leaf folder, a widespread pest of rice in Asia. I came to know this caterpillar during a sabbatical year at the International Rice Research Institute (IRRI) in the Philippines, one of the two homes of the famous "green revolution." IRRI had asked me to develop a method for testing rice varieties for resistance to the leaf folder. The goal was to breed any resistance found into the new, high-yielding varieties. In developing these new varieties of rice, IRRI had unknowingly "bred out" the natural resistance to insects characteristic of the many native rice varieties that had been grown in Asia for millennia.

Not long thereafter, the Central Agricultural Research Institute of Sri Lanka invited Gottfried Fraenkel, who had years before directed my Ph.D. research, to do the same thing. On his way to Sri Lanka, about 6 months after my arrival at IRRI, he stopped off to see me and asked if I had made any progress. With considerable self-satisfaction I handed my former "boss" a just-completed manuscript that described my method of testing for resistance to the leaf folder. Fraenkel went on to Sri Lanka and worked on other projects with the leaf folder, including one that showed how this insect folds a rice leaf to form the protective shelter in which it nibbles on the inner surface of the leaf.

Fraenkel and Faheemah Fallil wrote of the rice leaf folder, "Its characteristic behaviour is to spin a rice leaf longitudinally into a roll, by stitching together opposite rims of the leaf, and to feed inside this roll, leaving the epidermis on the outside of the roll intact." When forming a leaf roll, the caterpillar lies aligned with the length of the long, narrow leaf and rapidly swings the forepart of its body from side to side as it stretches a strand of silk, which issues from spinnerets near its mouth, from one edge of the leaf to the other. By repeating this in the same place as often as 100 times, it forms a thick band of silk anchored only at the opposite edges of the leaf. After completing a band, the caterpillar moves a short distance up or down the leaf to make another. This procedure is repeated until as many as 30 silken bands are completed. "A newly woven band," wrote Fraenkel and Fallil, "quickly becomes shorter by a process of contraction . . . which has the effect of bringing the points of attachment, usually the rims of the leaf blade, closer together . . . with each succeeding band, this distance becomes shorter until the leaf is completely rolled up."

Certain leaf-chewing caterpillars and beetles have evolved ways of coping with one of the many chemical defenses of plants, the flow of toxic sap into wounds made by chewing insects. Milkweeds are named for the sticky white sap (latex) that protects them from many leaf-eating insects. It is not only toxic but also sticky enough to gum up an insect's mouthparts. Stored under pressure, the latex flows rapidly along the branching leaf veins to the site of a wound. Beetles and caterpillars that specialize in feeding on milkweed stop the downstream flow of latex to their feeding site by severing veins that come from upstream, but as David Dussourd and Thomas Eisner reported, different insects accomplish this in different ways. Caterpillars of the queen butterfly, a relative of the monarch found in the southern United States, use two methods. Small caterpillars stem the flow of latex to their feeding site by cutting a surrounding, circular trench that severs the small veins. Large queen caterpillars use a different method, partially severing the midrib, the main conduit of latex, and eating the large and relatively latex-free downstream part of the leaf.

Many caterpillars and other herbivorous insects are camouflaged and hard to spot, but the leaves they damage could "blow their cover" by revealing their location to insect-eating birds. Convincing evidence shows that birds orient to damaged foliage in their search for insects. In Maine, Bernd Heinrich and Scott Collins found that foraging black-capped chickadees learned to focus their search for insects on trees with leaves damaged either artificially or by caterpillars. Later Jens Roland and his coworkers observed that pine siskins in British Columbia foraged for an average of over 4 minutes in apple trees with much visible leaf damage but only for about 0.7 minute in trees with little visible damage.

Some caterpillars cope with this danger by removing or disguising signs of their feeding, but others do not. In a survey of 43 species, Heinrich and Collins found that caterpillars which are presumably unpalatable to birds because they are bristly, spiny, or brightly colored (an advertisement of toxicity) make little or no effort to hide and are "messy feeders" that do not disguise or remove partly eaten leaves. They do not fear discovery because birds usually reject them on sight. But most camouflaged caterpillars are palatable and need to hide. They may disguise partly eaten leaves by trimming them to eliminate obvious ragged signs of feeding; they may eat all of a leaf before moving on; or they may clip the stems of partly eaten leaves, which then fall to the ground. Another strategy is to move away and leave damaged leaves far behind.

The larvae of certain beetles, moths, flies, and sawflies are leaf miners, highly specialized insects that literally mine within a leaf, tunneling between the upper and lower epidermal layers as they eat the soft tissues between. According to Stuart Frost in his book on insect life, they attack many species of plants, representatives of nearly all plant families, even some aquatic ones. Some feed on plants poisonous to vertebrates. One tiny leaf-mining caterpillar, for example, tunnels in poison ivy leaves. In a book on leaf miners, James Needham and his two coauthors wrote that this caterpillar "makes broad, tortuous, whitish, upper surface mines, that take on shapes as varied as those of water spilled on a smooth surface . . . Several larvae are at work together in nearly every mine."

Because a leaf's epidermal layers are translucent, the mine is clearly visible. Mines vary in form and are sometimes so characteristic of a particular species that the mine alone reveals the identity of the miner. Frost said that leaf miners "write their signature in the leaves." Some mines are long and narrow, often serpentine tunnels that wind widely through the blade of the leaf, and others are relatively straight tunnels. In either case, narrow tunnels are likely to widen noticeably from beginning to end, because the larva grows as it eats and forges ahead. Other mines, blotch mines, are wide caverns. Excavated in different and often specific areas of a leaf, they are of various sizes depending upon the size and species of the miner.

The contents of an abandoned mine reveal the history of its occupant. It may contain a series of molted skins of increasing size, marking a larva's progress as it grew and matured before molting to the pupal stage. The shed pupal skin may be at the end of the tunnel. Although some miners push their fecal pellets out of the mine, others leave them in the mine and tamp them down as they move on. But others may abandon a mine when it becomes too crowded with feces and move on to start a new mine in another leaf.

The leaf beetles—about 38,000 known species worldwide—all feed on the foliage, flowers, roots, or stems of plants. Larvae of the approximately 6,000 species of one group are leaf miners. In a recent issue of *Science*, Peter Wilf, Conrad Labandeira, and several coworkers described fossil ginger leaves from North Dakota and Wyoming with mines characteristic of those dug by these beetles. These plant fossils predate the earliest known fossils of the bodies of these beetles by about 20 million years, which places their origin about 70 million years ago, when flowering plants and insects were evolving and diversifying rapidly.

The minute, feathery-winged thrips are open pit miners of leaves. Just as human miners expose a seam of coal by stripping away the overburden of soil, thrips lacerate one of the epidermal layers of a leaf to expose the soft tissues beneath. The head of a thrips (this word is both singular and plural) extends downward as a long cone with a small opening from which three lancelike mouthparts can be protruded. The mouthparts do the lacerating, and the resulting slurry of cells is sucked up into the cone. Feeding by thrips damages the plant and causes a distinctive silvery sear on the leaf.

Many insects feed on roots, stems, or wood. Among the numerous root feeders is the northern corn rootworm. A native of the prairies of North America, this beetle fed on prairie grasses such as big bluestem before corn, which originated in Central America, was cultivated in North America. Adult northern corn rootworms feed on the silks of corn and the flowers of other plants, but as larvae they chew on the roots of corn. Long-horned beetles of several species feed on milkweed plants. Larvae chew on the roots and lower stem, and adults feed on the flowers or perhaps the leaves. Like most other insects that feed on milkweeds, ranging from aphids to monarch and queen butterflies, these beetles sequester milkweed toxins and when adult are conspicuously colored, red with black markings, to warn birds and other insectivores that they are poisonous. Long-horned beetles of the genus *Prionus* are much larger than milkweed beetles: up to 3 inches long, heavy-bodied, somberly colored in blackish-brown. In the larval stage they burrow through the soil from root to root of oaks and other hardwood trees. Heavily attacked trees, according to Frank Craighead, the author of *Insect Enemies of Eastern Forests,* may die gradually limb by limb.

The North American bean leaf beetle, which feeds only on beans and other legumes, has become a pest of one of our most important crops, the soybean, which originated in Asia. These beetles still feed on their native hosts, wild legumes such as tick trefoils, bush clovers, and cultivated lima and kidney beans, the last two gifts to us from the Native Americans. Adults chew holes in the leaves, and larvae feed on the roots, particularly on the nodules, characteristic of all legumes, which contain bacteria that fix atmospheric nitrogen and make this all-important nutrient available to their hosts and other organisms.

Research that I did jointly with Marcos Kogan, Charles Helm, and other entomologists at the Illinois Natural History Survey revealed how much of a head start these beetles give their offspring by where they lay their eggs. At the time it was known only that they lay their eggs in the soil somewhere near a host plant. But we wanted to know if the eggs are placed close enough to roots and nodules for the newly hatched larvae to reach them with a minimum of exhausting and perhaps fatal burrowing through the soil.

Locating an infinitesimal bean leaf beetle egg in the soil of a soybean field is harder than finding a needle in a haystack. But it can be done with an apparatus designed by William Horsfall to separate the tiny eggs of floodwater mosquitoes from samples of floodplain soil. Our samples, cores of soil weighing from 10 ounces to over 2 pounds, were cut in soybean fields with a tool like the one used to make holes in putting greens. They were first washed through three nested sieves, the bottom one a fine screen that retained the eggs and any soil or debris that was not caught by the coarser screens above. What remained on the fine screen was then washed into a narrow glass funnel filled with saturated brine. Sand and most of the debris sank to the bottom, and we discarded them by briefly removing the cork that plugged the stem of the funnel. A small amount of debris and eggs floated and were caught on a small, fine sieve, and then washed into a little porcelain bowl of tap water. The eggs were picked out with an eye dropper. This method was amazingly efficient. We recovered over 93 percent of the bean leaf beetle eggs we "seeded" into egg-free soil samples from cornfields.

Soil samples taken from different depths and at various distances from soybean plants showed that the great majority of eggs had been laid where roots and nodules were most concentrated. As Thomas Anderson and I found, the number of nodules and the combined biomass of roots and nodules is greatest within 3 inches of the stalk and at a depth of 3 inches or less.

All of the eggs were within 1.5 inches of the soil surface, and over 92 percent were within 3 inches of a stalk—73 percent within 1 inch, where over 85 percent of the biomass of roots and nodules is located.

The female long-horned beetles known as hickory twig girdlers prepare twigs for future occupancy by their stem-boring larvae in a most intriguing and seemingly "purposeful" way. As Frank Forcella reported, adult girdlers emerge from their host twigs in late September, about a month before the leaves fall. The females immediately begin to girdle twigs of hickories or other trees. Selecting a twig with a basal diameter of about a third of an inch, a female chews a narrow slot around the circumference of the twig's base. She cuts all the way through the bark and the phloem just beneath it but leaves the adjoining xylem largely intact. Cutting the phloem tubes prevents nutrients from flowing out of the girdled part of the twig, but since the xylem tubes are intact, water and nutrients from below can still enter and flow into the still living leaves on the girdled part of the twig. Consequently, the leaves continue to synthesize nutrients, which accumulate in the girdled part of the twig and increase its nutritional value for the larva that will ultimately bore in it. Girdling the twig just before leaf fall is advantageous because the leaves then salvage their nutrients by pouring them into the phloem, but instead of flowing to the roots these nutrients too accumulate in the girdled twig. After completely girdling the twig, the female, as Ephraim Felt explained, lays an egg or two in a puncture in its bark. The larvae spend the winter and the following summer in the twig. Most of the girdled twigs soon fall to the ground. If they were to decay before the following September, the larvae within them would be doomed. But hickory twigs, the usual choice of these insects, have chemicals in their bark that make them resistant to microbial decomposition for at least a year.

Among the many other stem-boring insects are the larvae of a great many beetles, most notably the metallic wood-boring beetles, bark beetles, and in addition to the hickory twig girdler, a multitude of other long-horned beetles. The moths are also well represented by stem-boring caterpillars. Outstanding among them are the carpenterworms, clear-winged moths, and some of the snout moths. The larvae of a few sawflies are stem borers, and the members of a family known as the stem sawflies burrow only in the stems of grasses and berries. The maggots of just a few true flies (order Diptera) have adopted the stem-boring way of life.

The raspberry cane maggot, the larva of a true fly, lives in the canes of both wild and cultivated plants. John Henry Comstock wrote in his classic textbook on entomology that this maggot "burrows in the new canes of black and red raspberries and blackberries and kills them. The eggs are laid on the young shoots in the spring. The larva bores into the pith of the shoot, and tunnels downward; when about half way to the ground, it girdles the wood beneath the bark. The part of the shoot above the girdle soon wilts, shrinks in size and droops over. The larva continues to burrow downward in the pith to the surface of the ground, transforms to a pupa without leaving its burrow in late June or early July; but the adult does not emerge till the following April." The new canes of the year will bear the next year's fruit, luscious berries that contain many tiny seeds. As anyone who has grown raspberries knows, robins and other birds gorge on the berries. They spread the seeds far and wide in their droppings. The destruction of new canes will, of course, decrease the plants' fruit production and thereby limit the expansion of their population.

The first carpenterworm (actually a caterpillar) I ever saw was in the trunk of a dead cottonwood I was chopping down. It was about 2.5 inches long and almost as big around as my little finger. So riddled with carpenterworm burrows was the tree that there was little doubt that it had been killed by these wood borers. Felt recalled that Asa Fitch, an early American entomologist, wrote that these insects "attack the stateliest oaks of the forest, probably ruining them in every instance." The carpenterworm "perforates a hole the size of a half inch auger, large enough to admit the little finger and requiring 3 or 4 years for the bark to close over it." Although oaks and cottonwoods are among the carpenterworm's favorites, it also attacks elms, locusts, willows, maples, and many other trees. The caterpillars, which take about 3 years to mature, pupate in large cells they excavate near the opening of a tunnel, first plugging the entrance with sawdust. Just before the adult emerges, the pupa squirms part way through the plug. The moth is large and has rather long, narrow wings with a span of about 3 inches. The eggs are laid in crevices in the bark, and the tiny, newly hatched caterpillars burrow into the tree immediately.

One of my great pleasures has been growing vegetables. I even raised some potatoes because it was so much fun to dig them, although it was easier and cheaper to buy them. I also planted several hills of pattypan squash, a favorite of my family and the neighbors. The first two years the squash plants thrived, but in the third year vines on a few plants were killed by squash vine borers, caterpillars of a clear-winged moth; and in the fourth year all my squash plants were killed by these borers. That year I saw many shed pupal skins protruding above the surface of the soil and also newly emerged, wasp-mimicking adult moths resting on the squash plants. According to Roger Friend's comprehensive account of this insect's life history, most of the eggs are laid on the underside of a vine at its base. The caterpillars bore into the vine and feed on the tissue that lines its cavity, usually causing it to wilt and die. When fully grown, the caterpillars leave the vine, burrow an inch or two into the soil, spin a cocoon, and in southeastern Ontario and the northern United States, where there is only one generation per year, they do not pupate and then emerge as adults until June of the next year. The squash vine borer, a native insect found east of the Rocky Mountains and south to Argentina, feeds on all of the squashes and pumpkins.

Bark beetles are among the most ecologically important insects in forest ecosystems. Some attack and weaken or even kill healthy trees, while others attack only weakened trees, killing them and beginning the recycling process that will ultimately return the tree's elements to the soil. There are many kinds of bark beetles, and among them they attack conifers, such as pines, spruces, cypresses, larches, and redwoods and many broad-leaved trees, including among others elms, hackberries, oaks, ashes, beeches, hickories, and cherries. Both the larvae and the adults, which tunnel at the interface of the bark and wood, are called engraver beetles because their intricate pattern of tunnels is etched both into the outer surface of the wood and the inner face of the bark—as is seen when a slab of bark is peeled from a tree. The pattern is feather-like. It begins with a long, straight egg gallery cut usually by a female—a male in some species—which enters the tree by burrowing through the bark. The eggs are laid one at a time and placed at more or less regular intervals along the egg gallery. Each newly hatched larva begins its own tunnel. At first it tunnels away from the egg gallery at a right angle, but as it burrows, feeds, and grows, its tunnel widens and usually curves away from its right-angle path. The burrowing destroys two tissues that are essential to the life of the tree, the cambium and the phloem. The cambium, the all-important growth layer, is only one cell thick but, like the phloem, forms a continuous cylinder that surrounds the trunk, branches, and twigs. As mentioned previously, the phloem, which lies between the cambium and the bark, is the conductive system that carries nutrients, mainly sugar formed in the leaves by photosynthesis, down to the root system for storage and redistribution. A branch or a whole tree dies if it is girdled by the destruction of a complete ring of phloem and cambium around its circumference. Bark beetles probably seldom girdle tree trunks, but they often destroy enough cambium and phloem to seriously weaken a tree.

The hickory twig girdler, which introduced you to the guild of stem-boring insects, is only one of a multitude of long-horned beetles—at least 35,000 known species worldwide and 1,200 in America north of Mexico. Without exception, all of these beetles are plant feeders. As larvae, they all burrow in plant tissues, some in herbaceous plants but most in the trunks, branches, or twigs of woody plants—often dead trees.

One of them is the Asian long-horned beetle, recently brought to the United States as stowaway larvae hidden in the wood of packing crates or cargo pallets from China. According to an article in the *Journal of Forestry* by Robert Haack and several coauthors, it was first found in the United States when a resident of Brooklyn, New York, reported that the Norway maples lining his street were riddled with large holes and that large black and white beetles were crawling on the trunks. A few years later another infestation of Asian long-horned beetles was found in the Chicago area, and just recently one has been found in Seattle and another in Austria. Since insecticides are all but useless against this wood borer, the current plan for its eradication is the destruction of all infested trees. But this plan is likely to fail because despite the cutting, chipping, and burning of infested trees the beetle has spread to nearby areas.

Newly emerged adult Asian long-horned beetles, wrote Haack and his coauthors, feed on the bark of twigs and mate on the branches or trunk of the host tree. Adults can cover several hundred yards in a single flight and, because they live 40 days or more, can travel considerable distances to find suitable trees. Egg-laying females gnaw through the bark to the cambium and then turn around to lay a single egg in the hole and cover it with a protective secretion. A female usually lays a total of 25 to 40 eggs. In New York, Asian long-horned beetles lay eggs from July to early November. Newly hatched larvae first feed in the region of the cambium, and then tunnel upward through sapwood and heartwood for from 4 to 12 inches, eating voraciously until they are fully grown, about 2 inches long. They pupate and emerge as adults in their tunnels and then gnaw their way to freedom, leaving circular holes about half an inch in diameter in the bark.

Asian long-horned beetles feed on broad-leaved trees of many species, among them maples, birches, poplars, willows, horse chestnuts, plums, black locusts, and elms. They readily attack healthy trees and can seriously

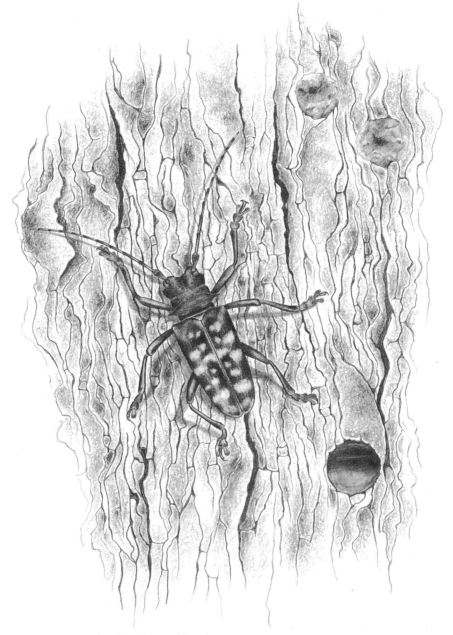

An Asian long-horned beetle sitting on the bark of a maple just above its emergence hole. Above it and to the right are two shallow pits in which others of its kind laid eggs.

weaken or kill them. The larvae, by boring in branches and the trunk, do most of the damage. Haack and his coauthors described how serious a threat these beetles are to our forests and shade trees: "Individual branches or entire trees can die if larval densities are high or if infestation continues for several years. Before a tree dies, however, heavily mined branches and stems commonly break, especially during strong winds."

Another long-horned beetle, which Craighead dubbed the ivory-marked beetle, comes close to holding the longevity record for insects. It bores in dead wood and also in healthy trees, among them oaks, hickories, birches, maples, and cypresses. It injures these trees, of course, but it is best known because it can survive for years after a tree is cut, the wood is seasoned, and even after it has been made into furniture. Adults of this lovely beetle, which may be as much as an inch long, not uncommonly emerge from furniture or woodwork in homes—hence its common name, furniture beetle. Writing in a book on insect records, Yong Zeng reported that one of these beetles emerged from a birch bookcase 40 years old.

How can this beetle survive without water for so long in dry, seasoned wood? The answer is that it survives on "metabolic water." The metabolism of a sugar or a fat yields water as a byproduct. The metabolism of a pound of fat, for instance, yields slightly more than a pound of water. The larva presumably digests cellulose, the major component of wood, breaking the cellulose down into its constituent sugars. It eats more wood and metabolizes more sugar than it needs for energy, dissipates the excess energy as heat, and retains the metabolic water to "quench its thirst."

Ecologists refer to insects and other animals that eat seeds as seed predators. Among them are many insects: some true bugs, beetles, ants, small wasps, flies, and moths. Just as a leopard or a praying mantis kills its prey, a seed predator kills a whole plant when it eats a seed. After all, the embryo in an acorn eaten by an acorn weevil could have become an oak tree, a giant of the forest. An insect that "parasitizes" a plant by eating its leaves or boring in a branch may or may not kill the plant, but a seed predator almost always does. Furthermore, as Daniel Janzen put it, a plant "knows" if its leaves or other vegetative tissues are being destroyed and can respond by deploying a chemical defense or grow replacement leaves. But the plant cannot "know" that its matured and already dispersed seeds, its offspring, have been destroyed, and therefore it cannot compensate for their loss. For these reasons, seed predators are especially effective at limiting the growth of plant populations. Seeds are usually the most nutritious part of a plant, a concentrated balanced diet in a compact package. As Frank Slansky and Antonio Panizzi observed, a seed must "be provisioned with enough energy and nutrients to produce sufficient roots, stem, and leaves to allow the developing seedling to achieve 'self-sufficiency.'" It is no wonder that seeds such as rice, corn, wheat, and other grains are the mainstay of the human diet.

The codling moth caterpillar, the worm in the apple, also burrows in several other fruits, including those of wild hawthorns. Soon after hatching from an egg laid on a leaf or a twig, the tiny caterpillar may enter a fruit from the side, but usually enters an apple through the little dimple at the blossom end of the fruit, the end opposite the stem. It then burrows to the core and often destroys the developing seeds. Another caterpillar, the pink bollworm, feeds in the fruits of cotton and other plants of the mallow family, okra, hollyhock, and the many species of the genus *Hibiscus*. Early in the season, these larvae burrow into flower buds of cotton but later in the season attack the developing fruits, the bolls, boring their way through the wall of the boll, the developing cotton fibers, and into the seeds.

The motive force of the seemingly inexplicable bouncing of a Mexican jumping bean is a seed-eating caterpillar, a relative of the codling moth, that lives within the "bean," which is actually a large seed of *Sebastiana*, a tropical member of the spurge family. The jumping of the seed is caused by the forceful impact of the caterpillar against the wall of the seed as it thrashes about. Walter Linsenmaier, in his beautifully illustrated *Insects of the World*, speculated that the jumping of the seed may serve the caterpillar. If the "seed is lying on the ground in a place unfavorable to the occupant, perhaps where there is too much sun, the caterpillar is able . . . to make the seed go leaping away."

Milkweed bugs suck nourishment from milkweed seeds with their piercing-sucking beaks. First, they liquefy the content of the seed, injecting it with their saliva, which, Vincent Wigglesworth notes in his book on insect physiology, contains enzymes that digest starch, proteins, and fats. Seeds are essential to milkweed bugs. If they are forced to feed only on vegetative parts of the plant, they grow slowly and often die. Until the tough and thick wall of the seed pod splits, their beaks cannot penetrate to the developing seeds, except where the walls are thinner at the pod's suture, the line along which it will split. When the pod begins to split open, they feed directly from the exposed mature seeds, which will eventually waft away in the breeze on their tufts of white floss. Milkweed bugs, as Carol Ralph noted, are gregarious and more likely to survive when they feed in groups than when they feed alone. Three or four of them may cooperate to exploit the same seed. According to the research of Jürgen Bongers and Wolfgang Eggerman, a lone milkweed bug cannot secrete enough saliva to liquefy the content of the seed so that it can be sucked out. But a group of bugs feeding together pool their saliva.

The larvae of the acorn and nut weevils, a small subgroup of the large family of snout beetles, live and feed in acorns or nuts. A snout beetle's snout is not a bundle of elongated mandibles and other mouthparts, as is the beak of a true bug or a mosquito. It is, rather, an extension of the head capsule ("skull") with a pair of small mandibles at its end. It may be broad and relatively short or very long and thin, as in the acorn and nut weevils. But in any case, the antennae, which in all insects always attach to the head, are on the snout, confirming that it is a part of the head capsule. The snout of the acorn weevil is very thin and longer than all the rest of its body. The female uses it as a substitute for an ovipositor. As Comstock described the process, an egg-laying female first bores a deep hole into an acorn, then turns around to drop an egg into this hole, and then turns back again to push the egg to the bottom of the hole with her long snout. Comstock said that an acorn weevil may insert her snout in the acorn up to the antennae, about three-quarters of the snout's length.

All of the world's approximately 1,300 species of seed beetles, family Bruchidae, spend their larval stage feeding within the seed of a flowering plant, most of them in the seeds of legumes such as acacias, locusts, mimosas, palo verde, clovers, beans, and peas. Others infest the seeds of a great many different plants of 24 other families, according to B. J. Southgate, of the Ministry of Agriculture, Fisheries, and Food in England. These seed beetles are small: the larvae of many grow to full size in a single seed, and the adults are usually less than two-tenths of an inch long. Several seed beetles have firmly established common names, such as bean weevil and pea weevil, that are misleading because these insects are not really weevils. Strictly speaking, only the snout beetles (family Curculianidae) are true weevils.

Some seed beetles glue their eggs one at a time to a single seed or to the outside of a seed pod. In an article on herbivorous insects, Arthur Weis and May Berenbaum reported that after a female lays an egg on a seed, she will—at least in some species—mark it with a pheromone that deters other

A female acorn weevil on a red oak acorn using her snout to drill
a deep hole in which she will deposit an egg.

females from laying eggs on the same seed. Other seed beetles lay their eggs
in small masses, usually on a single seed pod. In a desert in California,
Rodger Mitchell found, a seed beetle associated with the palo verde tree may
lay its eggs singly but sometimes stacks one directly upon another and covers both with a gluelike substance. The top egg protects the bottom egg
against drying out and against attacks by parasites, which together can kill
about 80 percent of single, uncovered eggs. Mitchell concluded that it is "advantageous for [the seed beetle] to put one egg on top of another, thus
sacrificing a second egg to double the survivorship of the first egg."

Newly hatched seed beetle larvae bore through the eggshell and directly
into a seed or first through the pod wall and then into a seed. Comstock's description of the behavior of the pea "weevil" tells the story: "when the larvae
hatch they bore through the pod into the young peas. Here they feed upon

the substance of the seed, which ripens, however, and in some cases will germinate when planted." Before molting to the pupal stage, the larva scores a circular groove in the seed coat, "leaving only a thin scale which is easily pushed away by the mature beetle."

Ever since people invented agriculture thousands of years ago in Mesoamerica, Mesopotamia, and the Nile Valley, they have been storing grain to tide them over the periods between crops. Seed-eating insects were quick to discover this bonanza and ever since have infested stores of grain ranging from a small box of popcorn kernels or a bag of bird seed in a home to a huge grain elevator—a prairie skyscraper—filled with tons of wheat, corn, soybeans, or other grains. The yearly loss may be as much as 50 percent in some countries, though it is much less in others. On average, about 10 percent of the grain harvested worldwide each year is lost to these insects.

Seed-eating insects may infest grain or grain products in your home. Since you can't apply insecticides to food, your only recourse is to examine every container of grain, cereal, flour, or other grain products in your home and discard them if they are infested. If you are thorough and if you meticulously clean kitchen cabinets and other places where infested material was stored, your home should be free of these insects.

My wife and I once had this problem. We saw small moths flying around the house, and in our pantry found their caterpillars infesting several packages of grain products that we seldom used. The wingspan of these moths wasn't much more than half an inch. The basal half of each front wing was off-white and the rest was coppery-red. They were Indian meal moths, which in the larval stage, Richard Cotton reports in his book on stored-grain pests, feed on whole grain, grain products of any kind, and almost any other organic substance ranging from dried fruit to powdered milk.

We did what I just recommended to you. But after a few weeks the moths were back. We again discarded infested material and cleaned our pantry thoroughly. But the moths and their caterpillars were soon back again. After the third time this happened, it finally dawned on me that these insects must have a refuge I didn't know about, an infestation somewhere else in the house. But where could it be and what could they be eating? We searched time and again but couldn't find an infestation other than the one in the kitchen. But one day my wife pulled out a drawer in a jewelry cabinet that she hadn't opened in a long time. Several Indian meal moths flew out, and there was the long-sought-for infestation. The caterpillars' feeding had riddled corn kernels strung on a leather thong to make a necklace. A few years earlier, my daughter Susan and I had visited a Native American pueblo near Santa Fe, New Mexico, where we bought my wife a gift of two necklaces of prettily dyed corn kernels.

All green plants, with the exception of mosses and algae, have two conductive tissues, phloem and xylem, which we can picture as plumbing systems that carry sap from one part of a plant to another. As you will remember phloem tubes reconduct nutrient-loaded sap down to the roots from the leaves, and xylem conducts nutrient-poor sap, almost nothing but water absorbed by the roots, up to the stems and leaves. Many insects with piercing-sucking mouthparts—some true bugs and most aphids, scale insects, and their relatives—ingest the nutrient-rich phloem sap, which is, on average, about 30 percent nutrients and only 70 percent water. But far fewer insects, among them spittle bugs and cicadas, drink the extremely dilute xylem sap, 99 percent water and less than 1 percent nutrients. Physically, of course, it takes much less effort for an insect to suck up phloem sap than xylem sap. Because it flows downward with the aid of gravity, a stream of phloem sap is under positive pressure, and therefore an insect with its mouthparts inserted into the phloem does not have to apply suction to receive the sap. But powerful suction is necessary to ingest xylem sap, which has negative pressure because it is drawn upward as water transpires from the leaves. Whereas phloem-feeding aphids have a small, weak pump in the head at the base of the beak, xylem feeders such as cicadas have much larger and more powerful pumps. Unlike the front of an aphid's head, the front of a cicada's head bulges forward prominently, forming a large internal chamber packed with powerful muscles that work its large suction pump.

Both phloem and xylem sap contain much more water than an insect can use. How do aphids and their sap-sucking relatives cope with the great excess of water they must ingest to get the nutrients they require? The problem is solved by a modification of the digestive tract known as the filter chamber. An insect's digestive tract is essentially a long tube consisting of three parts: the foregut extends from the mouth to the beginning of the midgut, which secretes digestive enzymes, absorbs nutrients, and connects to the hindgut, which carries waste to the anus. In a simple filter chamber, a length of the part of the midgut that adjoins the hindgut loops forward to contact the front end of the midgut, where it absorbs excess water from the ingested sap and passes it directly into the hindgut. Thus much of the water in the sap bypasses most of the midgut, and the result is a more concentrated solution of nutrients for the midgut to cope with. The excrement of these insects is the honeydew you became acquainted with previously.

🐜 🐜 🐜   Some insects can, to their own benefit, cause plants to develop tumor-like growths called galls. In *Bugs in the System,* May Berenbaum wrote that gall makers commandeer "the plant's hormonal system in such a way that the plant is induced to produce bizarre and unusual growths, which provide the insect with a place to live and with nice nutritious tissue on which to feed." Galls sap energy from the plant, slowing its growth and diminishing its ability to produce seeds. Among the gall-forming insects are some aphids, weevils, moths, flies, and wasps. In North America, reported Arthur Weis and May Berenbaum, about 1,700 insects are gall makers, and 70 percent of them belong to two families: the gall midges, flies of the family Cecidomyiidae, and the gall wasps of the family Cynipidae. (Alfred Kinsey, an expert on gall wasps, switched fields to write the famous "Kinsey Reports" on human sexual behavior.)

Well over 90 percent of the North American insect galls occur on flowering plants, most on various species of the rose, aster, and oak families. Galls can form on any part of a plant from the roots up to the leaves. Not only are gall makers generally host-specific, utilizing just one or a few closely related plants, but they also use only a specific part of a plant, such as the roots, stems, buds, leaf stems, leaf midribs, or leaf blades. Although galls come in various shapes and sizes, they are usually characteristic of the insect that induced them to develop. A gall-making species can be identified by its choice of host plant and the form and location of its gall.

The galls of aphids must have an opening through which the insects can escape because their piercing-sucking mouthparts could not break through the wall of a closed gall. The aphid-induced cockscomb elm gall, for example, an excrescence on the upper side of a leaf resembling the comb of a rooster, opens through a slit on the underside of the leaf. Other gall-making insects can gnaw through tough plant tissues. Their galls are closed. A typical closed stem gall has an inner chamber, in which the larval insect lives, lined by a layer of nutritive tissue on which it feeds. Surrounding the nutritive tissue is a thick layer of thin-walled undifferentiated cells, which is in turn surrounded by the stem wall, an inner layer of xylem and phloem, and an outer layer of epidermis.

Some common galls are conspicuous and likely to be noticed. Oak apple galls, caused by gall wasps, are globular, usually light brown, and large—as much as 2 inches in diameter. Ephraim Felt wrote in his book on plant galls that oak apple galls are "rather common on individual trees and occasionally are so numerous as to suggest a fair crop of fruit on an apple tree." They may hang from a leaf or consist of an entire deformed leaf, recognizable as such only because the gall arises from a leaf stem. Pine cone willow galls, induced by gall midges and always at the tip of a twig, look like miniature pine cones about an inch long. Hackberry nipple galls are small, only a sixth of an inch in diameter, but are conspicuous by their abundance—often a dozen or more of these rounded, nipple-like protuberances on the underside of virtually every leaf on a hackberry tree. The jumping plant lice that cause them, relatives of the aphids, escape from the upper side of the leaf via an opening in the gall.

In the winter you can easily find three types of conspicuous galls on dead stems of the previous summer's goldenrod plants. One of them, a rosette of deformed leaves about 2 inches across surrounding the stem, is caused by a tiny gall midge maggot that lives in the stem at the base of the rosette. It is known as the goldenrod rosette gall or the goldenrod bunch gall. The other two are large swellings of the stem. Spherical ones about an inch in diameter are caused by various species of goldenrod gall flies, which are picture-winged fruit flies and not to be confused with gall midges. In the winter, a central chamber is occupied by a large overwintering maggot, often used as bait by people who fish through the ice. In the autumn, it made an escape tunnel with an opening covered only by a thin layer of the gall's outer skin. In the spring the adult fly will shove its way through this thin cover. Many of these galls, especially if they are near a woodland, are pierced by large, ragged holes made by downy woodpeckers that pull out the maggots and eat them. Spindle-shaped goldenrod stem galls about an inch long are caused by caterpillars, several species of the same genus that attack different species of goldenrod. In the winter, these galls are unoccupied; the moth escaped in the autumn through a small round hole it made when it was still a caterpillar.

According to research done by David Hartnett and Warren Abrahamson, the spindle-shaped and spherical stem galls of goldenrod significantly limit

Galls common on goldenrod stems in the winter. Below the rosette gall are
a spindle-shaped stem gall with the caterpillar's escape hole and a nearly
globular stem gall with a hole through which a downy woodpecker
extracted the overwintering maggot.

both sexual reproduction by means of seeds and vegetative reproduction through the growth of new shoots from rootstocks. The gall insects caused plants to allocate resources to the growth of the gall that would otherwise have been used to produce seeds and rootstocks. On plants with galls, the share of the resources (measured as part of the biomass) allocated to the production of seeds can be decreased by as much as 43 percent.

Harvester ants—although they incidentally disseminate seeds—are probably the most important seed predators, and are often keystone species that largely determine the species composition of grassland or desert plant communities. By destroying seeds, harvester ants limit the populations of some plants. After Alan Andersen used an insecticide to destroy harvester ant colonies on experimental plots in a eucalyptus woodland in Australia, the number of new eucalyptus seedlings was 15 times greater than in similar plots from which the ants had not been removed. In a maize-growing ecosystem in Mexico, Stephen Risch and C. Ronald Carroll showed that by destroying seeds a harvester ant influences "which weed species become dominant and which go locally extinct, as well as the rates at which weeds recolonize areas." The ants decrease the abundance of weeds early in the growing season, thereby favoring farmers, because maize is most susceptible to competition from weeds when it is young. By preferentially harvesting grass seeds, the ant increases the ratio of herbaceous weeds to grassy weeds, again favoring the farmer, because herbaceous weeds are easier to eradicate than grasses.

Harvester ants and rodents compete with each other for seeds. As surprising as it may seem, competition from ants limits rodent populations, and competition from rodents limits ant populations. In a desert in Arizona, James Brown and two colleagues established four types of experimental plots: some from which only rodents were excluded, by trapping and fencing; some from which only ants were excluded, by the application of minute quantities of insecticide to the entrances of their nests; others from which both ants and rodents were excluded; and, finally, control plots from which neither ants nor rodents were excluded. There were more seeds on plots from which both rodents and ants had been excluded than on plots where both were present. When only rodents were excluded, the number of ant colonies increased by 71 percent; when only ants were excluded, the number of rodents increased by 20 percent. Furthermore, ants increased the diversity of plant species by harvesting mostly seeds of the more common plants and sparing the harder-to-find seeds of rarer plants.

Like harvester ants, all herbivorous insects may affect the succession of plants that recolonize a denuded area, such as an abandoned crop field, as it returns to its natural state. Walter Carson and Richard Root did experiments during the first three years of the succession in three abandoned crop fields in western New York. In each field they designated 10 pairs of plots. One randomly selected plot of each of the 30 pairs was treated with an insecticide once every 2 weeks; the other plot in a pair, the control plot, received no insecticide. Herbivorous insects, the dominant one a spittlebug, were abundant in the control plots but were virtually absent from insecticide-treated plots. Both the total plant biomass and the kinds of plants present were significantly affected by the herbivores. The ihivorous insects reduced the total plant biomass by about 31 percent, increased the number of different plant species present by about 9 percent, and so greatly reduced the abundance of meadow goldenrod that lance-leaved goldenrod became the dominant plant of the succession. Such effects of herbivores have ecological repercussions up and down the line. Limiting the plant biomass limits the total insect biomass that a field can support. But increasing plant diversity makes it possible for more different kinds of herbivores to inhabit the field. More species of herbivores makes possible an increase in the numbers of species of parasitic insects and predaceous insects and spiders. The end result is a considerable increase in diversity. And many ecologists believe that the stability of ecosystems tends to increase with their diversity.

This study leaves no doubt that herbivorous insects limit the growth of plant populations. But it considered only the first two levels of the food web: the nutrient-producing plants and the insects that feed on them. What are the effects of the third level, such as insect-eating vertebrates and predaceous and parasitic insects? Most insect-eating vertebrates, such as toads, lizards, or birds, feed indiscriminately on herbivores, parasites, and predators. How, then, do they affect the herbivores that control plants? Do they promote increases in plant populations by eating herbivores? Or do they increase herbivore populations by destroying the parasites and predators of herbivores? A few studies that include three levels of the food web—plants, plant-eating insects, and insect-eating lizards or birds—consider the plant-herbivore interaction in a more nearly complete ecological context. A study by Thomas Schoener and Catherine Toft and another by David Spiller and Schoener examined the effect on the herbivore-plant interaction of small

Anolis lizards ("chameleons") that prey on both plant-eating insects and insect-eating spiders. They found that some tiny islands, cays, in the Bahamas were inhabited by *Anolis* lizards while others were not—a natural experimental set-up for examining the role of these lizards in ecosystems. All of the islands had spiders that spin orb webs, which snare mostly flying insects. On islands without lizards, which eat spiders as well as insects, orb weavers were about ten times more numerous than on islands with lizards. On islands with lizards, there was only about two-thirds as much damage done to leaves by chewing insects as on islands without lizards. As Spiller and Schoener said, "lizards reduce the overall damage from herbivorous arthropods, even though they also reduce the numbers of web spiders which eat some herbivorous arthropods."

A similar experiment, done in Missouri by Robert Marquis and Christopher Whelan, compared the impact on trees of uncontrolled and bird-controlled insect populations. They covered white oak saplings with nets with a mesh too fine to admit birds but coarse enough to let in insects. Control groups of saplings were left uncovered. Comparing these two groups with a third group left uncovered but sprayed with an insecticide revealed a cascade of effects that went from the third level in the food web, birds, to the first level, the oak saplings. Birds reduced the population of plant-feeding insects by half; this, in turn, made possible a 33 percent increase in the aboveground growth of the saplings. Insectivorous birds limited the growth of insect populations in the community, and thereby significantly decreased the controlling effect of herbivorous insects on plant populations.

Just as insects transmit pathogens that cause diseases of animals, they also transmit viruses, bacteria, fungi, and nematode worms that cause diseases of plants. In their book on destructive and useful insects, Robert L. Metcalf and Robert A. Metcalf concluded that the destruction caused by diseases transmitted by insects "may rival in importance the destruction caused by their direct feeding." While many pathogens of plants are transmitted by wind, rain, mechanical injury, or other nonbiological factors, more than 400 known plant pathogens, about 250 of them viruses, are transmitted by insects, most often via the contaminated mouthparts of piercing-sucking insects such as leafhoppers and aphids. In fact, the Metcalfs wrote, these two groups of insects disseminate more plant diseases than all other kinds of insects combined. According to *Insects in Relation to Plant Disease* by Walter Carter, more aphids than leafhoppers transmit plant viruses; the green peach aphid alone is known to transmit over 150 different kinds of viruses.

The fungus that causes Dutch elm disease is transmitted mainly by an introduced species, the European elm bark beetle. The larvae pass the winter in the corky layer of the bark of dead or dying elms, metamorphose to the adult stage in spring, and then gnaw their way out, becoming contaminated with spores of the fungus if their host tree has Dutch elm disease. The hungry beetles fly to healthy trees to feed on living bark in the crotches of twigs, thereby inoculating the tree with any spores they carry. Since their larvae cannot survive in healthy trees, females burrow through the bark of dying trees to form egg galleries by tunneling parallel with the grain at the wood-bark interface. As many as 70 eggs are spaced along the two sides of a gallery. The larvae burrow and feed in the inner bark and the surface of the wood. The adult European elm bark beetles' habit of feeding on the twigs of healthy trees and laying their eggs in dying trees, which are likely to be infected with Dutch elm disease, creates a self-perpetuating cycle of disease transmission. Every healthy tree inoculated by a beetle that came from a tree infected with the disease will sicken and become a suitable breeding site for yet more bark beetles.

🐜 🐜 🐜 Every member of a species of plant or animal, wrote Peter Price in his book on insect ecology, is preoccupied with acquiring enough energy and nutrients to make reproduction possible, and an individual would be at an evolutionary disadvantage if it gave other species easy access to the store of energy and nutrients represented by the protoplasm of its body. Consequently, there is constant warfare between the eaten and the eaters. Prey animals evolve new defenses against predators, and the predators must then evolve ways of overcoming the new defenses, switch to other prey, or starve to death. Plants have their own ways of defending themselves against insects. Previously I described some of the amazing ways in which plants bribe ants to protect them against plant-eating insects, and how, when being munched on by herbivorous insects, they send out a chemical "cry for help" that brings parasites of the herbivores to the rescue. But other strategies in their repertoire of defenses are equally if not more amazing. Some plants have physical protection such as tough leaves, thorns, or hairs; most if not all contain chemical defenses such as toxic or repellent substances; and a few have even resorted to mimicry to warn away herbivores.

The leaves or stems of many plants—among them sycamores, mulleins, mouse-ear chickweed, forget-me-nots, and velvetleafs—are clothed with hairs that, at least in some species, have been shown to protect the plant against herbivorous insects. Almost all of the research demonstrating that leaf hairs ward off attacks by insects has been done with crop plants, but according to Donald Levin's article on plant defenses, it is a fair assumption that the results of these studies apply to many other plants as well.

Some varieties of soybeans have hairy leaves and others have smooth, hairless leaves. There is no doubt that leaf hairs protect soybeans against one of their worst enemies, misleadingly called the potato leafhopper—despite its common name, this little green insect, only an eighth of an inch long, sucks phloem sap from the leaves of over 100 species of cultivated and wild plants. A dense growth of hairs keeps it from reaching the surface of the soybean leaf to feed and to poke its eggs into the main veins. Soybean varieties with smooth leaves are massively attacked and seldom grow to be more than 8 inches tall. Hairy-leaved varieties suffer little damage, although the leafhoppers readily feed on them if they are shaved by an experimenter. They grow to be over 36 inches tall. Needless to say, only hairy-leaved soybean varieties are grown commercially. At the University of Illinois in Urbana-Champaign, Delmar Broersma and two coresearchers studied the protective value of hairy leaves. In one field trial, smooth-leaved plants had an average of about 31 potato leafhoppers per leaf, while hairy-leaved plants had an average of less than 4. The leafhoppers had a drastic effect on yield. Smooth-leaved plants produced less than 9 bushels of beans per acre while hairy-leaved ones yielded over 47 bushels per acre.

Both adult and larval cereal leaf beetles chew on the shoots and leaves of wheat, other cultivated grains, and wild grasses. In field trials, J. A. Schillinger, Jr., and R. L. Gallun found that the number of eggs laid on wheat leaves and the survival of the larvae decrease as the hairiness of the leaves increases. On a sample of a smooth-leaved variety they found 213 eggs; on a slightly hairy one, 74 eggs; on a moderately hairy variety, 38 eggs; and on a densely hairy one, only 4 eggs. The survival of larvae forced to feed on these varieties was also inversely related to the hairiness of the leaves. On smooth, slightly hairy, moderately hairy, and densely hair leaves, the survival rates were 92, 75, 60, and 20 percent, respectively.

A sanitation officer with the Austro-Hungarian army named St. v. Bogdandy described a folk remedy used in Albania and other Balkan countries to rid homes of bed bugs. In the evening, fresh leaves of the common garden bean are spread over the floor with their undersides facing up. In the dark of night the blood-sucking bed bugs leave their hiding places to crawl over the floor to beds occupied by people. In the morning, many bugs are found on the leaves, and are removed with the leaves and destroyed. Using this method, Bogdandy in one night trapped over 2 pounds of bed bugs in one room. Bogdandy did not explain why the bugs remained on the leaves. Later, Henry Richardson showed what happens: the leaves themselves do not attract bed bugs, but if a bug happens to walk over one, its legs are caught and held by tiny hooklike hairs on the leaf's underside. I have found no reports that these hooked hairs protect bean plants from herbivorous insects, but I have no doubt that they do.

In an interesting study, Lawrence Gilbert showed that certain species of passion fruit vines have probably won the evolutionary arms race with one of their arch enemies, caterpillars of the colorful, tropical heliconiine butterflies. He noted that one species of passion fruit vine was not attacked by these caterpillars while several others were. Close examination showed that the leaves of the vine that was not attacked were covered with sharply pointed, hooked hairs, while the leaves of the others were bare. The hooked hairs punctured the soft skin of the caterpillars, effectively snagging them and preventing them from moving. The following day snagged caterpillars were dead and dried out due to a combination of starvation and loss of blood. But caterpillars of another group of butterflies (family Ithomiidae) have found a way—one that requires cooperation between the members of a group—to avoid the sharp hairs that protect the leaves of their food plant, a member of the nightshade family. Beverly Rathcke and Robert Poole described how a group of these caterpillars, usually from four to six per leaf, spin on the underside of a leaf a hanging scaffolding of fine silk threads over which they crawl, avoiding the spines, to chew on the unprotected edge of the leaf.

Heliconiine butterflies usually lay their eggs on young tendrils of passion fruit vines, first visually inspecting them and not laying an egg if other eggs are present, thus sparing their offspring competition from other leaf-eating caterpillars. Many passion fruit species take advantage of the cautious egg-laying behavior of these butterflies. They grow convincing mimics of eggs on their tendrils and leaves, thereby dissuading female heliconiines from laying eggs on them. This form of mimicry has evolved independently several times, for, as Woodruff Benson and two colleagues noted in an article in *Evolution,* in different species of passion fruit vines, the false eggs are modifications of several different plant parts, such as modified nectar glands or even disposable flower buds that fall from the tendrils after growing to the size of a butterfly egg. Experiments by Kathy Williams and Lawrence Gilbert show that egg mimicry actually works. In one experiment they placed normal yellow eggs or eggs that had been dyed green on green tendrils of a species of passion fruit vine that does not have false eggs. Female heliconiines laid eggs on bare tendrils and those with green eggs, but seldom on tendrils with yellow eggs. The clinching experiment compared tendrils with natural egg mimics with tendrils of the same species from which the egg mimics had been removed. Females readily laid eggs on tendrils from which the egg mimics had been removed but laid far fewer on tendrils with intact egg mimics.

All, or at least the great majority, of the green plants, many of them without effective physical defenses, use chemical defenses to discourage attacks by plant-eating insects. These chemical warfare agents are called "secondary plant substances," secondary because—unlike the primary plant substances, such as enzymes, hormones, and the components of the protoplasm—they are not necessary to the physiological processes of the plant. According to Pierre Jolivet, a student of plant-insect interactions, about 30,000 secondary plant substances are currently known, and the botanist and biochemist David Seigler estimated that almost a thousand more are discovered every year. Most plants contain many of them—perhaps even a few thousand—but never all of them. Why are there so many secondary plant substances? One reason is the ongoing "evolutionary arms race" between plants and their enemies. If, for example, an insect becomes resistant to a plant's biochemical defense, the plant may counter by adding other defensive biochemicals to its arsenal, and the insect may respond by developing resistance to the new defense. In this way, more and more defensive chemicals evolve.

The secondary plant substances are a chemically diverse lot from 30 or more structural classes, and affect insects in many different ways. Some of them are not defenses—the scents of flowers, for example, are signals that attract pollinating insects—but most of them are an awesome array of defensive chemicals, including, among others, insect-killing poisons such as nicotine, pyrethrin, and Rotenone, and also chemicals that mimic insect hormones, such as juvenile hormone, which cause lethal developmental abnormalities by disrupting normal growth and metamorphosis. Numerous demonstrations of the effects of plant toxins on insects were summarized by

Gerald Rosenthal and May Berenbaum in their book on the interactions of herbivores and plants.

People use several plant toxins as insecticides. Nicotine, probably the first to be used in this way, comes chiefly from two plants, the common tobacco plant, *Nicotiana tabacum*, and *Nicotiana rustica*, which is grown mainly for its high nicotine content. Beginning in the late seventeenth century, people used water extracts of tobacco or tobacco smoke to kill aphids and other soft-bodied insects. Nicotine kills insects by blocking the transmission of signals in the central nervous system. Pyrethrin occurs in certain chrysanthemums, especially in the flowers. Around 1800 in Persia, now Iran, ground chrysanthemum flowers were found to kill lice and fleas. For several decades, the source of "Persian insect powder" was a closely guarded secret and it was sold at exorbitant prices. But by 1851, the secret was out, and pyrethrin was soon used worldwide to control many insects. Pyrethrin, or a synthetic variant of it, is still the active ingredient in household "bug bombs," and it is one of the natural insecticide used by growers of "organic" vegetables and fruits. It also kills by interfering with the central nervous system. The roots of certain shrubs yield another natural insecticide, rotenone, which kills by disrupting cellular metabolism. Ryanodine, used for the control of fruit pests and derived from the stems and roots of *Ryania*, a tropical shrub, kills by interfering with muscle contractions.

When Karel Sláma moved a laboratory colony of linden bugs, essential to his research, from Czechoslovakia to the Biological Laboratories at Harvard University, all 1,500 bugs died from developmental abnormalities. Some underwent extra molts but never became adults and others were "monsters" with a combination of immature and adult characteristics. These effects could have been produced only by a disruption of the bugs' hormonal systems caused by exposing them to excess juvenile hormone or some substance similar to it. Nothing like this had ever happened during the 10 years Sláma had maintained the colony in Prague. He and his Harvard colleague Carroll Williams found 15 differences in the way the colony was handled in Prague and at Harvard. One or more of them had to be the cause of the catastrophic failure of the colony. A systematic study eliminated all but one. "The source of the juvenile hormone activity was finally tracked down to exposure of the bugs to a certain paper towel (Scott, brand 150) which had been placed in the rearing jars. When this toweling was replaced by Whatman's filter paper, the entire phenomenon disappeared and all individuals developed normally."

At first Sláma and Williams thought that a chemical added to the towels was responsible, but other brands of American paper towels and tissues had the same extraordinary effect on the bugs. "Indeed, pieces of American newspapers and journals (*New York Times, Wall Street Journal, Boston Globe, Science,* and *Scientific American*) showed extremely high juvenile hormone activity when placed in contact with [the bugs]. The *London Times* and *Nature* were inactive, and so were other paper materials of European or Japanese manufacture." To make a long story short, Sláma and Williams finally determined that the substance with juvenile hormone activity, which was absorbed through the bugs' "skin," came from the balsam fir, an indigenous American tree used to make paper.

The molting and metamorphosis of insects are controlled by a complex of hormones, chemical messengers secreted by endocrine glands and carried to "target tissues" by the blood. When an insect has grown so large that it must molt its skin, the brain secretes a hormone that stimulates a pair of glands in the thorax to secrete another hormone, ecdysone, that triggers molting. At the same time, a pair of glands near the brain releases juvenile hormone, so called because it inhibits metamorphosis, causing a larva to remain a larva after the molt rather than metamorphosing to the pupal stage—or in the case of insects with gradual metamorphosis, such as linden bugs, it prevents them from going directly to the adult stage. When the insect is ready to become a pupa or an adult, the level of juvenile hormone in the blood is suppressed and metamorphosis proceeds. At this stage, exposure to superfluous juvenile hormone causes developmental abnormalities.

A number of plants produce substances that are similar in chemical structure to juvenile hormone and protect them from insects. In Malaysia, Yock Toong and two Americans discovered that a sedge, locally known as the grasshopper sedge, contains a juvenile hormone mimic and showed that it causes immature grasshoppers that feed on it to become abnormal adults. Not surprisingly, synthetic chemicals resembling juvenile hormone are being used as insecticides.

🐜 🐜 🐜 Most herbivorous insects feed on only one or a few closely related plants and will starve to death rather than feed on some other plant. In a book entitled *Insect Dietary,* Charles Brues put this insectan proclivity in perspective:

We ourselves are so accustomed to partake of a great variety of vegetable food according to the passing whims of our appetites, or with the turn of the seasons . . . The same is true very generally of our vegetarian farm animals whose grazing is not closely limited to particular forage plants except that the more succulent kinds are commonly most attractive and some conspicuously malodorous or otherwise unpleasant species may be generally avoided. We thus think of a mixed vegetable diet as quite natural for herbivorous animals and are startled to learn that at least one primitive marsupial mammal, the Australian Koala . . . lives continuously from generation to generation on the leaves of a particular species of eucalyptus . . . Among insects this kind of restriction to a single species of food plant or to a series of related species is by no means rare, and insects which do not show some such preference are the exception rather than the rule.

Although the degree of food plant specificity is a continuum, it is helpful to think of three broad categories. At one end of the spectrum are monophagous insects, which feed on only one or a few very closely related plants. The most widely known example is the silkworm caterpillar, which eats only leaves of mulberry and a few other plants of the same family. Other insects are oligophagous and feed on a wider variety of plants, which are usually members of the same family and contain a similar, if not identical array of secondary substances. For example, caterpillars of the familiar black swallowtail butterfly feed on celery, parsnip, parsley, dill, caraway, and many other members of the carrot family. Other insects, said to be polyphagous, feed on a wide variety of plants from different families. Many of the grasshoppers are highly polyphagous, and some, J. H. MacFarlane and Asgeir Thorsteinson discovered in their research, grow faster, gain more weight, and are more likely to survive if they feed on a mixture of several plants than on just one.

It seems intuitively obvious that in an ecosystem with hundreds or thousands of plant species, an insect can more easily find a meal if it is able to feed on all or almost all of those plants rather than on only one or a few of them. Why, then, are there so many insects that will feed on only one or a few plants, and why do even polyphagous species refuse to feed on many of the plants that surround them? The answer is that the continuing evolutionary arms race between plants and insects forces insects to specialize on just a few plants, those whose chemical defenses they have managed to breach. No insect can overcome all of the many different chemical defenses of plants. So all insects are bound to their particular host plants because they cannot cope with the chemical defenses of other plants. As Gottfried

Fraenkel and later Paul Ehrlich and Peter Raven argued, the unique flavors and odors of a plant, even those of its chemical defenses against other insects, have been adopted by the specialists that feed on it as chemical clues, "sign stimuli," by which they distinguish their own host plants from the many others whose defenses they cannot overcome.

From the evidence assembled above, it is clear that herbivorous insects are an essential controlling element of terrestrial ecosystems. Without them populations of many plants would "explode" and disrupt their ecosystem. But herbivorous insects, like plants and all other organisms, produce more offspring than are required to replace themselves, and therefore they are as prone to undergo population explosions as are plants, and could disrupt the "balance" of an ecosystem by seriously depleting or even eliminating populations of one or more plants. But as you will see next, herbivorous insects and all other insects have enemies that prevent their populations from increasing to troublesome levels.

CONTROLLING INSECT POPULATIONS

Imagine a tightly packed ball of fruit flies, the *Drosophila* of genetics fame, so large that it extends from the earth to the sun, 96 million miles. This huge mass would, as Donald Borror and his coauthors point out in *Introduction to the Study of Insects,* be the measure of the last of 25 generations if a single pair of these flies and their descendants reproduced, unlimited by deaths, for a year. (A female *Drosophila* lays about 100 eggs, and her offspring begin to reproduce in about 2 weeks.) The twenty-fifth generation would include about 10^{41} flies, the numeral 1 followed by a string of 41 zeroes. Only astronomers are accustomed to such huge numbers—one of them, Jim Kaler, told me that the weight of all the stars in our galaxy is only about 10 times greater, roughly 10^{42} pounds. Entomologists have, mostly for fun, made similar calculations for other insects. For example, one entomologist estimated that in a single summer a pair of house flies and their descendants could produce enough progeny to cover the earth to a depth of 47 feet. Harold Oldroyd repeated the calculations and reported in *The Natural History of Flies* that there would be only enough to cover Germany to that depth. But as he put it, "that is still a lot of flies."

Population increases of such magnitude are flights of fancy that can exist only in the mind. Nevertheless, they are compelling testimony to the immense potential for population increase common to all animals and plants. Such increases are possible because, as Thomas Malthus recognized in 1798, all organisms, including humans, produce more young than are likely to live, a form of insurance that increases the probability that, despite many hazards, some offspring will survive their parents. In biparental species, populations will remain stable if only two of these young live to replace their mother and father. But the survival of only a few extra offspring will ultimately result in a population explosion. Consider *Drosophila*. If 2 of each female's 100 eggs live to become reproducing adults, she and her mate will have been replaced and the population will remain stable from generation to generation, neither increasing nor decreasing. But if 4 of those eggs, only 2 extra, survive to adulthood, the population will double. If this trend continues, the tenth-generation descendants of only one pair will consist of 1,024 flies, and the twenty-fifth of over 3.5 million—not a ball 96 million miles in diameter, but still a lot of flies. And even after only five generations, the population will have increased by an ecologically disastrous 16 times. The obvious conclusion is that stability is possible only if on average 98 of each female's offspring do not survive.

Like plants, animals have two basic strategies for increasing the odds that their offspring will survive to pass their genes on to future generations, to breed prolifically or to breed in a restrained manner. The prolific breeders, such as the fruit flies, invest a minimum of resources and effort in each offspring but produce a great many of them in the hope that a few lucky ones will survive. The restrained breeders produce just a few offspring but give them a good start in life by investing in each one a large quantity of resources, sometimes just in the form of a large egg that contains an abundant supply of nutrients, but in other cases a long period of gestation and/or a great deal of parental care. An extreme example of a prolific breeder is the codfish, which may lay more than 6 million eggs but invests no more than a minimal quantity of yolk in each one. Mammals and birds are in general restrained breeders. People and elephants, for example, have only a few offspring during their lifetime but invest a great deal in each one of them, a long period of gestation followed by years of parental care.

Most insects are prolific breeders but a few are restrained breeders. The most famous of the latter—and probably the most restrained of them all—is the tsetse, the dreaded blood-sucking fly of Africa that transmits from wild animals to humans the pathogen that causes sleeping sickness. After being inseminated by a male, the female tsetse produces a single larva, a maggot, that she retains in a uterus-like organ for a gestation period of 9 to 10 days, during which time the larva, feeding on "milk" secreted into the uterus by a mammary gland, grows to a large size—sometimes, according to David Denlinger and his fellow researchers, weighing more than its mother. The mother is aided by powerful contractions of her abdomen as she gives birth to the fully grown larva in a shaded place on the ground. Almost immediately the larva burrows into the soil and soon it molts to the next stage of the life cycle, the pupal stage, which lasts 3 or 4 weeks and in which the metamorphosis from legless and wingless maggot to winged and highly mobile adult takes place. During her lifetime, a female produces very few offspring. The record, Denlinger told me, is 20, produced by a female protected and coddled in a laboratory. In nature females probably seldom produce more than 3 or 4.

Unlike the tsetse, the great majority of insects produce anywhere from a few dozen to many thousands of eggs. Brown-banded cockroaches, often pests in human habitations, are near the low end of the scale. During her usual lifespan of about 115 days, an average female lays about 150 eggs in 10 batches of 15, each batch enclosed in a protective capsule. A female tent caterpillar moth, having only vestigial mouthparts, cannot feed and lives for only a day, but nevertheless lays about 300 eggs in a single mass that is covered with a shellac-like coating and encircles a twig of the host plant, usually a wild black cherry tree, like a collar. Corn earworm moths, the adult stage of the fat caterpillars that we sometimes find under the husk at the tip of an ear of sweet corn, may survive for several weeks and lay from 500 to as many as 3,000 eggs, which they glue one at a time to fresh corn silks. A female wasp of the family Trigonalidae scatters about 10,000 minute eggs on the leaves of many different kinds of plants. They remain viable for several months. The larva of this wasp, Richard Askew notes in *Parasitic Insects,* is a parasite within the body of another species of parasite that lives within the body of a leaf-eating caterpillar. The trigonalid can get into a caterpillar's body only if a feeding caterpillar inadvertently swallows it when it is still in the egg stage. This is a long shot, so the female adult wasp lays many eggs that can survive in the open for a long time.

A few parasitic wasps have a way, unique among the insects, of greatly increasing their reproductive potential: polyembryony, the formation of multiple embryos from a single egg. Human reproduction is polyembryonic when a single developing embryo divides to produce identical twins, triplets, quadruplets, or even more little humans. Some of these parasitic wasps are always polyembryonic. One example is a tiny wasp that parasitizes caterpillars. The embryo from an egg of this wasp divides and redivides to form hundreds of identical embryos. Since a mother parasite may insert more than one egg into one of her victims, as many as 3,000 of her offspring can develop in one caterpillar, ultimately devouring all but the caterpillar's skin, and completely filling the skin as meat stuffing fills a sausage casing.

Population explosions of insects are common in monocultures of crop plants. In the Midwest, corn and soybean fields stretch as far as

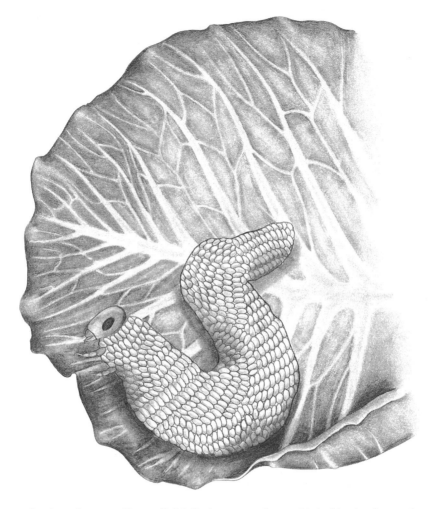

The skin of a caterpillar stuffed full of parasites that multiplied by dividing and
redividing before eating all of their host's inner tissues.

the eye can see, a huge and almost limitless banquet for insects. These
monocultures, wrote Mark Winston in *Nature Wars,* "are concentrated food
sources for organisms that previously foraged widely." In a natural ecosystem, an insect's preferred food plants, scattered among many other kinds of
plants, are difficult to find, and consequently population growth is checked,
but pest insects spread through a monoculture as a prairie fire sweeps
through dry grass.

Once in a great while, though, there are population explosions in natural ecosystems. The most monstrous and infamous of them are the outbreaks of migrating "locusts," as migrating grasshoppers are called, that occur in North Africa, the Near East, Australia, and elsewhere. Swarms of billions of flying locusts migrate for hundreds of miles and wreak havoc wherever they land. The Bible (Exodus 10:15), gives us a powerful and essentially accurate description of what an invasion of locusts is like. "For they covered the face of the whole earth; so that the land was darkened; and they did eat every herb of the land, and all the fruit of the trees . . . and there remained not any green thing, either tree or herb of the field, through all the land of Egypt." Swarms probably result from competition between nymphal locusts for the green plants they eat. When food is abundant and young locusts are not crowded together, they are in their nonmigratory solitary phase. When their population grows, they eat most of the available plants and crowd together on the remaining ones. The constant jostling from being in a crowd stimulates them to switch to their gregarious phase, which is anatomically and behaviorally distinct from the solitary phase. When the gregarious nymphs become winged adults, they fly off en masse in search of greener places in which to feed and lay eggs.

We cannot ignore the runaway growth of the human population. The world population of people, according to estimates published by Anne and Paul Ehrlich, was about 5 million when agriculture was invented about 10,000 years ago and about 400 million in 1650 c.e. when the first censuses were taken. In my own lifetime, the human population has grown at a frightening rate. When I entered high school it was about 2 billion. Now, less than 50 years later, it has tripled to over 6 billion, and in a few decades it will double to 12 billion. It is only a matter of time until, as Malthus warned, we outgrow our food supply and other resources.

But what kills the 98 superfluous fruit flies and the excess offspring of all other animals? The factors that put the brakes on the growth of animal populations—hazards from the victim's point of view—are many and varied. They can be physical. If there is a heavy rain, tiny, just-hatched grasshoppers are likely to become mired in mud and die when they try to make their way to the surface from an egg pod buried in the soil. Insects may be killed by a late frost in the spring or an early one in the fall. On small islands, flying insects are often blown so far out to sea by the wind that they perish before they can return to land. Consequently, many species on small islands are flightless, and thus safe from the wind.

Deaths may also be caused by biotic factors such as competition from other individuals for food or other resources. Host-specific beetles will not survive if they cannot find, among the many species of plants in an ecosystem, one of the few that evolution has designed them to eat. Unless a queen bumble bee can find a suitable cavity in the soil—often the burrow of a mouse or some other small mammal—soon after coming out of diapause in springtime, she will die without founding a colony. Much the same thing can happen to a honey bee queen that, with her accompanying swarm, has left the hive to found a new colony. In a woodland in Massachusetts, I saw a neat cluster of large, uncovered honey bee combs, tended by busy workers, hanging from the underside of a wide branch high in an oak. The workers had not found an acceptable cavity for a nest before the honey stored in their stomachs ran out, so they gave up and built their combs where the swarm hung from a branch. This unprotected colony perished when winter came.

Parasites, pathogens, and predators are very important in limiting the growth of insect populations. They include viruses, bacteria, fungi, certain plants, other insects, and insect relatives such as spiders, mites, and scorpions, as well as representatives of all the major groups of vertebrates: fish, amphibians, reptiles, birds, and mammals.

Parasitic and predaceous insects are essential to maintaining the balance of both natural and agricultural ecosystems. Several experiments show that populations of plant-feeding insects rapidly increase when their insect parasites and predators are artificially excluded. One of the early ecological experiments done in nature showed that both birds and ants significantly depress population growth of the western spruce budworm, a caterpillar that feeds on the needles of conifers. In a monumental effort, Robert Campbell, Torolf Torgersen, and Nilima Srivastava covered two 30-foot-tall Douglas firs, one of several of the budworm's food trees, with netting with a mesh too fine to admit birds but coarse enough to let insects through. Both birds and ants were excluded from another two trees, the birds kept out by netting and the crawling, wingless ants blocked from climbing the trees by a wide band of sticky material circling the base of each trunk. Two more trees, obstructed by neither netting nor sticky material, were controls accessible to both ants and birds.

Both birds and ants contributed significantly to the control of budworm populations, but they divided up the "territory" on a tree. Birds tended to feed in the top third of the tree and ants in the lower two-thirds. Nevertheless, only one of these two groups is sufficient to control budworm populations. If there were too few ants, the birds compensated for them, and if there were too few birds, the ants compensated for them. But when both birds and ants were excluded, budworm populations soared. In an earlier experiment, Campbell and Torgersen found that when ants were excluded, only 8 percent of the budworm pupae on Douglas fir branches were killed, but when ants had free access to branches, 85 percent of the pupae were killed.

In a stand of loblolly pines in the Kisatchie National Forest in Louisiana, Peter Turchin and two colleagues compared the survival rate of southern pine bark beetles protected from insect parasites and predators by mesh nets covering the trunks of pines with the survival rate of beetles exposed to their enemies on trunks that were not covered. Eggs of the beetle, which are laid under the bark, were more likely to survive when parasites and predators were excluded. Similarly, more adults emerged from under the bark when they were protected by nets.

The results of their experiment, which was repeated for 5 consecutive years, leave little doubt that the population cycles of southern pine beetles are caused by their enemies. The researchers' data on beetle survival show that populations of the beetles' enemies increased as their food supply, the beetles, increased. But, as is to be expected, the growth of parasite and predator populations lagged behind the growth of the beetle population. Not until a year after the beetle population had peaked did parasite and predator populations peak. After that the beetle population, and presumably that of its enemies, declined precipitously. Eventually, the beetles—largely freed of their enemies—began to increase again. Although the study stopped at that point, it can safely be assumed that the increase of beetles was followed by a corresponding increase of their enemies.

In 1998 Matthew Moran and Lawrence Hurd published the results of experiments that revealed a "cascade" of effects that flowed down the food chain from the carnivorous praying mantis at its top to the green plants at its base. Mantises, unlike the host-specific insect parasites of southern pine beetles, are generalists that will eat almost anything that moves—mostly insects and other arthropods. In an abandoned crop field overgrown with grasses and leafy plants, Moran and Hurd set up an array of experimental plots, each about 10 feet on a side and enclosed by a low fence of plastic

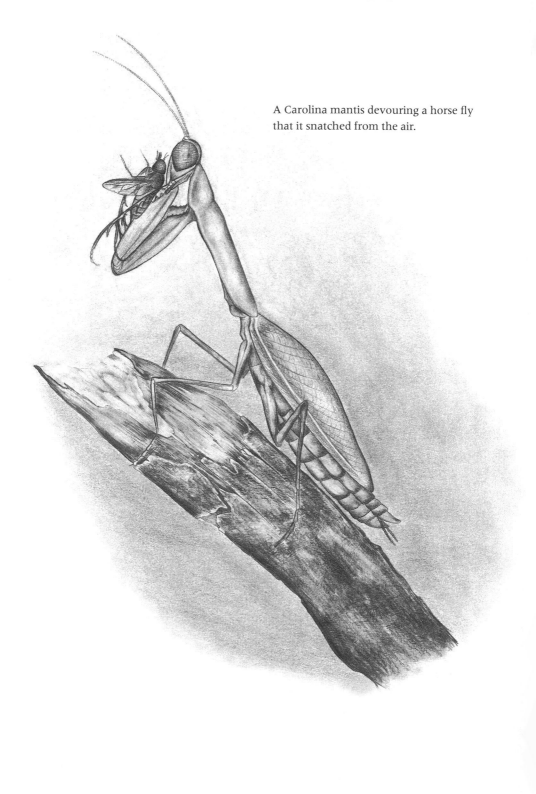

A Carolina mantis devouring a horse fly that it snatched from the air.

sheeting bearing an outer band of sticky material along all four sides. Crawling insects, including young mantises, were excluded, but flying insects and spiders floating on threads of silk freely entered from above.

None of the plots contained naturally occurring mantis egg pods, easily spotted in winter because they are large and conspicuous on the dead stalks of herbaceous plants. Early in May the experimenters placed two egg pods in each of the experimental plots, but placed none in control plots. A population of about 50 tiny mantises per square yard soon hatched from the pods. As the season progressed the mantis population declined, although quite a few almost fully grown but still wingless individuals remained when the experiment was stopped at the end of July. Then the number and biomass of all insects and plants were estimated. The results were unambiguous. In the plots with mantises, both the total count and the biomass of herbivorous insects was about 32 percent lower than in plots without mantises. A more telling statistic, the biomass of herbivorous insects per biomass of plants, was 45 percent lower where mantises were present. The cascade of effects extended to the plants; by decreasing the number of plant-eating insects, the mantises significantly benefited the plants, whose biomass was about 24 percent greater in plots with mantises than in plots without them.

Using one organism to control another has become a widespread way of managing both animal and plant pests. The first insect used as a biological control was probably a Chinese weaver ant related to the African weaver ant. These ants build nests by binding together living leaves of a tree with silk. The workers hold larvae in their jaws and use them to tie leaves together with silk that the larvae secrete. In an article in *BioScience,* H. T. Huang and Pei Yang cite a Chinese book of 304 C.E. that mentions using these ants to control insect pests of Mandarin orange trees. And in 1882, the Reverend H. C. McCook wrote:

> Through the courtesy of Rev. H. Corbett, a missionary of the American Presbyterian Board, at Cheefoo, China, I received a copy of the "North-China Herald," of April 4, 1882, containing an article by Dr. Magowan, of Wenchow, on the "Utilization of Ants as Grub-Destroyers in China." From this paper I quote the following sentences:
>
> "Accounts of the depredations of the coccids [scale insects] on the orange-trees of Florida, induce me to publish a brief account of the em-

ployment by the Chinese of ants as insecticides. In many parts of the province of Canton, where, says a Chinese writer, cereals cannot be profitably cultivated, the land is devoted to the cultivation of orange-trees, which, being subject to devastation from worms [caterpillars], require to be protected in a peculiar manner, that is, by importing ants from neighboring hills for the destruction of the dreaded parasite. The orangeries themselves supply ants which prey upon the enemy of the orange, but not in sufficient numbers; and resort is had to hill-people, who, throughout the summer and winter find the nests suspended from branches of bamboo and various trees. There are two varieties of ants, red and yellow, whose nests resemble cotton-bags. The 'orange-ant feeders' are provided with pig or goat bladders, which are baited inside with lard. The orifices of these they apply to the entrance of nests, when the ants enter the bags and become a marketable commodity at the orangeries. Orange-trees are colonized by depositing the ants on their upper branches, and to enable them to pass from tree to tree, all the trees of an orchard are connected by bamboo rods."

These weaver ants, voracious predators, destroy caterpillars and other large insects but spare the honeydew-producing Homoptera that supply a large part of their diet. When modern insecticides appeared, the use of these ants was largely abandoned. But now they are being used again, because insecticides create new pests that are resistant to the insecticides and whose populations explode because the insecticides kill the parasites and predators that normally keep their numbers down.

The first, the most famous, and one of the most successful of the biological controls established in modern times is the one responsible for the suppression of an insect, the cottony cushion scale, that was about to wipe out the infant citrus industry in California. The marvelous story of the conquest of this insect begins in 1868, when it was found on *Acacia* trees in Menlo Park in northern California. It spread rapidly, and by 1886 had reached the citrus groves of southern California with devastating effect. A few years later, after the scale had been stopped in its tracks, a grower remembered the catastrophe that had threatened him: "The white scales were incrusting our orange trees with a hideous leprosy. They spread with wonderful rapidity and would have made citrus growth on the whole North American continent impossible within a few years."

When in April 1887 Charles Valentine Riley, a brilliant scientist and the first chief of the U.S. Department of Agriculture's Division of Entomology, spoke at the California Fruit Growers' convention, he was asked to find a way of controlling the horrendous cottony cushion scale. In his book on biological controls, Paul DeBach wrote that Riley "stated his belief that the scale came from Australia, where it was harmless." Riley reasoned that parasites or predators controlled the scale in Australia and recommended that they be sought out and imported into California. He offered to send an entomologist to Australia, but feared that Congress would refuse to appropriate the necessary funds. (Riley was right. Congress had put a rider prohibiting foreign trips by employees on the appropriation bill for the Department of Agriculture. It was aimed at Riley himself, whose frequent junkets to Europe were made at government expense.) Riley suggested that the state of California in Los Angeles County appropriate the money. The citrus growers supported him, but no funds were available. Riley, however, soon circumvented Congress's ban on foreign travel. Albert Koebele, a Department of Agriculture entomologist, was attached to the American delegation to an 1888 international exposition in Australia—ostensibly to represent the State Department but actually to search for parasites and predators of the cottony cushion scale.

Koebele was successful beyond all expectations. The introduction into California of just one Australian predator, a ladybird beetle known as the vedalia, reduced the population of cottony cushion scales to insignificance. Koebele found the scale to be rare in Australia, so rare that most local entomologists barely knew that it existed, and local orange growers had never heard of it. Koebele first found a ladybird munching on a cottony cushion scale in a garden in Adelaide. He found more ladybirds elsewhere, and shipped many to California in 1888 and 1889. They were packed in wooden boxes with branches bearing scales for them to eat, and the boxes were kept on ice during the long journey. Some vedalias died en route, but during the last 2 months of 1888 and January of 1889, 129 that survived were placed on a caged tree in an orange grove in Los Angeles. (Another 385, most of them brought by Koebele when he returned from Australia, were placed in groves elsewhere.)

By early April of 1889, the caged vedalias had multiplied and eaten almost all of the cottony cushion scales on the tree. When the cage was opened, the beetles moved to nearby trees. Within 6 months of their initial introduction, the vedalias had virtually wiped out all the scales in the Los Angeles orchard and had spread to nearby orchards. By April 12, vedalias were being sent to other growers. As DeBach reported, "By June 12, two months after the cage was opened, 10,555 vedalias had been distributed to 208 different growers. By the end of 1889, the scale was no longer a threat anywhere in California." Populations of cottony cushion scales too small to cause damage are kept at this low level by the correspondingly small populations of vedalias.

Other biological controls, some also spectacularly successful, soon followed, for as DeBach wrote, the cottony cushion scale project "established the biological control method like a shot heard around the world." Among these spectacular successes was the virtual elimination of the coconut moth, also known as the Levuana moth, from Fiji, an archipelago of over 800 islands in the South Pacific. By 1925 the caterpillars of this moth, which feed on the leaves of coconut palms, had destroyed all or most of the coconut palms on several islands of the archipelago. It was feared that the moth would spread throughout Fiji and totally wipe out the production of copra, the dried meat of mature coconuts. In *The Coconut Moth in Fiji*, J. D. Tothill and two coauthors described the devastation that this insect had caused:

> coconuts on the islands under attack by *Levuana* turned from the normal fringe of shining, waving green fronds to a border of lifeless grey. Ovalau is roughly thirty-five miles in circumference, and every mile of shore-line has its quota of coconut palms. At the close of an outbreak not a single green coconut palm was to be seen, even single trees upon the slopes standing out as grey objects plainly marked from the surrounding greens of forest trees and grass.

Tothill and his colleagues found no coconut moths in southern Asia, the East Indies, or South Pacific islands other than Fiji. Nevertheless, they believed that they are not native to Fiji because they had no parasites in Fiji. I agree with them. It is unlikely that no parasite evolved to exploit an insect that had been present for hundreds of thousands of years or more.

Because they did not find coconut moths except in Fiji, Tothill and his group could not obtain parasites they knew to be adapted for attacking this moth. So they took another tack. They imported parasites of a related moth, *Artona*, in the hope that they would also attack the coconut moth, because most parasites are adapted to utilize several related species. *Artona* was uncommon, but they found a small population infested by several parasites, including the maggot of a tachinid fly in the genus *Ptychomyia*, at Batu Gajah, about 300 miles from Singapore in what was at the time the Federated Malay States. The tachinid maggots were in turn attacked by hyperparasites. It was, of course, necessary to prevent these hyperparasites from becoming established in Fiji lest they interfere with the biological control.

With great difficulty, they kept *Ptychomyia* alive on the long trip from Batu Gajah to Suva, the capital of Fiji. Today, living insects are easily and quickly sent from place to place by commercial airliner, but in 1925 there were no airlines. *The Coconut Moth in Fiji* relates the problems involved in shipping this parasite:

> A difficulty in the case of *Ptychomyia* was that on account of it being a tropical insect and having no hibernating period it could not be shipped in cold or even cool storage. A relay station in Queensland was out of the question on account of the risk of introducing *Artona* or *Levuana*, one of which would have had to be used as a food supply. The remaining alternative was shipment to Fiji in an active condition either by way of Australia or by direct freight boat. Via Australia was unsatisfactory for two reasons, namely, the danger of introducing *Artona* or *Levuana* and because shipment would have to be limited to the Australian summer months which might not coincide with *Artona* outbreaks. For practical purposes, therefore, it was expedient to attempt direct shipment. Aeroplanes were not available, and the cost of chartering either a fast naval vessel or merchant ship for the 4,000 mile journey would have been prohibitive. Our only hope was an occasional freighter, and it would therefore be necessary to keep *Ptychomyia* alive for not less than twenty-three days.

During the long journey, adult parasites would emerge from their hosts and would require fresh caterpillars in which to lay their eggs, and the caterpillars would have to have living coconut palms as food. *Ptychomyia* could survive only if accompanied by other members of its food chain. Large shipping cages were built and each was stocked with four or five coconut seedlings growing in cans. Most of the cages contained parasitized caterpillars, but others held unparasitized caterpillars that would be the hosts of the offspring of adult parasites that emerged en route. About 20,000 *Artona* caterpillars were placed on 85 young palms in 17 cages and sent by rail to Singapore, where they were transferred to a ship. The shipment arrived in Suva harbor 25 days after the parasitized caterpillars had been collected at Batu Gajah.

Three hundred and fifteen adult *Ptychomyia* survived and were placed in rearing cages with coconut moth caterpillars. The parasites multiplied and were soon released. Six months later they had spread throughout the area infested by coconut moths and were well on the way to controlling them. Only 3 years later, control was complete. The coconut moth had become so rare that visiting entomologists could not find it without help.

Some time before 1940, according to Donald Nafus, an entomologist at the University of Guam, a caterpillar that feeds on the buds, young leaves, flowers, and fruits of mango trees was unintentionally introduced on the island of Guam. No one knows where these invading insects, mango shoot moths, came from, but this species is widely distributed in tropical Asia and the Pacific islands. In its native range, it is seldom destructive—being at most a minor pest—because it is held in check by parasites and other natural enemies. But few parasitic insects on Guam attacked it, so mango shoot moths overran Guam and mango trees produced little or no fruit.

In 1986–87 parasites from southern India became established, a tachinid fly of the genus *Blepharella* and a tiny wasp of the genus *Euplectrus*. *Blepharella* females lay many tiny eggs on young mango leaves. They hatch only after being swallowed by a caterpillar. The parasitic maggot feeds on the caterpillar's internal tissues and eventually kills it. *Euplectrus* larvae are gregarious and live on the outside of the caterpillar as they suck its fluids. Before fastening her eggs to the host's body, the female wasp prepares the way for her offspring by stinging the caterpillar to arrest its development. These two parasites so greatly decreased the shoot moth population that by 1988 the yield of mango fruits was 40 times greater than it had been. Ecologically speaking, the reproductive rate of mangoes had been greatly increased.

🐜 🐜 🐜 As paradoxical as it may seem, modern synthetic insecticides have created scores of new insect problems in forests, orchards, and crop fields. What happens is that a plant-eating insect that has been too scarce to be damaging—but that is immune to the insecticide applied—undergoes a population explosion because some or all of its parasites and predators are not immune to the insecticide and are killed by it. Insecticide-induced pests of apples are a case in point. Insects and mites that had been at most minor and occasional pests in apple orchards vastly increased in number and became seriously destructive soon after DDT, the first of the modern "miracle" insecticides, became available in 1946 and was sprayed on apple trees. Among them were red-banded leaf rollers, caterpillars that chew on leaves and immature fruits, and many species of mites that suck sap from leaves, causing them to turn brown and die. The most important of these mites are the European red mite and spider mites. Up to seven additional insecticide applications were used to control just these mites. Some orchards were sprayed as many as 20 times in one season, including sprays for the usual pests such as the codling moth caterpillar.

In a book on insect pest management, William Luckmann and Robert Metcalf described an environmentally friendly and ingenious way of minimizing the use of insecticides by managing orchards so as to maximize the population of a predaceous mite, *Amblyseius fallacis,* that eats plant-feeding mites.

European red mites spend the winter as eggs on areas of rough bark on the underside of twigs high in the trees. Fallacis and spider mites overwinter only as mature females, hiding at the base of the trunk: under loose bark, among apple shoots, grasses, and other small plants, or hidden in debris on the ground. European red mite eggs hatch just before apple trees bloom; the tiny nymphs crawl to nearby apple leaves and feed, grow, and reproduce, often becoming numerous enough to damage the foliage so severely as to lower the yield and quality of the fruit. At about the same time, female fallacis and spider mites become active and begin to lay eggs at the base of the tree. The spider mites feed on almost any of the ground-cover plants, and fallacis mites prey on the spider mites. Fallacis and spider mites cannot survive unless a good stand of small plants is allowed to grow at the base of the apple trees. Both species steadily become more numerous, and ascend to the twigs and foliage of the trees later in the season. Then the fallacis mites control the European red mites, spider mites, and any other herbivorous mite on the trees.

Early in the season, before fallacis mites have ascended the tree, one application of a pesticide is usually required to suppress European red mites. But only this one spray is required if fallacis mites become abundant. Fortunately, there are several pesticides that do the job without killing fallacis mites, which have become resistant to them, and several insecticides control codling moths and other apple insects without harming fallacis mites. Resistant insects or mites are almost always a problem, but in this case the resistant predaceous fallacis mites are the solution to a problem.

From 1889 to 1945 the vedalia beetle did such a good job of controlling cottony cushion scales in California that this once all-too-abundant pest had become a rare insect. But in 1946 disaster struck in the guise of DDT, far too optimistically touted to be the weapon that would finally end humanity's war against pest insects. DDT certainly was a very efficient killer of insects, but it had such terrible environmental side effects, brought to the world's attention by Rachel Carson, that in 1972 it was banned in the United States, Canada, and many other countries. One of these side effects was the near extinction of bald eagles, ospreys, peregrine falcons, and other birds at the top of a food chain. This is a compelling story that should not be forgotten lest we make the same mistake again. The other side effect, the resurgence of pest insects, is yet another demonstration of the importance of parasitic and predaceous insects in preventing runaway population growth.

DDT caused a resurgence of the cottony cushion scale because it did not kill it but did kill vedalias and brought them to the brink of extinction. Paul DeBach, a professor of entomology at the University of California at Riverside, watched this disaster unfold. He wrote that many citrus groves "looked as if they were covered with snow because of the density of the cottony egg masses." Some trees, he reported, "were killed and many groves defoliated with subsequent loss of crop in the short two to three years before the use of DDT was voluntarily dropped or drastically modified."

In the late 1930s, I watched as Dutch elm disease denuded Bridgeport, Connecticut, of the stately American elms that shaded its streets and parks, even one old elm in whose shade George Washington had stood as he made a speech. By 1953, I was a graduate student in entomology at the University of Illinois. Dutch elm disease, moving ever westward, had arrived in Illinois in 1951, and together with another disease, phloem necrosis, was making a deadly sweep through the "forest" of American elms that shaded the campus and the twin cities of Champaign and Urbana. It was a sad sight, made even sadder by the ecological havoc wrought by an ill-advised effort to stop the spread of Dutch elm disease by spraying elms with DDT to kill the bark beetles that transmit the fungus that causes the disease from tree to tree.

The effort was futile. Chemical control did not work. By 1959, after 9 years of spraying, the campus had lost 86 percent of its elms, and only a few years later all were gone. Even worse, the application of DDT had two unforeseen, unwelcome, and all too obvious side effects. It virtually wiped out the robins on campus, and it "created" some new pests.

The concentration of DDT in the sprays was in itself too low to kill robins. Nevertheless, many dead and dying ones were found on the campus. One of these unfortunate birds has stuck in my mind all these years. While walking across campus to the main library, I came upon a dying robin lying on its back at the edge of a sidewalk. Its legs twitched as uncontrollable tremors—caused by DDT poisoning of its nervous system—shook its body. Jim

Sternburg's unpublished analyses of robin bodies and analyses by Roy Barker published later showed that the bodies of dead and dying robins—especially their brains—were loaded with lethal doses of DDT, more than could be accounted for by direct contact with the spray.

How did so much DDT get into their bodies? The answer is that robins were at the end of a food chain that concentrated DDT. When DDT-contaminated elm leaves fell to the ground, they were eaten by earthworms, including the big, juicy nightcrawlers that are a favorite of robins. The worms were immune to the insecticide, and they stored large quantities of it in their bodies, especially in their fat, and retained it through the winter. They were ecological time bombs. When the migratory robins returned the next spring, they ate many earthworms and accumulated a lethal dose of DDT. It didn't take many of these loaded earthworms to kill a robin. Barker calculated that a robin could acquire a lethal dose of DDT by eating only 11 contaminated nightcrawlers. In springtime, robins feed mainly on earthworms, and have been seen to eat as many as 12 in less than 20 minutes.

Spraying elms with DDT had another pernicious effect. By virtually wiping out their parasites, DDT caused population explosions of insects and mites that feed on elms and are immune to DDT. Before the spray program, they had been barely noticeable and were certainly not destructive. But after the application of DDT, aphids became so numerous that the sidewalks under elms were unpleasantly sticky with their honeydew. And by midsummer, the trees had a noticeable rusty cast, damage done by spider mites. The Putnam's scale, an insect that had been so rare that it was virtually unknown to Illinois entomologists, encrusted the leaves and twigs of elms and actually killed some. In 1954, the students in a course on insect ecology, of whom I was one, counted Putnam's scales on samples of elm twigs, taking 10 samples from trees that had been sprayed with DDT for 5 years and 10 from trees that had not been sprayed previously. The results were clear. DDT had caused an outbreak of Putnam's scale. On trees that had been regularly sprayed, we found 164 scales per sample, while on the others we found less than 9 per sample. Milton Tinker, a graduate student in the Department of Entomology, confirmed our results, leaving no doubt that there was a cause-and-effect relationship between spraying the elms with DDT and the population explosion of Putnam's scale.

🐜 🐜 🐜 Insects under attack by other insects have some amazing ways of avoiding their enemies or defending themselves against them. The complexity and ubiquity of these defenses is testimony to the universal and deadly threat from predaceous and parasitic insects. The victims have responded to this onslaught with many defenses, including running away, physically repelling parasites or predators, stinging them, warding them off with toxic or repellent biochemicals, and even suffocating parasites that make it into their bodies by enveloping them with layers of blood cells.

We don't think of caterpillars as having ears and responding to sounds, but at least some of them recognize as a warning the droning note of the beating wings of an approaching parasite or predator. In 1925 Dwight Minnich showed that caterpillars can hear. In response to clapping hands, human voices, or a violin being played gregarious caterpillars of the mourning cloak butterfly suddenly reared up in unison, raising their heads high. Their "ears" are fine hairs on their bodies. After Minnich singed these hairs, the caterpillars stopped responding to sounds. We now know that other caterpillars also react to the flight sounds of parasites or predators, defending themselves by rearing up, thrashing from side to side, biting, regurgitating, or by dropping to the ground on a silk thread.

In the 1970s Charles Hogue discovered that a large aggregation of caterpillars on a tree trunk in Costa Rica reacted when he shouted in their direction from about 32 feet away. "Each of the larvae," he wrote "responded to the sound of my voice at the same instant and in the same manner, a violent jerking of the anterior third of the body, so that the head, thorax, and anterior portion of the abdomen were arched upward or sideward. The caterpillars reacted only to sharp, high-pitched, and intense sounds; they did not respond to normal conversation." Hogue further tested this response by playing Strauss waltzes from a tape recorder in the immediate vicinity of the caterpillars; they responded similarly to the loud passages of the music. The approach of flying parasites, wasps or flies, elicited jerking movements that tended to fend off the attackers. This defensive behavior often worked but was not always successful. Nevertheless, attacking parasites experienced great difficulty in laying eggs on caterpillars and were sometimes deterred.

The effectiveness of physical defense is illustrated by the response of certain beetle larvae to parasitic wasps *(Bathyplectes)* introduced into North America as a biological control of the destructive alfalfa weevil. The wasps are more likely to succeed in inserting eggs into larvae of the relatively docile eastern strain than into the aggressively defensive larvae of the western strain. When a wasp touches a western larva, it, unlike its eastern counterpart, responds by rolling over. Kurt Volker and Robert Simpson described what happens: "A western larva stopped feeding as the parasite climbed on and initiated drilling. The larva then curled up (head to tail) while lying on its side, and rolled completely over while remaining curled. The movement was very rapid and usually one rollover followed another in a series of 4–5. This response almost invariably dislodged the parasite and the larva dropped from the sprig of alfalfa." In laboratory experiments conducted by Volker and Simpson, the wasps managed to insert an egg into 55 (76 percent) of 72 of the docile eastern larvae but only 24 (32 percent) of 73 western larvae. This difference may well be why these parasites have become established in eastern North America but not in the west.

Solitary digger wasps of many species excavate underground cells that they stock with bees, grasshoppers, or other insects as food for their larval offspring. The wasp larvae are parasitized by several flies and wasps that lay their eggs in the burrows. The diggers, noted Howard Evans, who is famous for his research on wasps, mislead the parasites by plugging the entrance to the burrow with soil before they leave to hunt for prey; some species even camouflage the plug with sticks, leaves, or other objects. Other diggers excavate false burrows near the entrance to the true burrow, leaving them unplugged. Unwitting parasites lay eggs in the dummy burrows; every one of these wasted eggs is one less threat to the digger wasp's offspring.

The larva of a tortoise beetle widely distributed in the Northern Hemisphere covers its back with a packet of shed skin and feces that deflects the attention of ants and probably other predators as well. This trash packet is carried on a long, freely movable, forklike appendage at the end of the abdomen. Thomas Eisner and his coauthors described the reaction of an attacking predaceous ant when the beetle larva thrust its fecal shield in the ant's face: "Confronted with the inanimate shield rather than the body of the larva, the ant was quick to lose interest in its quarry. It paused momentarily to palpate the shield and sometimes even to bite into it, but eventually the ant always abandoned the attack and walked away."

In an article called "The Workers' Bodyguard" published in 1968, Irenäus and Eleonore Eibl-Eibesfeldt described how small workers of a leafcutter ant "ride shotgun" to protect their larger sister workers from parasitic flies. With their scissor-like mandibles the large workers cut pieces several times their own size from living leaves and, joining a long column of their sisters, carry them back to the nest. The little parasites circle above the ants and dart in to lay an egg on the back of an ant's neck. The parasitic maggot burrows into its host's head and destroys the brain. "Empty-handed" large workers can usually deter these flies by snapping at them with their formidable mandibles, but they are too preoccupied to defend themselves as they cut out a piece of leaf and carry it back to the nest, holding it aloft like a big green parasol. At that point a small worker often comes to the rescue. She climbs onto the leaf piece, sits near its top, facing upward with her threatening mandibles wide open, and snaps at attacking flies, usually driving them off. More recently, experiments by Flavio Roces and Bert Hölldobler turned up convincing evidence that workers harvesting leaf pieces send out a "call for help" by stridulating, sliding one abdominal segment on another so a file on one rubs against a series of ridges on the other. The resulting vibrations, transmitted through the leaf on which the ants sit, are sensed by them through their legs. Small workers respond by climbing onto the leaf piece held by a large, calling worker.

Many insects are chemically armed against predators with internal toxins, stings, or repellent sprays or secretions. They advertise their defensive capability with conspicuous colors such as the orange and black of the toxic monarch butterfly, or with the piercing buzz of an angry wasp, or with some other warning signal that predators learn to heed after being sprayed, stung, or poisoned. Some harmless and perfectly edible insects escape predators by bluffing, mimicking the warning signals of well-armed insects. Experiments with birds and other vertebrate predators have demonstrated the efficacy of this double-barreled defense—warning and deterrent—and of mimicking the warning. Some predaceous insects and other arthropods learn to avoid stinging, toxic, or otherwise noxious prey, but there have been few studies to determine whether or not they are deceived by mimics of noxious prey.

Dorothy Feir and Jin-Shwu Suen found that milkweed bugs contain toxic chemicals, called cardenolides, that they obtain from the milkweed seeds from which they suck nutrients, just as monarchs, as Lincoln Brower, well known for his research on mimicry, reported, obtain cardenolides from the milkweed leaves they eat. The plant manufactures these chemicals to ward off insects, but some milkweed-eating insects, the milkweed bug and the monarch among them, have turned the tables on the plant. Not only can they tolerate cardenolides and feed on milkweeds, but they have appropriated the cardenolides as weapons against their own enemies. May Berenbaum and Eugene Miliczky, and more recently Todd Bowdish and Thomas Bultman, showed that the large Chinese praying mantis, introduced into the United States in 1896, learns to avoid toxic red and black milkweed bugs after a few attempts to eat them. Mantises readily captured milk-

weed bugs and held them in their forelegs as they chewed on them. But after eating part of a bug, they hurled it away, vigorously shook their front
legs, and then many of them regurgitated large amounts of orange fluid.
The mantises remembered their sickening experience for a surprisingly long
time. Two of them rejected milkweed bugs 3 weeks after they had last seen
one. Berenbaum and Miliczky later showed that they will also refuse to attack fake mimics of this bug, edible beetles painted black and red.

Quite a few insects and other arthropods discharge defensive chemicals
that discourage predators and parasites before they have a chance to mount
a physical attack. The defensive secretions of certain millipedes, centipedes, cockroaches, termites, aphids, beetles, flies, and ants are tenaciously
sticky—some also are toxic—and disable attackers by entangling their
mouthparts or legs. The larvae of some hover flies (family Syrphidae) eat
aphids and are thus subject to the attacks of ants that guard honeydew-producing aphids. When attacked by an ant, a hover fly larva arches over its
assailant and discharges a drop of sticky saliva onto its body. The ant immediately releases its grip on the larva and tries to rid itself of the entangling saliva. Centipedes have evolved two quite different versions of a sticky
exudate defense, which they use against spiders, ants, and other small predators. One group of centipedes, the lithobiids, discharges sticky threads from
legs on the rear of the body. Another group, the geophilids, discharges a
gluey liquid from glands on their undersides. Many species of cockroaches
also use the sticky exudate defense. The upper side of the hindmost segment
of the abdomen is covered with a sticky secretion that fouls the mouthparts of small predators such as centipedes and ants. Larval Mexican bean
beetles have soft bodies, but they are protected from ants, and probably from
parasitic insects too, by thin, multi-branched spines that cover their bodies.
When an attacking insect brushes against a Mexican bean beetle larva, the
fragile spines break and leak blood that contains a sticky substance that
gums up the intruder's body. The attacker then hobbles away and tries, often
in vain, to clean itself.

In southeastern Arizona, William Nutting and two coworkers studied the
defensive behavior of a tiny termite with the jaw-breaking generic name
Tenuirostritermes. Workers of this species go out, usually at night, to forage for
bits of dead grass and green leaves. Following trails they have marked with
an odorous pheromone, they march to their foraging site in columns that
may be more than 6 feet long, and are flanked on each side by guarding soldiers facing outward. About 20 percent of the individuals in a colony are sol-

diers, which are especially adept at immobilizing ants. The soldiers have a highly specialized and bizarre way of using a sticky secretion in defense. Their heads bear a long, forward-projecting snout from which a strand of sticky material, secreted by a huge gland in the head, is shot to a distance of more than three-quarters of an inch, about 6.6 times the length of the soldier's body. These soldiers are known as nasutes, from the Latin *nasut,* "large-nosed." In an article on these termites, Thomas Eisner and his coauthors described nasute soldiers as "little more than ambulatory spray guns."

In the tropics, social wasps are constantly threatened by ants that prey on the larvae in their nests. The paper nests of the South American vespid wasp *Mischocyttarus* are uncovered combs suspended upside down from twigs, leaves, or the eaves of buildings by long stems as thin as a thread. Although large predators such as beetles probably cannot crawl down the stem to reach the larvae in the comb—much as squirrels cannot crawl down a thin wire to get at a hanging bird feeder—ants are small and agile enough to navigate the threadlike stem. The wasps defend their brood against foraging ants by applying about once every 2 hours a repellent secretion from a tuft of hairs on the underside of the terminal segment of the abdomen. As Robert Jeanne wrote in an article in *Science,* "a worker moves to the top of the nest rubbing the tip of the [abdomen] over the nest surface, then approaches the nest stem headfirst, turns her body and reaches as high as she can . . . up the stem with the [abdomen] and rubs the stem for 1 or 2 seconds before rubbing down to the nest again." Jeanne showed that the secretion is very repellent to ants. He baited thin glass tubes with pieces of Brazil nuts pressed onto their tops, smeared some tubes with the wasps' secretion, smeared control tubes with a nonrepellent substance, and placed the tubes vertically where ants were abundant. Of 151 ants that made the attempt, 130 reached the tops of control tubes. But only after 1,154 ants had attempted the climb did 130 make it to the top of a tube smeared with the wasp repellent.

The large, heavy-bodied, slow-moving, and flightless lubber grasshoppers of the southeastern United States confront attackers with a noxious, frothy exudate accompanied by a warning display. Although most other grasshoppers are camouflaged when seen against their usual background, the bright yellow and black lubber is flagrantly conspicuous even in repose and when not threatened by a predator. At the sight of an intruder, a lubber rises up on its legs and reveals its small but brilliantly crimson hind wings. If the intruder does not back off, the lubber forcibly discharges, with a loud hiss, a brown froth from two of the porelike openings to its respiratory system. Insect-eating birds are visibly frightened by the warning display and abandon any thought of making the lubber a meal. From the boardwalk of the Anhinga Trail in Everglades National Park, I once saw a purple gallinule walk slowly over floating water lily pads, pecking at almost any insect that was in reach, but ignoring a big lubber grasshopper that crossed its path—the gallinule stepped right over it, probably because it remembered a previous disagreeable encounter with one of these insects. Ants, important enemies of grasshoppers, have poor eyesight and are not likely to perceive a lubber's warning display, but they are markedly repelled by the grasshopper's froth, which smells strongly of carbolic acid.

Among the insects that discharge repellent chemicals as a spray, which some of them may shoot to a distance of several feet, is the large and relatively heavy-bodied, warningly colored, two-striped walkingstick of the southern United States, which has two large glands in its thorax that secrete a highly irritating liquid that can be sprayed from two openings just behind the head. According to an article in *Science* by Thomas Eisner, the spray is discharged instantly in response to "mild traumatic stimulation," as when a leg is pinched with forceps. Both glands or only one of them discharges, depending on whether the walkingstick is stimulated on both sides or only one side. Eisner noted that its "marksmanship is precise." The spray always drenched the tip of the offending forceps. If he tapped the insect on its back, both glands aimed their spray straight up; if he pinched its rear end, the two glands discharged directly to the back; if he touched both antennae with a heated probe, two jets of spray shot to the front; if he pinched one leg, only the gland on that side discharged, invariably hitting the leg that had been pinched. Ants and ground beetles that bit walkingsticks were sprayed and were always instantly repelled, rushing away to clean their bodies.

In 1972, Eisner described the bombardier beetle's potent defense. The bombardier, warningly colored with bright blue and orange, has at the tip of its abdomen a specialized organ that prepares and sprays an obnoxiously repellent fluid. This organ has one reservoir that contains hydrogen peroxide and another that contains chemicals called hydroquinones. If a threatening predator ignores the bombardier beetle's warning coloration, the beetle raises its abdomen and turns it toward the intruder. At the same time, the contents of the two reservoirs are passed into a third chamber in which they undergo an explosive chemical reaction catalyzed by certain enzymes, producing benzoquinones and spraying these irritating chemicals toward the predator with an audible pop and at the temperature of boiling water. Attacking toads are sprayed in the mouth and show obvious signs of distress, gaping wide and rubbing their tongues against the ground. The bombardier's spray is known to have a similar effect on certain spiders, and it is safe to assume that it affects insects and other predaceous arthropods similarly.

Even if a parasite gets past its victim's chemical or other external defenses and manages to insert an egg into its blood-filled body cavity, all is not lost. Some insects have an internal backup defense called encapsulation. In this process, graphically described by Michael Strand and Louis Peck, certain mobile blood cells recognize the parasite as a foreign object, move to it, and flatten themselves against its body. Then they release into the blood substances that attract more blood cells, which also adhere to the parasite and each other, forming a thick capsule that completely covers the parasite and kills it by cutting off its access to oxygen.

🐜 🐜 🪰 These are just a few of the many ways in which natural selection has shaped insects to defend themselves against predators and parasites. How then, in the face of such formidable defenses, do predators manage to subdue their prey and how do parasites succeed in infesting their hosts? There are many answers to these questions. To start with, not all insects have totally effective defenses and some have virtually no defenses. Generally speaking, defenseless species survive because they produce so many offspring that a few are likely to escape the notice of their enemies. Many offspring are sacrificed so that a few can survive. These "sacrificial" individuals are the victims of generalist predators or parasites that have not evolved specialized counter-defenses for dealing with well-defended insects. Most predators are opportunistic generalists that, like spiders or praying mantises, take any insect that happens to come along. Among the few specialist predators is a fly (*Stomorhina*) that as a larva eats only eggs in the underground egg pod of a locust. "As soon as a locust has completed laying a batch of eggs and withdrawn its ovipositor from the ground," wrote Richard Askew, "the female *Stomorhina* moves to the place and lays its own eggs in the still-soft fluid froth that covers the top of the egg pod." He believes that the strong-flying adult flies probably follow migrating swarms of their victims.

Some parasites are also generalists. For example, the larvae of a certain tachinid fly *(Compsilura concinnata)* are known to be internal parasites of at least 200 species of caterpillars and of insects of two other orders. But many parasites are specialists that attack only one or a few closely related insects.

The larvae of all known species of the tiny flies of the genus *Cryptochetum* are internal parasites of members of only 1 of the 16 families of scale insects, the giant scales and ground pearls. The parasitic larvae have no mouthparts, but absorb nutrients from a scale's blood through their skin. Larvae of all of the big-headed flies, the Pipunculidae, are internal parasites of leafhoppers, spittlebugs, and insects of related families. The species of different genera specialize on a particular host family: some parasitize only leafhoppers, others only planthoppers, others only spittlebugs, and yet others only treehoppers. Adult females of most species pounce on an immature host and insert an egg into its abdomen. But those that parasitize spittlebugs lay their eggs in adults, presumably because the immatures are covered and protected by a "spittle" nest composed of hundreds of tiny bubbles.

That parasites utilize only one or a few closely related hosts reflects the evolution of ever more specialized counter-defenses to overcome the many different defenses against potential parasites. Different insects have evolved different defenses, and no parasite that I know of can overcome them all. There has been an escalating arms race between host insects and parasites similar to the one between plants and herbivorous insects. Evolution has driven parasites to specialize on insect hosts whose defenses they can overcome with counter-defenses. "In the face of competition from other parasite species," wrote Askew of this evolutionary process, "the polyphagous parasite will have a higher survival rate on those hosts where it is a more successful competitor, and it will therefore become more and more adapted to parasitizing an increasingly restricted host range."

There are many countermeasures against the defenses of a host. For example, using her exceptionally long ovipositor, the female *Megarhyssa* breaches what is probably the major defense of her horntail host, the thick layer of wood that covers it and that cannot be penetrated by parasites with shorter ovipositors. Using her 3-inch-long ovipositor, about twice the length of her body, she drills through solid wood to place an egg in the tunnel a horntail has burrowed in a dead tree.

Encapsulation often works against generalist parasites, but parasites that specialize on one or just a few hosts have evolved countermeasures that prevent their host from encapsulating them. One complex and exceptionally interesting countermeasure, discussed by Mark Lavine and Nancy Beckage, involves a partnership between the parasite and a virus. This virus, known as a polydnavirus, suppresses the host's immune system, which if not suppressed would trigger encapsulation by means that have yet to be discovered. As the parasite egg passes through its mother's reproductive system, it is bathed in a fluid that contains the virus, and is thereby coated with the virus that will make possible the survival of the larva that will hatch from it.

Although it seems improbable that creatures as small as insects could affect the population size of vertebrates, some as large as moose and bison and even most of the smallest ones considerably larger than the largest insects, there is abundant evidence—some of it discussed on the following pages—that they are very important in helping to avert ecologically disruptive population increases of birds and mammals.

9

CONTROLLING VERTEBRATE POPULATIONS

A clipper ship, its tall masts with white sails taut in the wind, is in my view the most comely of ships and one of the most beautiful creations of our species. When the clipper *Lightning* reached Australia in 1859, it was no doubt a welcome sight, but Australians were soon to wish that it had never docked. It brought with it the root cause of an immense ecological catastrophe, a shipment of 24 wild European rabbits from England. Huge hordes of their descendants would soon overrun the southern half of Australia, eating everything green in sight and turning huge tracts, including valuable grazing lands, into deserts. Thomas Austin, whose name is still anathema in Australia, released these rabbits on his extensive holdings in the state of Victoria, hoping they would multiply and provide new targets for sport shooting.

The rabbits more than lived up to their proverbial reputation for rapid multiplication. In a comprehensive book on the biological control of the European rabbit in Australia, Frank Fenner and Francis Ratcliffe chronicled the wildfire spread of these rabbits, which soon became Australia's most destructive pest. By 1865, only 6 years after their introduction, 20,000 rabbits had been killed on Austin's estate, and they abounded on neighboring properties. Fenner and Ratcliffe wrote, "When the population increase had gained momentum, the rabbits proceeded to spread northward and westward with a speed which we believe to be without parallel in the whole history of animal invasions." They crossed the state of New South Wales from south to north, breeding as they went, at the incredible rate of 70 miles per year. They ultimately became established on more than half of Australia's 3 million square miles, throughout most of its temperate region, but they were only sparsely distributed in or were absent from its tropical areas.

This rapid dispersal caused much apprehension, especially among ranchers. "While the rabbit advanced," wrote Fenner and Ratcliffe, "it was pre-

ceded by reports (growing ever more alarming) of its destructiveness in those areas where it had become established and had built up numbers. With no experience and little biological understanding to guide them, the authorities tried to cope with the rabbits in various ways that were ineffective and sometimes very costly." The building of several thousand miles of expensive barrier fences to stop the advance of the rabbits is testimony to the people's realization that they were faced with an impending environmental disaster. The longest of these fences crossed the state of Western Australia, extending more than 1,100 miles from near Hopetoun on the south coast of the Indian Ocean almost due north to the opposite coast at the southern end of the Eighty Mile Beach near Port Hedland. The fences did not stop the advance of the rabbits. Some were built too late and it was impracticable to maintain them well enough to exclude all rabbits.

In 1950, Australia instituted a biological control, the introduction of the myxoma virus of rabbits, to alleviate the horrendous damage done by these feral European invaders. The disease caused by the virus, myxomatosis, was first discovered in South America, where it causes relatively mild symptoms and is seldom fatal to the native rabbits of the genus *Sylvilagus*. These rabbits and the virus evolved together and over the millennia reached an accommodation that let both survive. The virus became less virulent to its host and the host became resistant to the virus. Such evolutionary accommodations between pathogens and their hosts are to be expected. After all, the pathogen will inevitably go extinct if it kills all of its hosts. But the myxoma virus did not occur in Europe, and European rabbits (genus *Oryctolagus*) were not resistant to it. In South America, the virus, transmitted by mosquitoes to domestic European rabbits, caused a severe disease that was nearly 100 percent fatal. Similar situations have involved humans. When Native Americans contracted measles—a disease new to them—from European explorers, they often died, although Europeans, long associated with this disease, suffered milder symptoms and were far less likely to die. This is not to say, however, that coevolved diseases, insect-borne or not, do not help to limit population growth of their hosts. Even though they have long been associated with the myxoma virus and usually survive an infection, rabbits of the Western Hemisphere are sometimes killed by it and when infected by it may develop tumors and other symptoms that can interfere with reproduction.

With almost no exceptions, the myxoma virus can be transmitted from a sick rabbit to a healthy one only by the bite of a blood-sucking insect. In the parlance of medical entomologists such a transmitter of pathogens is called a vector. The myxoma virus flourished in Australia because after it was intro-duced—unaccompanied by a vector—it found capable and widespread na-tive vectors in place and ready to transmit it from rabbit to rabbit. The virus had little or no difficulty adapting to these vectors because it required no more than simple mechanical transmission. If the sucking mouthparts of a mosquito or black fly were contaminated, the virus was injected when the vector pierced a rabbit's skin.

The virus spread rapidly. Fenner and Ratcliffe said: "by the end of the 1952–53 season myxomatosis, by natural spread and with man's assistance, had reached virtually every part of the continent in which rabbits are found." The high death rate, well over 90 percent, decreased the population enough for predators such as dingoes and foxes to keep it in bounds.

"The great overall reduction in Australia's rabbit population as a result of myxomatosis must be classed," wrote Fenner and Ratcliffe, "as an ecological event of the first magnitude." During the almost 100 years that rabbits in-fested Australia, they caused major shifts in the balance of plant and animal populations. Their near extermination largely reversed these shifts. Con-trolling them eased their tremendous grazing and browsing pressure on her-baceous plants and the seedlings of shrubs and trees. Restored pastures ac-commodated more sheep and cattle, and although no data were collected, there can be no doubt that populations of native plant-eating marsupials also rebounded. Conversely, populations of dingoes, foxes, eagles, and other predators that had benefited from the plethora of rabbits presumably de-clined.

As was to be expected, after the virus and the European rabbit were brought together for the first time, they began to coevolve toward a more benign relationship. Within a few years, rabbits were becoming resistant and less deadly strains of the virus were beginning to appear. As Elizabeth Finkel reported, within 4 years the effectiveness of the virus had dropped from 99 percent to 70 percent and has since continued to decline. In 1995 another disease of rabbits, caused by the calicivirus, was introduced and has so far been stunningly successful.

The transmission of disease-causing organisms by insects or other arthropods was unproven and the very idea scorned until the last quarter of the nineteenth century. Malaria—the name means bad air—was thought to be caused by noxious miasmas rising from swamps. Yellow fever, which virtually halted the construction of the Panama Canal, was said to be the result of exposure to contaminated bed linens or clothing. It wasn't until 1878 that the English parasitologist Patrick Manson first proved that a pathogen can be transmitted by an insect. While working in China, he found that the nematode worm that causes filariasis in humans is transmitted by a mosquito. A well-known and spectacular symptom of this disease is the gross swelling of limbs and other body parts known as elephantiasis. (The scrotum and penis of one victim were so enlarged that he had to rest them in a wheelbarrow when he walked.) It wasn't until 1893 that the Americans Theobald Smith and F. L. Kilbourne took the next step by showing that the cattle tick is the vector of the protozoan that causes cattle fever. In 1897, the English physician Ronald Ross, with the help and guidance of Manson, demonstrated that bird malaria is transmitted by mosquitoes. After the American pathologist and bacteriologist Walter Reed showed that yellow fever, a viral disease of both humans and New World monkeys, has a mosquito vector, the concept of the transmission of pathogens by insects was well established.

The effects of myxomatosis on rabbits in Australia and, as you will see below, of bird malaria on the native birds of Hawaii prove that insects—mainly mosquitoes in these instances—can have devastating effects as transmitters of disease-causing pathogens. Our feelings about birds and mammals are much more compassionate than is our attitude to insects. Birds are beautiful and sing sweetly, and cats, rabbits, and other mammals are charming and cuddly. Therefore, we tend to look upon diseases and parasites of these sympathetically viewed creatures as lamentable afflictions that cause pain and suffering. And so they do. But birds and mammals, like all other organisms, produce more offspring than are needed to replace themselves, and so their populations are as likely to increase out of bounds as are insect populations. Thus disease and parasites are important and necessary factors in the equation that controls the "balance of nature" by limiting the growth of vertebrate populations that, left unchecked, would seriously disrupt or even destroy their ecosystems and thereby make life difficult or impossible for themselves and many other animals and plants.

There are many insect-borne diseases, caused by various viruses, bacteria, protozoa, or nematode worms, that attack mammals, birds, reptiles, and even amphibians such as frogs. Insects and other arthropods that suck blood—but not all of them—are often the vectors of the pathogens that cause these diseases. Ticks, which are all blood feeders, are the most important of the noninsect arthropod vectors. Many mites are blood suckers, but are less likely to be vectors of pathogens than are ticks or insects, because a mite is much less likely to move from one animal to another, usually spending its entire life on the same bird, mammal, or reptile. A tick, in contrast, usually spends each of its several life stages on a different animal. Many of the blood-sucking insects are gadabouts, likely to feed on several hosts during their lifetimes. Among them are certain true bugs, fleas, and above all many flies, including mosquitoes, biting midges, black flies, horse and deer flies, and relatives of the house fly such as the tsetse, stable, and horn flies. Lice, as you will see below, are by no means gadabouts, but some switch hosts often enough to be significant vectors.

There is another way in which insects, most of them not blood suckers, transmit pathogens. As Gerald Schmidt and Larry Roberts reported in their book on parasitology, dogs and many other carnivorous animals become infected with a nematode by eating a dung beetle or a vertebrate animal that has eaten a dung beetle. The nematodes reproduce in the vertebrate, releasing eggs into its intestines. The beetles swallow eggs when they feed on the droppings of an infected animal, and the nematode goes through part of its development, but not to the point of sexual maturity, in the body of the beetle. If the dung beetle is eaten by a carnivore, the juvenile nematodes burrow into the wall of its aorta and after about 3 months migrate into the wall of the nearby esophagus to complete their development. About 3 months later, the worms lay eggs, which pass through the digestive system and are expelled in the feces. Given a hungry dung beetle, the cycle will begin again.

Relatively little, Robert Harwood and Maurice James write in *Entomology in Human and Animal Health,* is known about the diseases of wild animals— except when they affect domestic animals or are "reservoirs" for pathogens that are transmitted to humans. Among diseases in the latter category, to mention just a few, are eastern equine encephalitis, caused by a virus transmitted by mosquitoes from birds to humans and horses; the infamous bubonic plague, caused by a bacterium transmitted from domestic rats and wild rodents to people by fleas; at least one form of African sleeping sickness, caused by a trypanosome (a protozoan) transmitted by tsetse flies from antelopes and other wild mammals to humans; the well-known Lyme disease of

North America, caused by a bacterium transmitted from birds and small mammals, such as white-footed mice, to humans by immature deer ticks; and several mosquito-transmitted pathogens that cause malaria in people or birds.

🦟 🐜 🐛 Bird malarias are caused by many different pathogens of the genus *Plasmodium,* single-celled amoeba-like protozoans small enough to fit inside a red blood cell. (Human malarias are caused by four other species of the same genus.) Unlike the myxoma virus, plasmodia have a complex and obligatory relationship with their mosquito vectors. In the mosquito they are present as male and female cells that reproduce sexually and bear "spores" (sporozoites) that a feeding female mosquito may inject into a bird along with her anticoagulant saliva. Once in the bird, the sporozoites go through two asexually reproducing stages. First the sporozoite enters a liver cell and divides to form many merozoites, which are released into the blood, enter red blood cells, and then divide into many individuals, the precursors of the sexually reproducing stage, that are released into the blood plasma when the cell bursts and may be ingested by a mosquito taking a blood meal—thus another cycle begins.

The unintentional introduction of a mosquito that transmits bird malaria had a calamitous effect on the native birds of Hawaii. In Australia, a pathogen, the myxoma virus, was brought to an existing population of native vectors, while in Hawaii a vector was brought to a pathogen that had always been there in the winter months in ducks, plovers, and sandpipers that migrated from North America every autumn. But the malaria pathogen had never spread to birds that are permanent residents of the islands. It remained locked in the bodies of the migrants, because in Hawaii there were no vectors to transmit it. Mosquitoes of any sort were unknown in the Hawaiian Islands until their inadvertent introduction in 1827 by a whaling ship, setting the stage for the transmission of malaria from the North American migrants to Hawaiian land birds. Especially grievous is the extinction of many of the honeycreepers that still survived when the first Europeans arrived. This family evolved on Hawaii and occurs nowhere else in the world. The ancestor of the honeycreepers was a species of finch that arrived on these volcanic islands on an unknown date some time in the 1 to 5 million years during which the islands were rising above the sea. The descendants of these birds underwent an evolutionary radiation that produced nectar drinkers, seed crackers, fruit eaters, and insect eaters which variously sought their prey by gleaning leaves and twigs, by creeping on the bark of trunks and large branches, and even by digging into wood with a short, stout lower mandible and picking insects from their burrows with a long, thin, curved upper mandible.

When the native Hawaiians arrived on the islands about 1,400 years ago, they found 85 or more species of honeycreepers. They exterminated at least 35 of them to make the luxurious, colorful cloaks, each made of the feathers of as many as 10,000 birds, worn by their chiefs. Early in the nineteenth century European explorers and naturalists found many native birds still present. But by the late nineteenth century, the land birds of the Hawaiian Islands were declining. This decrease was partly the result of habitat destruction, the clearing of native forests from the lowlands and lower slopes of the mountains by European and American settlers. But malaria and another mosquito-borne disease, birdpox, were also taking their toll. In 1902, H. W. Henshaw, an ornithologist, reported that "dead birds are . . . found rather frequently in the woods on the island of Hawaii, especially the iiwi and the akahani [= apapane]."

In an article in the September 1995 issue of *National Geographic,* Elizabeth Royte wrote of the current plight of the native Hawaiian flora and fauna. Of the 50 species of honeycreepers still extant when the first Europeans arrived, only 21 remain and 14 of them are endangered. She relates that birds literally fell from the trees during recent outbreaks of avian malaria, and explains how feral pigs, at least 100,000 of them, exacerbate the malaria problem in the few remaining forests of native vegetation. The pigs knock over giant tree ferns and hollow out the trunks to eat the starchy core. Pools of rainwater in these hollows are important breeding sites for the larvae of mosquitoes that transmit bird malaria.

The native Hawaiian birds are far from the only birds that suffer from malaria. There are many species of the malaria parasite, and they occur in many different birds—even penguins from South Africa and New Zealand—and they may ultimately be found to occur in virtually all birds except for those that live in areas, such as Antarctica, where there are no mosquitoes. In North America, malaria-causing plasmodia have been found in the Canada goose, various ducks, Forster's tern, the ruffed grouse, the eastern screech owl, and many songbirds such as the robin, bluebird, cliff swallow, tree swallow, gray catbird, red-winged blackbird, cowbird, house sparrow, common yellowthroat, orange-crowned warbler.

🦟 🦟 🦟 Now and then we hear news reports of an epidemic among humans of a mosquito-borne viral encephalitis somewhere in the United States or Canada. These can be debilitating and sometimes fatal diseases, especially in the very young and the elderly. The most frequently occurring have been St. Louis encephalitis, eastern equine encephalitis, and western equine encephalitis—the last two so named because they affect horses as well as humans. In 1999, another type of encephalitis made a big splash in the news. The West Nile virus, never before seen anywhere in the New World, suddenly appeared in New York City, sickening 62 people and killing 7. Since then, West Nile virus has spread to many more states. The first sign of this disease in an area is often a dead crow found lying on the ground.

West Nile and the others are basically viruses of birds and have been found in scores of species, the West Nile virus already in more than 60 North American birds. All four of these viruses are transmitted from one bird to another by the bites of mosquitoes. Most of the time, epidemics in urban humans start after the virus is spread from the wild birds that are its usual hosts to such urban birds as pigeons and house sparrows. The urban mosquitoes that feed on these city birds occasionally attack people and transmit the virus to them. But they do not transmit it from human to human. Epidemics can be stopped by using insecticides to control these bird-feeding vectors, which are relatively uncommon compared to the hordes of pest mosquitoes that regularly bite people but are not disease vectors. So don't think that a vector-control program has failed just because you are still being bitten by mosquitoes.

There are essentially two ways to use these insecticides: the first is to try to kill widely scattered, flying, adult mosquitoes by broadcasting insecticides from airplanes or fogging machines that cruise the streets, and the second is to target the relatively few pools of organically contaminated water in which the larvae are concentrated. Fogging is highly visible and thus a good public relations ploy, but according to the medical entomologist Robert Novak, it is seldom effective. But treating the larval breeding sites is usually very effective. In urban areas, these sites are often along streets in the catch basins below the grates that cover the openings to storm sewers. The basins hold water that, to the benefit of mosquito larvae, is contaminated with decaying grass clippings and the other organic debris on which grow the bacteria and other tiny organisms that are the larvae's food. One of these small basins may support hundreds of larvae that will as adults disperse over many city blocks. As military tacticians know, it's better to bomb a cannon factory than to go after cannons that have been widely dispersed.

🐜 🐜 🐜 Even if they don't transmit a disease, blood-sucking insects, ticks, or mites may kill an animal outright or weaken it enough to limit its reproduction or even altogether prevent it from producing offspring. Injury to the host may be caused by the excessive loss of blood or even by the impairment of normal behavior by frequent annoyance and irritation. The debilitating effect of blood feeders is exacerbated when it is combined with other stresses such as disease, a scarcity of food, or unfavorable weather.

Lice are external parasites that spend their entire lives on the body of a bird or a mammal, where they mate, lay their eggs, feed, and grow to maturity. The evolution of the louse life style was facilitated by these insects' gradual metamorphosis; the requirements of the nymphs are virtually identical to those of the adults and they are not encumbered by an immobile and helpless pupal stage that could not hold onto the host. Long ago in evolutionary time, they lost their wings as an accommodation to living in plumage or dense fur; their bodies became flattened; and their legs were adapted for holding onto the host. They can move from one host to another only when two hosts are in close bodily contact, as when a mammal suckles her young, when a bird broods her nestlings, or when adults mate. (The last situation probably provides the most frequent opportunity for the infamous pubic or crab louse of humans to switch hosts. It and a similar one that infests gorillas are the only two species in the genus *Phthirus*.) In the laboratory, lice do best at a temperature close to that at the surface of the host's body. For example, eggs of the chicken louse hatch after 3 to 5 days in an incubator at about 99°F. (The internal body temperature of a chicken is about 107°, but the temperature at the body surface is much lower.) Slightly lower temperatures prolong the hatching period to 14 days, and nymphs soon die if the temperature is lowered to 90°F. Lice abandon their host when it dies. When the clothing of Thomas à Becket was removed on the day after his murder in 1170, a horde of body lice seethed over his haircloth underwear.

There are two quite different groups of lice: the biting lice, which have chewing mouthparts and which I will discuss later, and the sucking lice, equipped with mouthparts for piercing skin and sucking blood. The sucking lice are nicely adapted for holding onto their hosts. Their bodies are flattened

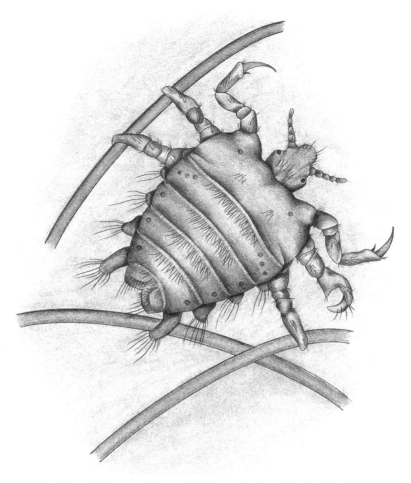

A crab louse holding tightly to pubic hairs. Note the claw that
opposes a "thumb" on one of the leg segments.

and their legs adapted for grasping. They press themselves flat against the
host's skin and hold so tenaciously to hairs that they are seldom torn loose
by scratching and almost never fall off.

All 500 or so sucking lice live only on mammals, but not on the most
primitive mammals, the egg-laying echidna and duck-billed platypus; or on
any of the marsupials, such as opossums, kangaroos, or koalas. But the
world over they parasitize many of the other mammals, among them cattle,
giraffes, camels, peccaries, goats, mice, moles, dogs, river otters, even seals
and walruses. Sucking lice are mostly host-specific; many live on only one

or a few closely related species. For example, the crab louse, the head louse, and the body louse infest only humans, and all of their close relatives live on monkeys or apes.

Blood, the only food of sucking lice, is a poor source of the B vitamins, required by insects and all other animals. How, then do they survive? Sir Vincent Wigglesworth, long the dean of insect physiologists, reported that sucking lice, other blood-sucking insects, and insects that eat such vitamin-poor foods as wood or plant sap harbor in their bodies friendly microorganisms, such as bacteria, which provide them with vitamins. In nymphal lice, the microorganisms are housed in special organs associated with the digestive system. In adult male lice the microorganisms degenerate, but in females they migrate to the ovaries, are incorporated in the eggs, and are passed on to the next generation.

Robert Harwood and Maurice James, eminent medical and veterinary entomologists, noted that while quite a bit is known about the effect of lice and other parasites on people and domestic animals, far less is known about their effect on wild mammals. But it seems safe to assume that sucking lice have more or less the same effect on wild and domestic animals. In 1965, according to the entomologist C. Dayton Steelman, lice infesting cattle, sheep, swine, and goats caused a combined loss of about $97.8 million—an amount over $520 million in today's dollars. Although expressed in dollars, this figure reflects an ecological reality: deaths and, to a significant degree, losses in weight gain and milk production, all of which limit population growth. In a U.S. Department of Agriculture bulletin, H. E. Kemper and Harold Peterson describe the impact of lice on cattle in biological terms.

> The irritation caused by the parasites is evidenced by the efforts of animals to obtain relief by rubbing and scratching. Some of the animals in grossly infested herds have large areas of skin that are partly denuded of hair and are raw and bruised from rubbing against posts and other objects. The lowering of the vitality of animals affected with lice and the general unthrifty condition produced by these parasites often result in an increased percentage of deaths among cattle during unfavorable seasons.
>
> Calves, young stock, and old, weak, poorly nourished cattle suffer most from the ravages of lice. Heavily infested calves do not . . . gain weight normally during the winter and often remain stunted until spring, when most of the lice are shed with the old hair coat . . . Mature

cattle in full vigor suffer less from infestation with lice; nevertheless if they become very lousy they will not gain weight and there will be a loss in the production of either meat or milk.

🐜 🐜 🦟   A very few insects with complete metamorphosis have, despite their helpless pupal stage, become permanent residents of the bodies of mammals, as are the lice, which have gradual metamorphosis and have no pupal stage. The sheep ked, sometimes inappropriately called the sheep tick, is a wingless, ticklike, blood-sucking fly that, even though it has a pupal stage, is a permanent resident in the fleece of a sheep or the hair of a goat. A wingless adult ked could not find and board a sheep if it spent its maggot and pupal stages somewhere other than on the sheep—perhaps in dung or carrion, as do most other maggots and as did the maggot ancestor of the ked. Boarding a sheep would be a difficult climb for a wingless adult, and at any rate, the flock of sheep would probably have moved on by the time a free-living maggot grew to adulthood. Sheep keds have solved this problem by not having free-living larvae. Like tsetses, they retain a larva within their body, nourishing it with milk secreted in the "uterus," an en-larged area of the common oviduct, and giving birth to it when it is fully grown and ready to become a pupa. During her lifetime, a female sheep ked may give birth to 10 or 20 maggots. The mother glues each newly born maggot to the hairs of the sheep. The helpless and immobile pupa can remain on the sheep because of the unique way in which it and the maggots of all the higher flies, as contrasted with the more primitive flies such as mosquitoes and gnats, prepare for pupation. Before it pupates, the carrot-shaped maggot assumes an oval shape and darkens and hardens its soft white skin. Only then does the pupa separate itself from its hardened larval skin. But it does not shed its larval skin, called a puparium. It re-mains within it, protected by it as a moth pupa is protected by its silken cocoon. The adult fly bursts its way out of the puparium with a large bladder that protrudes through an opening at the front of the head when inflated with blood and pushes off a detachable cap from the front of the puparium. After emergence, the bladder is withdrawn into the head and the open-ing closes.

🐜 🐜 🦟   Blood-sucking insects that fly, mainly mosquitoes and biting flies, can have surprisingly great deleterious effects on the growth, reproduction, milk production, and even survival of mammals. As anyone who has been in the north woods in spring knows, black fly females—but not males—are fierce biters that can drive you out of the woods as hundreds or even thousands swarm around you. Their serrated mandibles are uniquely modified for cutting through skin. They overlap like the blades of a scissor and snip rather than pierce like a hypodermic needle, as do the mouthparts of other blood-sucking insects. Black fly larvae live attached to a rock or some other object under flowing water, their gills absorbing dissolved oxygen from the water, and their mouthparts filtering tiny creatures, algae, and decaying organic matter from the current. They pupate in a cocoon attached to a submerged object. It takes the adult fly only about a minute to escape from the cocoon and rise to the surface in a bubble.

Black flies are usually abundant, and unusually massive outbreaks can seriously weaken or even kill domestic birds and mammals. In Canada, according to Steelman, yearly outbreaks from 1944 to 1948 killed over 1,100 farm animals, reduced the milk production of dairy cattle by about 50 percent, and seriously reduced the weight gain of beef animals. In 1923, 16,000 farm animals died in Romania because of a massive black fly outbreak, and in 1934, an outbreak killed 13,900 domestic animals in Yugoslavia, Bulgaria, and Romania. Today, insecticides protect livestock from black flies and so the catastrophic losses of earlier years are rare. But wild animals receive no such protection.

On a birding trip in May 1999, I drove to Point Pelee in Ontario, one of the great birding places of North America. At the entrance to the park a sign warned that the flies were bad that day. And so they were. They drove me from the park. Stable flies, relatives of house flies and similar to them in size and appearance, were everywhere and bit my legs and ankles right through my socks. The bites (both sexes suck blood) were painful because the piercing beak—a makeshift modification of the sponging beak of the house fly—tears its way through the skin rather than piercing gently and usually painlessly, as do the mouthparts of a black fly or mosquito. Stable flies breed in piles of wet straw or other rotting vegetation, at Point Pelee in thick windrows of aquatic vegetation that the waves of Lake Erie toss up on the beaches.

Stable flies can be very injurious to animals, not only because of the blood loss they bring about but also because of the nervous behavior caused by the pain of the bites. Animals under attack, often by hundreds or even thousands of these flies, stamp their feet continuously to dislodge them. According to Steelman, in 1965 this fly caused a loss in cattle production in the

United States of $142 million, well over $750 million today. Only one stable fly per cow per day reduces milk production by 0.7 percent. Only 70 flies per day cut milk production in half—probably enough to starve or at least stunt the growth of a nursing calf or of a fawn or a bison calf. According to "Earthweek: A Diary of the Planet," a column produced by the *Los Angeles Times*, stable flies—the very same species that so viciously pestered me at Point Pelee—have recently been so horrendously numerous in Tanzania's Ngorongoro Crater wildlife park that they killed 6 lions and seriously injured 62 others.

Horn flies, named for their habit of clustering at the base of an animal's horn, take blood meals mainly from cattle. They are closely related to stable flies but are only about half the size. Heavy infestations may number several thousand per animal. They breed in cattle droppings, laying eggs on the surface of a cow pat almost as soon as it is dropped. Within 2 minutes the dung no longer attracts females. One of my colleagues commented, "When the tail goes up, the female fly darts out." On average there are about 150 larvae per dropping. They hatch from the eggs in about a day, become fully grown in 3 to 5 days, pupate in the dropping or nearby soil, and emerge as adults about a week later. As is the case with stable flies, both sexes suck blood. But they cause greater losses—almost a billion dollars in one year—probably because they are more numerous on cattle and because they breed in such close association with them.

On an insect collecting trip to Texas in 1960, a fellow entomologist, Bill Downes, and I camped on a deserted Gulf of Mexico beach on the Bolivar Peninsula. Late in the afternoon we made a fire of driftwood, heated our meal of canned food, and then watched the sun go down over the wide marsh on the landward side of the peninsula. When the sky began to darken, we saw a large, dark cloud rising from the marsh. At first we thought the marsh was on fire and the cloud was smoke. But we soon realized that we were seeing a vast swarm of countless millions of salt marsh mosquitoes leaving the marsh in which they had lived during their aquatic larval stage. The females, the blood-sucking sex, were searching for a blood meal to provide the protein they require to develop eggs. Since the evening breeze was blowing the mosquitoes towards us, we quickly smeared ourselves with insect repellent. We soon learned that mosquitoes—and surely other blood-sucking insects, too—need not transmit disease to be severely injurious. Shortly thereafter the mosquitoes arrived and plagued us almost beyond belief. Hundreds hovered just in front of our faces; some bit despite

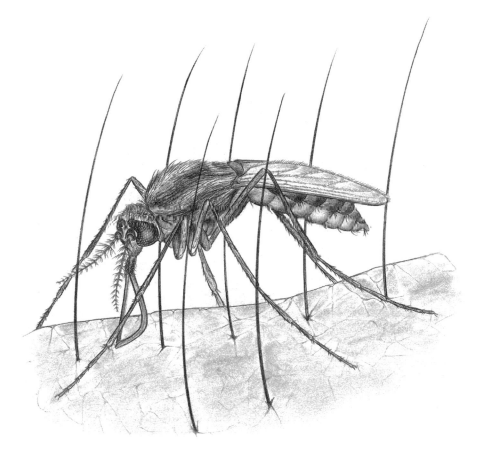

A female mosquito sucking blood from a human arm
through the tube that pierces the skin.

the repellent, and in the brief period during which we removed our boots before taking refuge in our screened tent, scores of them landed on our repellent-free ankles and tortured us with their bites. These mosquitoes attack all of the wild and domestic mammals of the area. Steelman wrote, "In the Texas coastal marsh, cattle-raising is uneconomical because mosquito attack reduces feeding time while the cattle are fighting mosquitoes and causes loss of blood from mosquito bites." He also reported that young and weak cattle have been killed by mosquito attacks.

In *People of the Deer,* Farley Mowat described the great swarms of mosquitoes that torment people, caribou, and virtually all other mammals and birds of the tundra in the Canadian Arctic. But even where they are not astronomically abundant, mosquitoes take their toll of blood from mammals, birds, and even reptiles and amphibians, and thereby limit their reproductive potential.

Few people have a first-hand acquaintance with parasitic insects that live within or on the outside of the body of an animal as they devour skin and flesh. Virtually all of these parasites are maggots, the larvae of flies. Although people are seldom infested, wild and domestic animals often are. In the Americas, the most important of the external flesh-devouring maggots is the screwworm. Maggots that are internal parasites are collectively known as bots or warbles. Broadly speaking, there are four groups: the human bot *(Dermatobia)* of South America; bots that live in the nasal cavities and sinuses of grazing mammals; warble flies, which live beneath the skin of rodents and even-toed hoofed animals, such as swine, deer, bison, cattle, giraffes, and sheep; and finally, the bots that live in the stomachs of odd-toed hoofed grazers and browsers such as horses, zebras, tapirs, and rhinoceroses.

The bots known as ox warbles or cattle grubs (family Oestridae), wrote F. C. Bishopp and his coauthors in a U.S. Department of Agriculture bulletin, hatch from eggs glued to the hairs on a cow's body, burrow through the skin, and then wander through muscles and other tissues until they reach their host's back 5 or 6 months later. There they cut a breathing hole through the hide and form a large pus-filled boil. From 5 to 10 weeks later, when they have grown to the size of the last joint of a person's thumb, they drop to the ground, pupate in the soil, and emerge as flies about a month later. (Ox warbles are not musical. Their name is akin to the obsolete Swedish *varbulde,* a boil.) These bots are a significant drain on the host. Animals plagued by bots have evolved countermeasures, and the bots have responded by evolving behaviors that overcome these countermeasures.

Some warbles, such as the bomb fly that parasitizes cattle in Europe and North America, are very open about laying their eggs. The females of this species, named for the threatening way they approach a potential host, buzz loudly as they repeatedly "dive bomb" it in order to glue eggs to its hair. One female may lay as many as 800 eggs during her lifetime, placed one to a hair, mainly on the lower legs but also on the belly. A cow that is the target of these bombing runs raises its tail high in the air and rushes about wildly until it evades its tormentor by running into water or a brushy area. Another bot, the heel fly, is much more circumspect when placing its eggs on a cow, a behavior that probably evolved to counter the cows' wild and sometimes successful attempts to avoid an egg-laying fly. The sneaky heel fly lands in the shadow of a cow and crawls unobtrusively over the ground to glue its eggs to hairs on the heels of a standing cow or the belly of a recumbent one. In addition to cattle, ox warbles regularly parasitize bison, and probably some other wild hoofed grazers and browsers such as deer and elk. Several species of warble related to the ox warble attack caribou, deer, goats, and sheep. The warble fly of both wild and domestic caribou (reindeer) occurs throughout the range of its host in northern Eurasia and in northern areas of Canada and the United States.

Flies of a different family (Cuterebridae) are common warble-like parasites of wild New World rabbits and rodents such as field mice, wood rats, squirrels, and chipmunks. Females lay their eggs in the host's haunts. The larva enters the host through its skin or a body opening and migrates to a place under the skin, where it forms a boil much like that of an ox warble. The late William Horsfall, a medical entomologist, noted that in Canada in 1955 about 50 percent of chipmunks and in Wisconsin in 1958 about 20 percent of cottontail rabbits were infested. Because they are large, the maggots seriously injure the host. Some species lodge in the groin of male chipmunks and rats and often sterilize them. Among them is a chipmunk-infesting species that is aptly named *Cuterebra emasculator.*

The human bot of Central and South America, another cuterebrid, has the ultimately sneaky way of getting its eggs onto the body of a person, some other mammal, or even a bird. The female does not approach the host directly, but instead catches a blood-sucking fly, mosquito, or, rarely, a tick, attaches a dozen or more eggs to its body with a quick-drying glue, and then releases it unharmed. When and if the egg-carrying intermediary lands on a warm-blooded animal, the bot larvae immediately pop out of the eggs, drop to the skin, and burrow in. This fly, often known as the tórsalo, infests a wide range of hosts: humans, monkeys, cattle, swine, cats, dogs, horses, mules, sheep, goats, birds, including toucans and antbirds, and probably many other wild birds and mammals of the forest. The larva does not wander in the body. It lives just beneath the skin at its point of entry, forming a boil-like lesion with a breathing hole. After about 6 weeks, the mature larva drops to the ground and burrows into the soil to pupate.

Some parasitic maggots, the stomach bots (Gasterophilidae) and the nose bots (Oestridae), live only in a natural cavity of the host's body. Larvae of the various species of the former group live in the stomachs of horses, zebras, rhinoceroses, and elephants. The behavior and life history of the cosmopolitan horse bot (*Gasterophilus intestinalis*—the genus name means "lover of stomachs" in Greek) have been well studied because it injures horses. Females lay about 1,000 eggs, tightly attaching each one to a hair, mainly on the backs of the knees, where horses often lick themselves. The embryos are ready to hatch after about 5 days, and are stimulated to do so by the warmth of the licking tongue. The tiny maggots burrow into the tongue and later pass into the stomach, where they firmly attach themselves by their mouth hooks and extract nutriment from the tissues of the stomach and absorb it from the stomach's contents. They remain attached throughout the summer and winter, but the following spring, when fully grown, nearly an inch in length, they pass out with the horse's droppings, and pupate in the soil or dry droppings. In 3 to 5 weeks the adult bot fly emerges. A horse's stomach may contain hundreds of bots, which cause digestive disorders, may obstruct the passage of food from the stomach to the intestine, and may lead to secondary infections by bacteria. Occasionally they even cause death.

Nose bots live in the nasal passages and sinuses of a wide variety of large plant-eating mammals. One or more species occur in parts of Eurasia, Africa, Australia, and the New World, and among them parasitize sheep, goats, pigs, warthogs, camels, horses, zebras, hippopotamuses, deer, elk, caribou, and wildebeests, hartebeests, and other antelopes of Africa. These flies have probably flourished as long as mammals have existed; kangaroos, among the most primitive of mammals, have a species of their own, which lives in the windpipe rather than in the sinuses. In an article in *Smithsonian,* John Ross described a "fist-size mass of yellow, pulpy maggots" that he saw seethe and writhe when a Khanty reindeer herder of northwestern Siberia exposed the sinuses and nasal passages of a reindeer he was butchering. The incidence of infestation by this parasite may be as high as 100 percent in some herds, and many young animals are killed, with reduced milk production by infested mothers probably a contributing factor. The sheep nose bot also parasitizes goats and wild relatives of sheep. In the summer and autumn, flying females dash at sheep to deposit active larvae, which hatched in the mother's birth canal, in or near the nostrils. The larvae immediately move up the nasal passages into the sinuses and attach to the membranes. The next spring, fully grown and almost an inch long, they crawl out through the nostrils or are sneezed out by the sheep, pupate in the soil, and emerge as adults from 3 to

6 weeks later. As many as 350 maggots have been found in the head of a single sheep. In New Mexico, 90 to 95 percent of adult sheep have nose bots, an average of about 20 per animal. Both the adult flies and the larvae seriously interfere with the lives of their hosts. When adult flies threaten, sheep stop feeding and run away or shake their heads, stamp their feet, and huddle together holding their noses to the ground. The maggots cause inflammation, a copious nasal discharge, sneezing, labored breathing, and sometimes giddiness or "blind staggers."

Black flies, mosquitoes, sand flies, and other winged blood feeders that attack birds are just transient visitors, staying only long enough to get their fill of blood. But other parasites—most of them blood feeders, too—are more closely associated with their hosts. Biting lice, and certain other louse flies and mites, remain permanently on the body of their avian hosts. Others live permanently in bird nests and, for the most part, make contact with the bird only to take a blood meal. Among them are fleas, some maggots, certain relatives of the bed bug, and some louse flies and mites.

The true bugs of the family Cimicidae are called bed bugs, a misleading name because only 2 of the 89 known species associate with humans. These 2 species occasionally attack birds that associate with people, such as house sparrows and pigeons. But there are 26 other bed bugs that are exclusively or almost exclusively associated with birds, hiding in or near their nests during the day and coming out at night to suck their blood.

Because all bed bugs are wingless and cannot move very far, they assure continued association with their meal ticket by parasitizing only mammals or birds that have a permanent dwelling or reliably return to the same place after an absence. Sixty-one species parasitize bats, which return to roost in the same cave every day. No bed bug attacks primates other than humans, presumably because humans are the only primates that inhabit permanent dwellings. Most birds build temporary nests, used for a season and then abandoned, but some reuse the same nest in succeeding years. The 26 bed bugs that attack birds parasitize only species that return to the same nest or nest site year after year.

A bed bug aptly known as the swallow bug infests nests of the highly colonial cliff swallow of North America. Cliff swallows build gourd-shaped nests of mud pellets in protected places—under an overhanging rock ledge, a bridge, or the eaves of a building—and usually return to the same site year after year. When the swallows migrate south in the fall, the bugs remain behind for the winter in nests or crevices near a nest. When the swallows return in the spring, the bugs resume their attacks. (If the birds don't return, the bugs can survive without feeding for up to 3 years. With luck—for the bugs—the site may be reoccupied before the swallow bugs starve to death.) At night the bugs suck blood from adult swallows and nestlings, and during daylight they hide in the nest or nearby crevices. By treating some nests with an insecticide, which killed the bugs but did not harm the swallows, and comparing them with a control group of nests that were not treated, Charles and Mary Brown showed that nestlings from treated, parasite-free nests were large and healthy. Those from untreated, heavily infested nests were small and sickly, and many of them died.

Most louse flies, sparsely hairy, flattened, and leathery, are blood-sucking parasites of birds. Members of a small but geographically widespread family (Hippoboscidae), they vary widely in their commitment to the birds they parasitize. Some have functional wings throughout their adult lives and fly from one bird to another. Although winged females could fly off, they usually remain with the same host throughout their lives and mate only with males that come to the host. Males, to the contrary, often fly away from their host, presumably looking for more females to inseminate. Other louse flies can fly but shed their wings after they locate a host and will spend the rest of their lives on it. A few species have only vestigial wings or no wings at all and can only crawl from within a nest to an occupying bird.

Like the sheep ked, louse flies facilitate their parasitic way of life by not having a free-living larval stage. Instead, the female retains a maggot in her pseudo-uterus, nourishes it with a milklike secretion, and gives birth to it when it is fully grown and ready to molt to the pupal stage. The newly born maggot forms a puparium just as do the sheep ked and other evolutionarily advanced flies such as house flies, nest flies, and stable flies.

Many louse flies that parasitize birds that reuse the same nest year after year—such as swifts and bank swallows—deposit their puparia in the host's nest. Here the pupa spends the winter, and in the spring the adult emerges from the puparium and sucks blood from a returning bird. Many louse flies that infest birds that reuse nests lack wings or have small wings useless for flying. They need not fly to find a host; a host will usually come back to them. But louse flies that infest birds that build a new nest every year do have wings, because they must fly to locate a new host. They place their puparia on the ground, where they are hidden by fallen leaves and other organic debris during winter.

At Algonquin Park in Ontario, Gordon Bennett examined 84 species of birds and found louse flies on 56 of them, all songbirds (mostly thrushes, warblers, and sparrows) except for saw-whet owls and 4 species of woodpeckers. His data are based on 6,448 birds that he captured and examined during the summers of 1957–1960. About 16 percent were infested.

🕷 🕷 🕷 You may have seen large, shiny, green or blue flies sipping nectar from flowers, especially the yellow umbrella like clusters of wild parsnip, a common roadside weed. Most of these insects are blow flies, which are carrion feeders as maggots. But among them may be a related shiny blue fly of the genus *Protocalliphora,* which in the maggot stage inhabits birds' nests and is a bloodthirsty parasite of nestlings. These nest flies occur worldwide. A dozen species or so are the only North American flies that suck blood in the larval stage.

Nest flies lay small batches of eggs—often only 10 to 20 per batch—in nests with partly grown nestlings, apparently attracted by the odor of the young birds. They never lay in nests that are empty or contain only eggs. Within a day, the sluglike maggots hatch and immediately burrow into the nesting material. Like leeches, hungry maggots crawl onto the nestlings, usually at night, and hold on with a sucker at the front end of the body as they drink blood for about an hour. When satiated, they burrow back into the nest. After taking several blood meals over a period of about 12 days, they are fully grown and transform to the pupal stage in the nesting material. Adults emerge 2 or 3 weeks later.

To judge from what is now known, it seems that all or almost all North American birds that build a nest are parasitized by some species of nest fly. Although the flies do not specialize on a particular species of bird, they do have other preferences, such as for cup nests or nests in cavities. In 1991, Catherine Rogers and her coworkers observed that in Ontario a nest fly that associates with tree swallows, which nest in natural cavities in dead trees, also infests nests of a wide variety of unrelated birds that also nest in cavities: screech owls, downy woodpeckers, great crested flycatchers, house wrens, European starlings, and eastern bluebirds.

Nest fly infestations are common. Tree swallow nests are usually infested,

72 percent in one series of observations and 82 percent in another. A nest is likely to be infested by 10 or 20 maggots, but there may be many more. Ninety-three percent of 127 mountain chickadee nests examined in Modoc County, California, in the 1980s by Clifford Gold and Donald Dahlsten harbored anywhere from only 1 to over 100 nest fly maggots each. In Ontario, Rogers and her colleagues found 167 in a tree swallow nest with only three nestlings. In California, a chestnut-backed chickadee nest contained 273 maggots, and almost 13 percent of the mountain and chestnut-backed chickadee nests examined contained 100 or more maggots. The record seems to be 373 in a nest of a black-billed magpie in Europe.

Rogers and her colleagues found that even "heavily parasitized nestlings do not necessarily suffer unusually low growth rates or high morality." Nevertheless, the loss of blood by a nestling may be very great. The evolutionary biologists L. Scott Johnson and Daniel Albrecht suggested that although blood loss to nest flies has little or no effect on nestlings, the real burden may be borne by the parent birds. The argument is that the parents give parasitized nestlings more food to compensate for the loss of blood, and that the extra energy expended in foraging for this food could have been used to raise more offspring and thereby increase the parents' evolutionary fitness.

Fleas are wingless, blood-sucking parasites of mammals or birds. According to *Insects of Australia,* compiled by the Division of Entomology of the Commonwealth Scientific and Industrial Research Organization, there are about 2,390 known species of fleas in the world. Most are parasites of mammals, mostly rodents, but 132 species are associated with birds. The number of bird species attacked greatly exceeds the number of species of bird fleas. Although some fleas are somewhat specific in their choice of hosts, many others are generalists that parasitize many different kinds of birds. In Great Britain, for example, the hen flea attacks about 75 species of birds in addition to chickens.

Almost all fleas parasitize hosts with a permanent home, such as the nest of a bird or the lair of a mammal, where the larvae can live and find food. Humans, the only primates with permanent homes, are the only primates infested by fleas. But, as is almost always the case in matters biological, there are a few exceptions to this generalization. In *The Ecology of Ectoparasitic Insects,* Adrian Marshall reported that the larvae of certain fleas that parasitize cattle or other hoofed mammals live on the ground where their hosts pasture. The flea *Uropsylla tasmanica* is probably the only flea that is a parasite during its larval stage. In Tasmania and Australia the anatomically specialized larvae of this flea burrow into and feed on the skin of their marsupial hosts.

Fleas are among the few parasitic insects with complete metamorphosis that are associated with the host throughout both their larval and adult stages. The bird fleas, like the bed bugs, reestablish contact with a host from one generation to the next by parasitizing only birds that return to the same nest for several years in succession. Larvae and pupae are always in the nest itself. Adults also live in the nest, but from time to time board the host to take a blood meal. The different behaviors of larval and adult fleas are reflected by the advice, beguiling to nonentomologists, that Miriam Rothschild and Theresa Clay offered to flea collectors in their engaging book *Fleas, Flukes, and Cuckoos:*

> The simplest way to collect bird fleas is to take a nest from which the fledglings have recently flown and to keep it in a cardboard box or linen bag. Providing the nest is damped periodically, the larval or pupal fleas continue to develop in the debris or rubbish in the bottom, and in due course hatch out. It is a more lengthy process to collect them off the bodies of their host. Less than one bird in ten harbours fleas, and then generally only one or two specimens at a time . . . The maximum number recovered from a bird is 25 specimens from a house-martin, a species of swallow that does not occur in the Western Hemisphere. On the other hand, no less than 4,000 have been bred out of a single martin's nest.

A few species of fleas break the general rule that adults do not remain permanently on the host. Among them are the females of the sticktight flea, which occurs in the southern United States on wild birds and domestic poultry, particularly on the combs and wattles of chickens. Before mating, both male and female sticktight fleas hop about much as do other fleas. Soon after the female has taken a blood meal, she attaches herself to the skin of the host with her mouthparts and copulates with a male. Robert L. Metcalf and Robert A. Metcalf described an infestation of females on a chicken as "clusters of dark brown objects about the face, eyes, earlobes, comb, and wattles made by hundreds of small, flattened fleas that have their heads embedded in the skin so that they cannot be brushed off." Gravid females expel their eggs forcibly, propelling them for some distance, thus assuring that they fall from the bird and, with good luck, land in its nest or roosting place, where the larvae will find food. Little is known about the effect of the sticktight flea on wild birds, but the Metcalfs said that young fowl are often killed, and that infested hens lay fewer than the usual number of eggs.

Other bird fleas hop about quite actively and lay their eggs in the host bird's nest. The larvae that hatch after a few days resemble tiny caterpillars but have neither eyes nor legs. With their chewing mouthparts, they ingest organic debris in the nest, including—and probably very important in their diet—the feces of adult fleas, which consist mainly of blood that has passed through the flea virtually undigested—perhaps through the larva's own mother or father. Fully grown larvae pupate in tiny silken cocoons usually festooned with particles of debris.

In many species of fleas , including at least some bird fleas, the adult sheds its skin in the cocoon, but delays its emergence from the cocoon until some cue alerts it to the presence of a host in the nest. (Some species can survive in the cocoon without food for as long as a year.) The cue is usually vibrations caused by movements of the host but can also be a sudden increase in the temperature or carbon dioxide level. If the host is absent—often for months at a time in the case of migrant birds—there will be no blood donor for a newly emerged flea. The adult is better off if it remains quiescent in the cocoon and conserves its energy until a host appears. Gottfried Fraenkel, one of the world's great insect physiologists, demonstrated to his classes the fleas' response to vibrations by banging on a table on which he had gently placed a covered dish containing cocoons of the Oriental rat flea. The hungry fleas, previously undisturbed for weeks, almost immediately came out of their cocoons and jumped about frantically in search of a host that wasn't there.

The nesting behavior of their hosts has shaped the life cycle and behavior of fleas that parasitize birds, strongly linking them to their respective hosts. The relationship between the sand martin (known in North America as the bank swallow) and the flea that parasitizes it is a well-understood example of the closeness of this link. In Europe sand martins are parasitized by the flea *Ceratophyllus styx* and in North America by *Ceratophyllus riparia*. These fleas are host-specific, almost never found on other birds. Robert Lewis, a noted expert on fleas, recently told me that these supposedly different fleas actually belong to the same species.

The highly colonial sand martins raise their offspring in a nest of feathers and grass placed at the end of a long burrow they dig in a steep-sided bank,

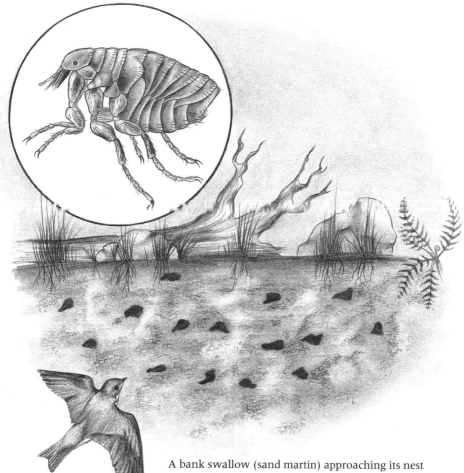

A bank swallow (sand martin) approaching its nest in a burrow in a bank of earth. The nest is infested by fleas like the one in the inset.

usually in a streamside bluff or in the clifflike face of a sand pit dug by people. Most burrows are between 24 and 26 inches long. North American sand martins migrate to South America for the winter and Eurasian ones winter in either Africa or southern Asia. In the spring all return to the same nesting site they occupied the previous year. Sometimes they dig a new burrow but more often they reuse an old one. No fleas are on migrating swallows in the fall or winter. They stay behind in the temporarily unoccupied nests as mature adults poised to emerge from their cocoons at a moment's notice when

properly stimulated the following spring. Then they board a recently returned swallow that enters the burrow or move to the burrow's entrance—
often in large aggregations—and try to get on any swallow that approaches.

In a series of simple but elegantly designed experiments, David
Humphries of the University of Aston in Birmingham, England, figured out
how sand martin fleas exploit their hosts. In the spring, the fleas coordinate
their emergence from the cocoon and the resumption of activity with the arrival of the swallows returning from the south. How do they manage this?
Humphries reasoned that they must become active either in response to
some change in the weather, probably the seasonal increase in temperature
that coincides with the arrival of the swallows, or in response to vibrations
caused by a swallow when it inspects the old nest. He then did two simple
but incisive experiments that determined which of these two factors actually
activates the fleas. The first experiment, done in late autumn after the swallows had left, showed that a mechanical stimulus is sufficient to activate the
fleas. He disturbed nests in several burrows that had been occupied the previous summer "with a long flexible cane bearing on its tip a stiff tassel of
rope." Fleas began to appear at the entrances to these burrows within 2 days
but, as he expected, none appeared at similar burrows that he did not disturb. The second experiment, done in the spring, eliminated weather factors. Before the first swallows arrived, he blocked with thin twigs the entrances to several burrows that had been occupied the previous spring. Birds
could not enter but fleas could exit. After the swallows returned, many fleas
appeared at the openings of unobstructed burrows that had been entered by
swallows, but none appeared at burrows blocked with twigs. The inescapable conclusion is that the fleas emerged after being disturbed by a swallow,
and that neither the warmth of spring nor other climatic factors—certainly
not blocked by a few twigs—had any effect. No more than a brief disturbance sufficed. A swallow that examined a nest and then left activated the
fleas as effectively as one that stayed on to occupy the nest.

Humphries next determined what caused the fleas to move to the entrance of the burrow from the nest at its other end. He hypothesized that
they responded to the light from the burrow's entrance. Such responses
are often phototactic, instinctive orientations to light. In laboratory experiments, he found that sand martin fleas do indeed respond phototactically. In
a horizontal cardboard tube with a light at one end they moved toward that
end of the tube but reversed direction if the light was shifted to the other end
of the tube. Other experiments showed that the fleas are negatively phototactic for the first day after they emerge from the cocoon but are positively

phototactic thereafter. In nature, the brief initial period of negative photo-taxis keeps the fleas within the nest long enough to mate with other recently emerged fleas or to make contact with a returning swallow that might have been temporarily absent. But if the burrow remains unoccupied, the switch to positive phototaxis leads the flea to the entrance, where it has a good chance of boarding an exploring swallow.

Sand martins often hover within 3 or 4 inches of a burrow entrance for as long as 30 seconds. They may then leave, or enter the burrow. Even a swallow that briefly hovers at the entrance and then leaves could transport fleas to another burrow that it will occupy later. Fleas waiting at a burrow entrance use their powerful legs to leap onto any bird that hovers nearby. Humphries's next experiments showed that the instinctive responses of the fleas, honed to a fine edge by natural selection, are exquisitely suited to their dispersal via swallows that come close to but do not enter a burrow.

He was first alerted to this hitchhiking when he saw fleas leap out into space when he cast a shadow on the burrow entrance. He suspected that this seemingly inappropriate behavior was a response to what the fleas perceived as the shadow of a hovering swallow. A simple experiment confirmed his suspicion. He used a small, square black card mounted on a stiff wire to imitate a hovering bird. When he held the card in front of several burrows, 85 of 103 fleas leaped to intercept it. In control experiments, using the wire without a card, not one flea of 7 aggregations totaling 88 individuals jumped. Clearly, the fleas were stimulated to leap by the shadow of the card, and would presumably be similarly stimulated by a hovering sand martin. Their behavior is even programmed to avoid futile responses to swallows or other birds that flit past quickly without pausing. They simply delay their response, not jumping until 1 to 6 seconds after the shadow appears.

In a similar experiment, he determined the effect of the distance of the bird from the burrow entrance. When the model, a piece of hardboard cut to the size and shape of a swallow's silhouette, was held about 3 inches away, 54 of 75 fleas leaped to intercept it, and 39 actually hit the model. When it was held about 6 inches away, fewer fleas leaped, and only one of them hit it. Fleas that miss the bird fall to the bottom of the bank, but are not doomed. They crawl up the bank and position themselves for another leap from the entrance to a burrow.

🦟 🦟 🦟 You have already heard about the sucking lice, all of which are external parasites of mammals. Now we come to the biting lice, most of which parasitize birds, although a few prefer mammals. Like sucking lice, they are small and flattened, and their legs are modified for grasping, but unlike the sucking lice, their mouthparts are basically of the chewing type, always including functional mandibles that in all but a few species are used only to eat solid food. Those few exceptional species consume blood. Most of them just lap up blood that oozes out when their mandibles lacerate the skin of a bird or puncture the quill of a developing feather. But the great majority of biting lice feed on loose scales of skin, other organic debris, and the barbules of feathers—especially down feathers—which they cut into bite-sized pieces. Feathers, like hair, fingernails, and hooves, are composed of keratin, which few animals can digest because it is resistant to most digestive enzymes. Vincent Wigglesworth, an expert on insect physiology, says that clothes moths, hide beetles (Dermestidae), and biting lice seem to be the only insects that have enzymes that can digest this unusual protein.

In his authoritative book *Parasitic Insects,* Richard Askew pointed out that "of all the insects, lice are the most completely committed to parasitism." A biting louse will probably spend its entire life, from egg to adult, on the body of one bird, and its descendants for several generations may stay on that same bird. Of course, some of these descendants must transfer to another bird, usually to an individual of the same species. If some do not do so, their family line will end with the inevitable death of their host. Transfers from one host to another are possible when two birds are in close bodily contact, as when gregarious birds such as starlings and blackbirds roost shoulder to shoulder. But the dispersal of biting lice probably peaks during the birds' breeding season, when copulations are frequent and later lice can easily move from parents to nestlings.

Sometimes winged louse flies transport bird lice to a new host, a form of hitchhiking known technically as phoresy. The louse crawls onto a louse fly and holds on tightly. If the fly lands on a bird the louse gets off and makes itself at home if the bird is an acceptable host, which is likely to be so because the flies usually prefer to stay with the same species of bird. Considering only biting lice and louse flies, at least 405 instances of phoresy are recorded in the scientific literature. These passenger-carrying flies were collected from many different birds, ranging from bitterns and egrets to sparrows and orioles. A surprising proportion of louse flies transport lice. Figures cited in the literature include 22 percent of a sample of 180, 43.5 percent of a sample of 156, and 25.5 percent of a sample of 55. A fly usually carries only 1 or 2 lice, but sometimes more; the record is a fly with 31 lice clinging to its body.

Some biting lice live happily on several unrelated species of birds, but many species are finicky about their hosts and can survive on only one or a few closely related birds. A particularly interesting example of host specificity involves the lice of the cuckoos. Most of them (including all North American species) build nests and care for their offspring. But others, about 40 percent of the 143 species, mostly Eurasian, African, and Australian ones, dump their eggs in the nests of other birds, abandoning them to the care of unwitting foster parents. None of the many different kinds of lice that infest the numerous foster parents exploited by the common cuckoo of Eurasia ever establish themselves on the parasitic cuckoo nestlings. Cuckoos deceive the foster parents of their young but do not deceive the lice that infest the foster parents. According to Rothschild and Clay, in England the common cuckoo is parasitized by only three lice of the aptly named genera *Cuculoecus* ("inhabitant of cuckoos"), *Cuculicola* ("dweller on cuckoos"), and *Cuculiphilus* ("lover of cuckoos"). Lice of these genera infest cuckoos all over the world, but never infest other birds.

To a tiny louse, a bird's body is a veritable world. Different species of lice split up the available territory on a bird by occupying different regions of its body. As the Russian parasitologist V. A. Dogiel noted, several species of bird lice share the bodies of glossy ibises. One lives only on the feathers of the head and neck; another only on those of the breast, belly, and sides of the abdomen; and two or more species on the wing and tail feathers. Theresa Clay noted that a species is precisely adapted to cope with life on the particular part of the bird it occupies. Lice that live on the wings or body below the neck are built to escape the preening bill of their host—elongate and greatly flattened so they can hide in the groove between two barbs of a feather, and sleek and agile so they can escape the preening bill by moving rapidly. They place their eggs in rows in the groove between barbs, firmly sticking them in place with a cement-like substance. Protected by the raised edges of the barbs, they are not likely to be dislodged by a preening bird's bill. Lice that live on the head and neck are short and not much flattened—somewhat round-bodied and not designed to hide in a groove. They do not hide their eggs but lay them singly or in small clumps at the base of a feather. On the head, they cannot be reached by the bill; they can only be reached by the claws, which are much less adept at removing parasites.

Most mites, many of which are parasites, are tiny; the major exception is the ticks, which—although many think of them as being a separate group—are just big blood-sucking mites. Between 20,000 and 30,000 species of mites have been discovered and named, but experts estimate that as many as 500,000 are still unknown to science. Mites inhabit virtually any ecological niche you can imagine. About a third of the known species are parasites that probably infest most terrestrial animals from insects to humans.

Many associate with birds, living on their bodies or in their nests. Some nest inhabitants, notably some members of the family Dermanyssidae, visit the nest's residents to suck blood. Some permanent residents of the bodies of birds are external parasites that suck blood or eat feathers, and others are internal parasites that live within the bird, often in developing feathers. Some that live in the plumage are benign scavengers that eat only bird dander and organic debris. But some others are predaceous mites that roam through the plumage preying on parasitic and scavenging mites. Many parasitic mites occur on only a single species of bird or on a few closely related ones. Such host specificity is probably widespread, but the picture is still incomplete because so many species remain undescribed and because little or nothing is known about the natural history of most of the species that have been described.

In the family Dermanyssidae, we see a spectrum of behaviors, a transition from blood-feeding nest inhabitants that only occasionally visit resident birds to permanent external parasites that spend their whole lives on the body of one bird. Except for a few species that are pests of poultry, little is known about the many dermanyssid mites that parasitize other birds. But what is known about the species that attack poultry surely tells us something about those that attack wild birds. Mites in the latter group probably inflict similar damage and limit population growth, as do poultry mites, by significantly slowing development, reducing egg production, and sometimes even killing a bird.

Chicken mites, common almost all over the world, also parasitize several species of wild birds, report J. P. Linduska and Arthur Lindquist, experts on arthropod parasites of wildlife. In chicken houses, these tiny mites hide in crevices or under debris during the day, but at night come out and crawl onto a chicken to suck blood, their only food. They lay eggs in their hiding places. After they hatch, the mites molt three times as they mature to the adult stage in only 7 to 10 days, their growth fueled by large blood meals they take just before the second and third molts. Females lay their eggs over a period of several weeks, taking blood meals all the while. The northern fowl mite, another pest of poultry and also a dermanyssid, is much more closely associated with its host. It also feeds on blood, but usually spends its whole life on a bird's body and lays its eggs on the down feathers. It attacks many wild birds, among them North American kingbirds, robins, sparrows, blackbirds, grackles, and meadowlarks.

The feather mites, a motley group defined by their habits rather than their scientific classification, or taxonomy, do not suck blood but eat feathers, horny layers of the skin, and organic debris in the plumage. Except for penguins and cassowaries, some birds of every order are known to be infested by feather mites, but we do not yet know how many different kinds there are or how many species of birds harbor them. Many mites have adopted this way of life, members of at least 19 different families, and there are certainly many hundreds—probably many thousands—undiscovered. Ultimately, acarologists (students of ticks and mites) will probably find that most birds are inhabited by one or more species. Many kinds of feather mites live on the surface of a feather, but some live within the quill. Quill-inhabiting species invade new, developing feathers, feed on their inner pulp, and grow to adulthood as the feather matures. Adult females leave the quill of the mature feather and disperse to lay eggs on developing feathers on the same bird or on a young bird brooded in the nest or under the wing of its parent.

Although a number of insects parasitize birds, few prey on them in the way that a coyote catches and devours a rabbit. But on rare occasions, praying mantises sitting in ambush at feeders that dispense sugar water captured and fed on hummingbirds. A few species of ants prey on birds, mammals, and other vertebrates. Various reports of native fire ants preying on quail chicks were summarized by Bernard Travis in 1938. These ants sometimes kill newly hatched chicks, but more often they crawl into an egg that has just been pipped, sting the ready-to-hatch chick to death, and tear it to pieces that they carry back to the colony. Travis reported that of 2,456 nests of bobwhite quail observed near Tallahassee, Florida, in 151, 6.2 percent, some or all of the chicks had been killed by these ants.

The nomadic army ants of Central and South America and the driver ants of Africa are the most ravaging of the ants that attack and kill birds. Army ants do not, as portrayed in horror stories, threaten whole villages as they advance on a mile-wide front. But the unexaggerated truth is impressive enough. Swarms of these ants, anywhere from 150,000 to 700,000 strong, advance on a front that is often 60 feet wide. Raiding swarms move out from their bivouac early in the morning. In a book on army ants T. C. Schneirla wrote that huge raids of *Eciton burchelli,* the most studied of the army ants, "bring disaster to practically all animal life that lies in their path and fails to escape." They capture mainly forest insects and other arthropods for food, but may also kill larger animals. Schneirla goes on to say, "I have seen snakes, lizards, and nestling birds killed on various occasions; undoubtedly a larger vertebrate which . . . could not run off, would be killed by stinging or asphyxiation. But lacking a cutting or shearing edge on their mandibles, unlike their African relatives the 'driver ants,' these tropical American swarmers cannot tear down their occasional vertebrate victims."

The driver ants—so named because they drive all living creatures before them—have an even greater limiting effect on population increases of birds and other vertebrates. In some ways they are similar to army ants. Both are predators that send out huge raiding parties to kill prey and carry it back to the colony. But unlike army ants, drivers dig nests in the soil, and their colonies are usually much larger, consisting of as many as 20 million individuals. Like army ants, they are nomadic, moving to new hunting grounds after they have stripped an area of prey. Unlike army ants, driver ants have sharp mandibles with which they cut their prey into small pieces that they bring back to the nest. In an 1845 letter to J. O. Westwood, secretary of the Entomological Society of London, the Reverend Thomas S. Savage, M.D., described the activities of driver ants in West Africa. The letter, published in the 1847 volume of the *Transactions of the Entomological Society of London,* reported that driver ants "fiercely attack any thing that comes in their way, 'conquer or die' is their motto . . . They are decidedly carnivorous in their propensities. Fresh meat of all kinds is their favorite food . . . They attack . . . lizards, snakes, etc. with complete success. We have lost several animals by them . . . monkeys, pigs, fowls, etc."

Savage goes on to describe how drivers "butchered" a fowl:

The feathers were *pulled out,* sometimes one, two and three ants would be seen tugging most lustily at one, but I am inclined to think that the largest feathers were extracted by lacerating the flesh at their root, though I was not able to decide this point fully . . . [cutting up the prey] began at the beak, and was gradually extended backward. The neck being half stripped, they then began the work of laceration at the eyes and ears. It was some time before any visible impression was made, but at last . . . deep cavities appeared, and muscles, membranes and tendons were reduced and borne off to their habitation.

Parasites can be very injurious. According to the ornithologists Larry Clark and J. Russell Mason, seabirds have abandoned their nesting colonies because of rapidly increasing tick populations. Young cliff swallows from nests infested by ticks called swallow bugs were small, sickly, and often died, while those from bug-free nests were larger, healthy, and much more likely to survive. The effect of blood-sucking mites on young starlings was more subtle; there was no immediately obvious effect, but birds from infested nests were anemic and therefore less able to supply their wing muscles with enough oxygen to support sustained flight. If parasites decrease their host's evolutionary fitness, natural selection will favor any behavior of the host that limits the damage done by the parasites.

Dogs, coyotes, and foxes scratch and nip to get rid of fleas, but the fleas usually save themselves by hopping away on well-muscled legs. A tongue-in-cheek story—one of the tallest of the tall tales—tells how the sly fox outwits his fleas. First, the fox finds a little ball of milkweed fluff and, holding it in his teeth at the very tip of his snout, ever so slowly wades into a pond. As he goes deeper and deeper, the fleas hop ahead to escape the rising water. Finally, as the last dry place on the fox's body, his snout, is about to submerge, the fleas hop onto the milkweed fluff. The fox then releases the fluff, and it floats away carrying the flummoxed fleas. Although this is pure fiction, mammals really do have effective, although usually more prosaic, ways of fighting their insect enemies.

Monkeys, apes, and primitive people groom each other, picking off nits and lice. The cape buffalo of Africa keep off biting flies by wallowing in mud that coats their bodies. Elephants chase off pestiferous insects by tossing sand and dust onto their backs. Horses twitch their skin when bothered by flies, disturbing the flies and making it difficult for them to settle down and stab their mouthparts into the skin. In Africa, the lashing tails of wildebeests drive away or even swat and kill blood-sucking flies such as tsetses. Many other mammals, including horses and the North American bison, have similarly busy tails that serve the same purpose. Knowing a good thing when they see it, Africans drive away tsetses and other biting flies with a hand-held whisk that is usually made of the tail of a wildebeest.

Mammals and iguanas have formed mutually beneficial associations with

certain birds that eat ticks and other parasites they pluck from their compan-
ions' bodies. In the Galápagos Islands, native mockingbirds visit land iguanas
to eat ticks. As Keith Christian put it in an article in *Auk,* when a mockingbird landed on its back, the iguana "assumed a cooperative posture . . . , raising itself off the ground as high as possible on all four legs and remaining
motionless while the mockingbird picked ticks off its body." In Africa, there
are two closely related species of birds, the yellow-billed and red-billed oxpeckers, that perform a similar service for large grazing and browsing mammals. Oxpeckers, wrote Hugh Cott in *Looking at Animals,* associate with all
the large plant eaters except the elephant, and are commonly seen on rhinoceroses, elands, giraffes, buffaloes, zebras, and warthogs. Most animals
tolerate the oxpeckers as they run all over their bodies like woodpeckers on
a tree—up and down the legs, under the belly, or on the head to check the
ears and nostrils for ticks or insects. Even when the animal gallops, they
hold on with sharp curved claws. Their main diet is ticks, but they also eat
blood-sucking flies, and probe open sores to feed on blood. In short, they obtain all their food from the mammals with which they associate. Cott reported that of 58 oxpeckers examined in Tanzania, 55 had ticks in their
stomachs—a total of over 2,000.

🕷 🐜 🪰 The ways birds control the lice, mites, and other creatures that
infest them are even more wonderful than the fictional method used by the
fox to rid itself of fleas. Birds use naturally occurring insecticides they obtain
in two ways. As you read earlier, some rub their feathers with ants, anoint-
ing themselves with formic acid or other insecticidal biochemicals secreted
by the ants. Many birds that reuse their nests year after year incorporate in
them the fresh foliage of plants that contain insecticidal chemicals.

Clark and Mason found that starlings, which may repeatedly nest in the
same cavity, incorporate fresh, green foliage in their nests; males put it in an
empty nesting cavity even before nest building begins. These cavities are
likely to harbor overwintering mites from the occupants of the previous
year. A survey of 137 North American songbirds by Clark and Mason revealed that a third of them put fresh plant material in their nests. But songbirds that nest in cavities are almost six times as likely to use fresh plant material as are birds that build a new nest in the open every spring. Hawks,
eagles, and related raptors build nests in the open, but 30 of 49 species surveyed reuse their nests from year to year, and 22 (73 percent) of those 30 in-

corporate fresh plant material in their nests. Cavity-nesting red-breasted nuthatches do not put foliage in their nests, but the noted ornithologist Arthur Cleveland Bent reported that they repeatedly smear the entrance to the cavity with pitch, starting when nest building begins and continuing until the first young have flown. In northern woods they use balsam fir and spruce pitch; farther south, they use the pitch of pines. Pitch is the source of turpentine, a mixture of strong-smelling, volatile compounds that are repugnant to insects.

There is convincing evidence that plants used by these birds actually contain biochemicals that deter or kill parasites. Clark and Mason found that the green plants starlings incorporated in their nests were not randomly chosen, but constituted only a small subset of the many plants that grew nearby, consisting mostly of those that contain volatile biochemicals that kill or deter parasitic arthropods. Among them are wild carrot (Queen Anne's lace) and fleabane, which are both known to be insecticidal, the latter having a traditional common name that reflects its insecticidal nature. S. Sengupta of the Zoological Survey of India reported that every one of 120 house sparrow nests in a suburb of Calcutta contained leaves of the neem (margosa) tree. Neem has been used as an insecticide in India for centuries, and today extracts of neem are sprayed on cotton and other crops to prevent damage by various leaf-feeding caterpillars and sap-sucking whiteflies. There is no doubt that wild carrot foliage controls blood-sucking mites in starling nests. In an experiment done outdoors with naturally occurring starling nests in cavities, Larry Clark continually removed all green foliage from one group of nests and periodically added wild carrot foliage to another group. The results are astonishing. Clark wrote, "By the end of the first breeding attempt in May, nests protected with wild carrot foliage contained an average of 3,000 mites, whereas nests without plant material contained an average of 90,000 mites . . . By the end of the second breeding attempt in June, nests with wild carrot foliage added contained an average of 11,000 mites, while nests with no green plants contained approximately 500,000 mites."

🐜 🐛 🦟 Dale Clayton's experiments with rock doves (feral pigeons) showed that bird lice which eat feathers can destroy quite a bit of a bird's plumage. Although rock doves do not ant, they regularly preen to remove lice from their feathers. By placing a harmless metal bit in their bills, Clayton prevented rock doves from preening for various durations of from 12 to 34 weeks. Louse populations increased progressively in accordance with how long the birds were prevented from preening, and the total weight of the doves' plumage decreased as louse populations increased. Comparing louse-laden birds with louse-free birds, Clayton found that a large population of lice reduced the average total weight of plumage by 23 percent in one experiment and by 28 percent in another. Birds are warm-blooded. Their metabolic processes produce heat that keeps their bodies at a constant temperature at all times. Because feathers are insulation that lessens the loss of heat from the body, the loss of so much plumage results in a great loss of heat and thereby forces birds to expend more of their limited supply of energy to maintain their body temperature, energy that could be devoted to raising offspring.

A bird heavily parasitized by mites, lice, or other insects will probably be too weak to assume its fair share of parental care and will also be a source of parasites that will infest its spouse and its offspring. Even if only females give parental care, as do grouse and other fowl-like birds, a parasitized father may still be a liability if he lacks genes for personal hygiene—genes that may be present in a less parasitized male and be passed on to his offspring. It would certainly be advantageous for females to recognize and reject heavily parasitized males as mates.

During the summer of 1982, I was happily doing research at the University of Michigan's Biological Station—affectionately called the bug camp by local residents—on beautiful Douglas Lake in northern Michigan. The surroundings were wild and lovely and my research on mimicry was going very well. William Hamilton and Marlene Zuk were at the station that summer and had just formulated the new and exciting hypothesis that the combs and wattles of roosters, and the air sacs of male grouse, not only are "ornaments" to entice females, but also are highly visible barometers that reveal the state of a male's health and his parasite load. The Hamilton-Zuk hypothesis received strong support from the results of experiments with sage grouse done by Margo Spurrier and her coworkers at the University of Wyoming. Sage grouse mate on leks, sites where many males congregate to strut and posture as they compete for visiting females. They fluff up their feathers, fan their tails, and inflate two large, bare-skinned, yellow air sacs on either side of the throat. Linda Johnson and Mark Boyce had already shown that females are more likely to choose louse-free males than males infested with biting lice. Then Spurrier and her colleagues demonstrated that females assess the parasite load of courting males by inspecting the condition of their air sacs. Biting lice cause conspicuous hematomas, blood-red spots, on the air sacs. Spurrier and her coworkers found that these red spots are clues females use to recognize lousy males, proving this experimentally by showing that females rejected louse-free males whose air sacs had been marked with spots by a red felt-tipped pen.

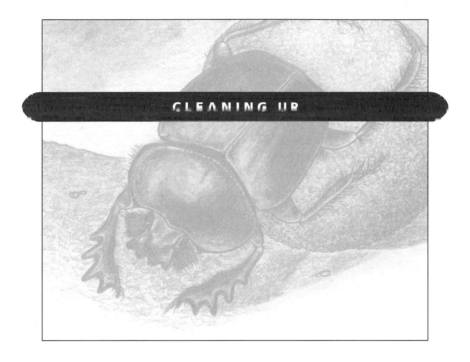

CLEANING UP

10

RECYCLING DEAD ANIMALS

All living things must sooner or later die, and after death the molecules of their bodies must be returned to the soil and the atmosphere so they can be reused by new life over and over again. (Statistics tell us that your body and mine are almost certain to contain at least one molecule that was once in the body of Julius Caesar.) The Bible addresses this inevitable cycle of life and death. After Adam and Eve ate of the apple and fell from grace, Jehovah expelled them from the Garden of Eden, condemning them to live in the real world: "In the sweat of thy face shalt thou eat bread, till thou return unto the ground; for out of it wast thou taken: for dust thou art and unto dust shalt thou return" (Genesis 3:19). But how is a dead person or some other creature returned to the soil? The Book of Job (24:20) has a biologically realistic answer: "The womb shall forget him; The worm shall feed sweetly on him." This worm is not really a worm at all. It is an insect, more specifically a maggot, the larval stage of a flesh fly.

We know that death must ultimately come to all people as to all organisms. Jean Henri Fabre, the great French entomologist and writer, said, "The animal, with its humbler destiny, is spared that apprehension of the hour of death which constitutes at once our torment and our greatness." But only through death is renewal possible. The cycle of life and death is a prerequisite for the evolution that produced and continues to produce the beautiful and ever-changing tapestry of life that includes us. If there were no death, there could be no evolution. Death makes room for new life, and only in new life, the offspring of animals and plants, can the usefulness of new genetic mutations be tested by natural selection. It eventually eliminates harmful mutations, but often molds beneficial ones into new anatomical, physiological, and behavioral characteristics that make it possible for animals and plants to adapt to ever-changing environments. If organisms did

261

not change, life could not cope with new challenges; it would stagnate and soon come to an end. Organisms will have to cope with climatic changes as global warming proceeds, or perish. Cold-adapted animals will have to survive in a warmer climate or shift their range northward. Either option is likely to require new adaptations. An analogy with the evolution of human cultures makes a similar point. New generations contribute new ideas, and there is always the chance that a new generation will produce a challenger to William Shakespeare, Charles Darwin, Ludwig van Beethoven, Marie Curie, or Albert Einstein.

We think of dead animals, dung, and dead plant matter as waste to be disposed of. But to bacteria, fungi, earthworms, vultures, hyenas, jackals, and tens of thousands of insects, the carcass of an animal, a pile of dung, or a heap of plant detritus are valuable food resources. Indeed, paleontologists and entomologists speculate that the first insects on earth were omnivorous scavengers that exploited both dead plants and dead animal matter. To this day, some scavengers, also known as decomposers, are omnivores that feed on both dead plants and dead animals, but others have become specialists that utilize only one particular type of "waste." Decomposers are inevitably preceded by, associated with, or succeeded by bacteria, the universal decomposers, the ultimate recyclers. Scavenging insects enormously facilitate and accelerate the role of these beneficial microbes by chewing and thereby fragmenting dead organic matter.

Flies, as we learn from the Bible, are very important decomposers, but they are far from the only insects that eat dung, dead animals, and dead plants. Some or even all members of 18 of the 32 orders of insects are scavengers. At least some species of 8 of these 18 orders are particularly important ecologically: The wild relatives of the wingless silverfish live in forest litter and feed mainly on decaying vegetation. Silverfish are the innocuous little insects with three long tails that are seen darting around in buildings. Springtails, not considered to be insects by most entomologists but rather a group unto themselves, are tiny, wingless, jumping creatures. Many live in the soil or leaf litter feeding on decaying organic matter, and usually make up for their small size by being exceedingly abundant. Most cockroaches are omnivores that eat decaying plant and animal matter. Later, you will meet a social species that nests in rotting wood. All termites live in complex societies, and all either eat dead plant matter or use it as a medium on which to cultivate fungi. They are by far the most important decomposers of dead plants on earth. Some of the 350,000 known species of beetles are decomposers that feed on carrion, dung, or dead plant matter—even solid wood. They are among the most important of the insectan recyclers. The caterpillars of a few moths feed on plant detritus, and a few others—some of which have become household pests—eat hair and the dried-out remnants of tissues on skeletonized animal carcasses. According to Bert Hölldobler and Edward O. Wilson in their monumental book *The Ants,* ants are probably the most abundant insects on earth. Some, such as fire ants, collect flesh from carcasses of mammals or other vertebrates, and others feed on dead insects. A few of the latter live in the nests of other ants and feed on their rubbish; some live near the nests of other ants and collect dead insects discarded by their larger neighbors, including corpses of the neighbors themselves. Certain flies, mainly blow flies and flesh flies, are the most important decomposers of carrion. And the larvae of many other flies are important decomposers of dung.

In *The Glow-Worm and Other Beetles,* Fabre, using commonplace examples, expressed the world's need for carrion eaters:

> Beside the footpath in April lies the Mole, disemboweled by the peasant's spade . . . The passer-by has thought it a meritorious deed to crush beneath his heel the chance-met Adder; and a gust of wind has thrown a tiny unfledged bird from its nest. What will become of these little bodies and so many other pitiful remnants of life? They will not long offend our sense of sight and smell. The sanitary officers of the fields are legion.

Practically all of the carrion-feeding insects are fly maggots, ants, adult and larval beetles, or certain caterpillars. A few others are minor players; even some adult butterflies get into the act by sucking liquid from carcasses. Bacteria compete with insects for dead flesh, and almost everywhere, some vertebrates—such as vultures and opossums in North America—compete with scavenging insects and may even consume a dead animal before insects can find it. But carrion-eating insects are not completely shut out by either their large or their minuscule competitors; they are abundant everywhere. Some insect scavengers fight back. Certain carrion-feeding maggots, for example, secrete an antibiotic that kills bacteria, and burying beetles hide small animal carcasses from mammalian scavengers.

🐜 🐜 🐜 A white-footed mouse's misfortune was good luck for a pair of burying beetles. After it died, its body was food for a brood of larvae the beetles would care for together. Like other members of its species, this mouse was big-eyed and nocturnal, lived in a brushy area at the edge of a woodland, and fed chiefly on seeds and insects. It was of average size, about 6 inches long—including 3 inches of whiplike tail—and weighed about three-quarters of an ounce, just the right size to be handled by a pair of burying beetles. One evening, as the mouse chased a ground beetle, it was attacked by a young screech owl. The inexperienced owl slashed it with its sharp talons but lost its grip. The wounded mouse crawled under a bush, but died before sunrise. Luckily for the burying beetles, an opossum searching for food nearby did not find the dead mouse. But some time during the dark of night it was found by the burying beetles.

Carrion-eating mammals such as opossums, skunks, coyotes, and even black bears compete with burying beetles for the bodies of small dead mammals and birds. But experiments done in the field by David Sloan Wilson and Julie Fudge showed that these beetles usually find and bury small bodies before they are discovered and eaten by scavenging vertebrates. At the University of Michigan Biological Station, Wilson and Fudge—at intervals from May to August—laid a cumulative total of 778 fresh bodies of dead house mice on the ground in a hardwood forest. A length of dental floss was tied to the hind legs of each mouse to help locate it if it was buried. Within 24 hours 95 percent of the bodies had been discovered, and of those 94 percent were found by burying beetles and only 6 percent by scavenging mammals

The 75 known species of burying beetles (genus *Nicrophorus*, family Silphidae) are, Michelle Scott reported, confined to the Northern Hemisphere and are most diverse and numerous in its cooler areas. All have three things in common. They care for their young, raise them on the carcasses of small mammals or birds, and bury the carcasses—some just under the leaf litter and others to the amazing depth of 2 feet—to protect them from competitors such as other burying beetles or other carrion-feeding animals. But they vary in their choice of habitat and breeding season. Different species occupy fields and meadows, coniferous forests, or hardwood forests. Some breed early in the spring, and some not until summer.

Male and female burying beetles come together when both are attracted to the same dead animal by its odor or when a male emits a sex pheromone that he may broadcast from a carcass suitable for breeding or from a place where there is no carcass. It is to a female's advantage to respond to a male's signal and to mate with him even if he has not found a dead animal. She will obtain sperm that she can store for several weeks. Thus if she finds a suitable carcass, she can raise a brood of larvae on her own even if she is not joined by a male. A male that finds a small dead animal not occupied by a female emits his sex-attractant pheromone. If other males happen to locate the same carcass, he tries to chase them off. As Anne-Katerina Eggert and Josef Müller wrote, excluding rival males is "essential to the maximization of a male's reproductive success." He will father more of his consort's offspring if he keeps other males away.

A small dead animal may be interred by a lone female, in some species by a cooperating group of several beetles, or, as in the case I describe here, by a female and a male together. If the carcass is on hard ground, the beetles drag it to softer ground, where digging is easier. They bury it by tunneling back and forth beneath it. As they bring more and more soil to the surface, the dead animal sinks down into the ever-deepening hole and is finally covered with soil. As the burial proceeds, the pair remove hair or feathers, shape the carcass into a ball, and cover it with anal secretions that kill bacteria and thereby delay decomposition. All the while the beetles destroy the eggs and maggots of flesh flies and kill or drive off other burying beetles. David Sloan Wilson and W. G. Knollenberg think that mites which hitch rides on burying beetles may help to protect the carcass by piercing blow fly or flesh fly eggs. According to Eggert and Müller, after the burial, the beetles create an open space around the carcass, in which they can move around freely, by constantly forcing their way all around the ball of carrion, thereby compacting the soil.

The female lays from 4 to 30 eggs in the soil near the carrion, the number laid being dependent on the size of the carcass. The pair mate frequently and constantly remove soil particles and fungal growths from the carrion ball and moisten it with anal and oral secretions. When the larvae hatch, they crawl to the carcass, probably guided by a combination of its odor and sounds the adults make by stridulating. The adults ingest flesh from the carcass and then regurgitate predigested carrion to the larvae. As Edward O. Wilson put it, the beetles interact with their larvae much as a mother bird interacts with her nestlings. As a parent approaches them, they rear up and "make grasping motions with their legs in what appears to be begging movements." In response, the parent spreads its mandibles and feeds each larva in turn. As the larvae grow in size, they beg less often and more often feed on their own. The mother stays with her offspring until they are ready to pu-

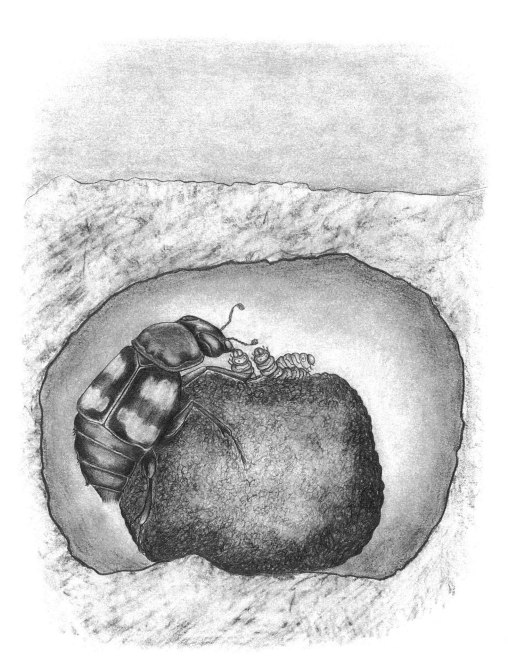

A burying beetle feeding regurgitated carrion to larvae in their nest.

pate, but the male leaves a few days earlier. Except for bones and bits of skin, all the carrion is gone by the time the larvae burrow into the soil to molt to the pupal stage. In a few weeks they emerge from the soil as adults.

In 1834, the French entomologist Jean Lacordaire proposed that burying beetles can reason, an erroneous conclusion based on a second-hand story about an eighteenth-century German naturalist who tried to mummify a dead frog in the sun. To protect it from burying beetles, he put it on top of a stick thrust into the soil. When the frog fell to the ground, burying beetles found and buried it, thereby undermining the nearby stick and causing it to topple. Lacordaire assumed that the beetles had purposely undermined the stick to cause the frog to fall. This "wonder of nature" was finally shown to be false by famous experiments done by Fabre. First, he hung a dead mole upside down from a raffia fiber tied to the top of a vertical stick so that just its head and shoulders touched the ground at the base of the stick. Burying beetles found the mole, and began work by burying its head and shoulders; finally they buried the rest of the body when it fell after their digging dislodged the adjacent stick. Fabre believed that they had unknowingly undermined the stick because its base was so close to the body of the mole. He then hung another dead mole from the top of a stick he had thrust into the soil at an angle. Again the head and shoulders rested on the ground, but the base of the stick was at some distance from the body. Beetles buried only the head and shoulders, because they did not dislodge the stick, which was stuck into the ground well away from where they were digging.

🐜 🦟 🪰 Although burying beetles take an occasional snack from the carcass of a large animal, they play a negligible part in its decomposition, never trying to bury it or to raise larvae on it. Among the most important recyclers of dead animals larger than mice or chipmunks are the maggots of several species of blow flies (Calliphoridae) and flesh flies (Sarcophagidae), which may appear on a carcass within minutes or hours of death. The larvae and adults of several beetles arrive later. A succession of different species appears until the carcass is skeletonized, culminating with the arrival of caterpillars of certain small moths and certain beetles, both of which eat hair and the dry remnants of tissues. Some of these latter insects, among them the clothes moths, have become household pests that damage furs and woolens.

The importance of maggots as recyclers of carcasses from the size of a squirrel to the size of a moose can hardly be underestimated. Fabre quoted

Carolus Linnaeus, the founder of taxonomy, who wrote that "three flies consume the carcass of a horse as quickly as a Lion could do it." Linnaeus's statement seems to be a gross exaggeration until we factor in the enormous reproductive potential of flies. On average, three female blue bottle flies (blow flies) lay a total of about 900 eggs. If the weather is warm and all of their offspring survive, after about 22 days, they will have well over 20 million grandchildren in the form of fully grown maggots. Assuming they weigh about 100 milligrams each, they will in the aggregate weigh about 445 pounds. If about 50 percent of their food intake is used for growth, then they would have eaten almost 900 pounds of carrion to attain their final weight, more than enough to dispose of the body of almost any horse. Przewalski's horse, once the wild horse of Asia but now extinct in the wild, weighs about 800 pounds. The largest species of zebra, Grevy's zebra, weighs a bit over 900 pounds. And keep in mind that these weights include the bones, which neither maggots nor lions eat.

The decomposition of an animal's body, a continuous process effected by a succession of insects and microbes, begins within minutes after death and ends when the corpse is reduced to a bare skeleton. A number of entomologists have studied the succession of insects that contribute to the decomposition of various dead animals, among them pigs, sheep, dogs, cats, rats, and guinea pigs. But for practical reasons that I will explain much more is known about the decomposition of human bodies. Nevertheless, under similar conditions the sequence of insects that appear on the decomposing bodies of all these animals are similar although not identical. The species of insects involved vary with the geographic locale, but they have similar roles and are usually closely related to their counterparts in other locales.

M. Lee Goff, a professor of entomology at the University of Hawaii, is a forensic entomologist, a scientist who uses his knowledge of carrion-feeding insects to help solve crimes, especially murders. Knowing the time of death often makes it possible for the police to tie a suspect to a murder. Forensic entomologists can, among other things, accurately determine the time of death of a corpse that has been exposed for days, weeks, or even months, a corpse too old to be dated by forensic medical examiners. Goff's recent book, *A Fly for the Prosecution,* which will fascinate even the squeamish, details the progressive decomposition of a human body. Although the process is continuous, he has, for the sake of convenience, divided it into five broadly defined and intergrading stages.

The fresh stage begins at the moment of death. There are few changes in the body's external appearance; it looks like "an immobile, unconscious but living person." Nevertheless, if the corpse is outdoors, blow flies, especially blue bottles—but depending upon the season and the environment, also green bottles and black blow flies—may find it within 10 minutes of death, probably by means of odors not perceptible to people. They lay masses of eggs, which will hatch in less than a day, within wounds or natural body openings. As soon as the maggots hatch, they begin to feed. Flesh flies are considerably larger than blue bottles, and appear at about the same time or, according to Kenneth Smith in his book on forensic entomology, somewhat later with a second wave of arriving flies. Unlike blow flies, flesh flies lay newly hatched maggots rather than eggs. The mother retains the eggs within her body until the embryos are fully developed. They hatch in the "birth canal" just before they are laid, and she wastes no time in depositing her tiny larvae on a carcass. As Fabre put it: "She is surprisingly quick about her work. Twice over—buzz! Buzz!—the tip of her abdomen touches the meat, and the thing is done." Goff says that some females may "not even land on the corpse, instead squirting a stream of tiny larvae directly into a body opening or wound while still flying." Between them, the blow flies and flesh flies are very important in the decomposition process. Without them, a carcass decomposes very slowly and retains its form for months. If flies are present, 90 percent of the available soft tissue on the carcass is gone within 6 days. Not long after the flies arrive, predaceous insects begin to appear at the corpse, such as wasps that may capture adult flies on the wing and ants that carry fly eggs or maggots back to their colonies.

The bloated stage begins when the abdomen starts to swell with the gases of decay produced by bacteria breaking down tissues in the body. With death, the body's defenses against its usual bacterial inhabitants no longer function, and they spread out and proceed to feed on the inner organs and tissues. During this stage, there are still many blow fly and flesh fly maggots on the body, and eggs and maggots of the familiar house fly become abundant. Blow fly, flesh fly, and house fly maggots do not have mandibles and cannot ingest solid foods, but they do have sharp, recurved "mouth hooks" with which they can scrape and lacerate. When these maggots feed, they slash the flesh with their mouth hooks and flood it with protein-digesting enzymes secreted in their intestines and released with their feces. According to Ilkka Hanski of the University of Helsinki, the maggots ingest their meal after these enzymes and the bacteria have predigested and liquefied the flesh. "As the population of maggots increases" wrote Goff, "the corpse becomes even more attractive to various predatory species of beetles, ants, and wasps, as well as to parasites of the maggots and their pupae." Among the more frequently seen predaceous beetles are the small black hister beetles, which feed on small insects, including maggots; the rove beetles, which eat maggots and are recognizable by their short wing covers; and species of the family of checkered beetles, which are predators, except for two that I will tell you about later. There are also burying beetles, which, of course, feed on both carrion and maggots.

The decay stage begins when "the maggot masses feeding externally . . . and the bacteria internally finally break the skin of the corpse, allowing the gases to escape and the corpse to deflate." Maggots remain abundant until about the middle of this stage. When they are fully grown, they leave the body to burrow in the soil nearby and molt to the pupal stage, as did the many maggots that came before them. (H. B. Reed, Jr., reported that the soil near and under a dog carcass in this stage of decay had been thoroughly pulverized by the digging of hordes of maggots and other insects.) During this stage the number and variety of both carrion-eating and predaceous beetles increase, and by the end of this stage they have become the predominant insects on the body.

Early in the decay stage, a very unusual maggot, the larval form of a hover fly specifically known as the drone fly, can usually be found in pools of liquid in the carcass of almost any dead animal. It is known as the rat-tailed maggot, because it has a thin tube as long as the rest of the body growing from its back end like the tail of a rat. It is a snorkel, a breathing tube that extends to the surface of the liquid to reach the oxygen of the atmosphere, allowing these maggots to occupy ecological niches in which other animals would die of asphyxiation, pools of polluted fluids that lack dissolved oxygen because of the processes of decay. The adults look like honey bees and probably gain some protection from insectivorous birds by their mimicry of a stinging insect. The ancients, mistaking drone flies for bees, believed that honey bees are generated from the rotting carcasses of animals. In Judges 14:5–9 we read that some time after he had killed a lion, Samson returned and found that there were "bees in the body of the lion, and honey. And he took it into his hands, and went on, eating as he went."

Toward the end of the decay stage, two closely related checkered beetles

(Cleridae) appear on the body. Unlike the other members of their family, most of which eat insects, these two species of *Necrobia* feed on remnants of dry flesh that cling to the skeleton. They have become household pests known as the red-legged and the red-shouldered ham beetles. Robert L. Metcalf and Robert A. Metcalf wrote that both adults and larvae feed on smoked meats such as ham and bacon. They do not distinguish their natural diet from pieces of dried meat hanging in a larder.

By the end of the decay stage, little remains of the body except skin, cartilage, and bone; the body has been reduced to 20 percent or less of its original weight. Now begins the post-decay stage, characterized by beetles and caterpillars that eat hair and the last scraps of dry tissues. Hair, like fingernails and feathers, is composed of keratin, an unusual protein that few animals can eat because it is resistant to the usual digestive enzymes. Vincent Wigglesworth wrote that only three kinds of insects can digest keratin: the feather-eating biting lice, larval and adult carpet, or hide, beetles (Dermestidae), and larval clothes moths (Tineidae). The beetles and moths surely lived on dry carcasses millions of years ago, long before humans evolved. Like the ham beetles, they did not invade our dwellings until several thousand years ago, when we learned to preserve meat by drying it, to wear and sleep on furs, and, more recently, to weave woolen clothing and carpets. After these last scavenging insects have stripped the last edible bits from the skeleton, only about 10 percent by weight of the corpse is left.

As the body of an animal decomposes, wrote Goff, organic fluids of various sorts seep into the earth beneath. For some time after all but the bones are gone, when the carcass is in the final stage of decomposition, these fluids nourish a community of many different soil-dwelling creatures; in terms of numbers and biomass, the most important are mites and springtails. These last decomposers persist for a few years, but eventually the fluids and every other vestige of the body will be broken down to its component molecules. Even the bones disappear, gnawed by rodents, and consumed by various insects such as checkered beetles and, according to Kenneth Smith, even certain tropical termites. Nothing remains except sweet, rich soil. The garden soil that you crumble in your hand contains molecules that once composed the bodies of living animals, perhaps some from a mouse that burying beetles consumed only a few years ago, some from what was left of an elk after a pack of wolves devoured most of it a few hundred years ago, and even some from the body of a mammoth killed over 10,000 years go by hunters of the Clovis culture.

Before the young English poet Rupert Brooke died in the First World War, he wrote:

> If I should die, think only this of me:
>   That there's some corner of a foreign field
> That is for ever England. There shall be
>   In that rich earth a richer dust concealed;
> A dust whom England bore, shaped, made aware,
>   Gave, once, her flowers to love, her ways to roam,
> A body of England's, breathing English air,
>   Washed by the rivers, blest by suns of home.

🐜 🐜 🐜 In the United States and Canada, according to Goff, there are currently only 15 people who are routinely involved in forensic entomology. With their knowledge of the natural history and developmental rates of carrion-feeding insects, they often determine the time of death of a body with amazing precision, even of one that is several weeks old and little more than a skeleton. The forensic entomologist identifies the insects associated with a corpse and notes their developmental stage. Blow flies, for example, may be represented by eggs, newly hatched maggots, fully grown maggots, or pupae. A record of conditions during the days or weeks preceding the discovery of a body is obtained from a nearby weather station. With knowledge of how the weather—especially air temperature—affects the growth rate of the species of blow fly in question, a forensic entomologist can determine the time of death to within a few hours. For example, at 71.6°F the eggs of black blow flies hatch after 16 hours and become fully grown maggots after 3–4 days. If eggs and no later stages are present on an exposed corpse, death cannot have occurred more than 16 hours previously. Similarly, if fully grown maggots and no later stage are present, death occurred about 3–4 days before the discovery of the body.

An actual case recounted by Wayne Lord involved an unidentified body found along a highway in central Indiana in mid-July. The fully clothed body was almost skeletonized, and only a little soft tissue remained. An autopsy showed that death had been due to natural causes. The body was to be identified by searching the missing persons files for the period during which the decedent could have died. It was, therefore, important to establish the time of death. Solely on the basis of the appearance of the remains, medical examiners concluded that death had occurred 2 to 3 months before the corpse had been found, indicating that the missing persons files should be searched for the period from mid-March to mid-May. It would have been a futile effort. A forensic entomologist found living puparia of the black blow fly with the corpse, indicating that it could have been exposed for no more than 35 days. A more detailed entomological analysis revealed that death had occurred about 30 days before the discovery of the body. When the remains were finally identified, it was found that the decedent had last been seen hitchhiking near the death scene 31 days before the body was discovered.

In 1984, the skeletonized remains of a toddler were recovered from a shallow grave on the slope of Koko Head Crater, Oahu, Hawaii. On the body Goff found an assemblage of arthropods characteristic of the post-decay stage of decomposition. Among others, there were hide beetles, red-legged ham beetles, hister beetles, and various species of mites. He determined the time of death by comparing this assemblage of species with the insects and mites that he and his students had found during the sequence of decomposition stages in controlled studies of pig carcasses. His "analyses of this assortment of species gave a preliminary estimate of 51 to 76 days between death and the recovery of the specimens from the remains." He then refined this estimate by looking more closely at "the patterns of activity and occurrence of stages of insect life cycles." Adult hide beetles were present but larvae were not, although their molted skins were there. In his decomposition studies in similar habitats, the last hide beetle larvae had been observed 51 days after death. But how long after the child's death had the larvae metamorphosed to the adult stage? The molted skins of these beetles are fragile and decompose rapidly, but those that Goff found were in good condition, leading him to estimate the postmortem interval to be about 52 days. The child's father, who was later found guilty of her murder, admitted that she had been killed 51 days before her body was found.

There are other applications for forensic entomology. Goff discusses using maggots as indicators of drugs or poisons that may have been in the tissues of a corpse. The flesh of the corpse may be gone, but maggots that fed on the flesh may still be present and are likely to contain traces of any drugs or other chemicals that may have been in the dead body. Analyses showed that this is the case with cocaine, tranquilizers, insecticides, and several other chemicals. Even the empty puparia of flies contain traces of drugs that were in the body.

Robert Pickering, a forensic entomologist working with an archaeologist, found a completely different application for forensic entomology. He found traces of puparia attached to pottery figurines found in tombs in western Mexico that were 2,000 or more years old. Fully grown maggots had left the corpse in the tomb and some had formed their puparia on nearby grave goods such as ceramic figurines. Faked figurines are common but have no puparia attached to them. Thus the presence of puparia indicates a genuine figurine, although their absence does not necessarily indicate a fake.

Improbable as it may seem, an ever-increasing number of physicians use certain blow fly maggots to cure otherwise untreatable infections, such as osteomyelitis, a deep-seated infection of bone that may occur in diabetics. Sterile maggots are placed in the wound and allowed to do their thing—which is to feed on decaying tissues. They are much better at cleaning a wound than a surgeon with a scalpel. A surgeon must cut away some healthy flesh to excise all infected flesh, but the maggots, fussy feeders that eat only flesh infected with bacteria, leave healthy flesh virtually untouched. In an article on maggot therapy, Ronald Sherman pointed out that the maggots have three major effects: they liquefy and remove infected tissue; they hasten wound healing; and they disinfect the wound. As long ago as 1935, William Robinson showed that the maggots flood the wound with an antibiotic substance, allantoin, a waste product of the metabolism of protein.

Maggot therapy probably dates far back into prehistoric times. As Sherman and two coauthors wrote in an article on the medicinal use of maggots, it was used by Australian aborigines, the hill peoples of northern Burma, and possibly the Mayans. William Baer, a physician, told of a surgeon in Napoleon's army who was much troubled by the recurrent appearance of maggots in his patients' wounds. He did everything he could to get rid of them, even though he wrote, "Although these insects were troublesome, they expedited the healing of wounds by shortening the work of nature, and causing the sloughs to fall off." Baer also quoted J. F. Zacharias of Cumberland, Maryland, a physician in the Confederate army during the Civil War: "During my service in the hospital at Danville, Virginia, I first used

maggots to remove the decayed tissue in hospital gangrene and with eminent satisfaction. In a single day they would clean a wound much better
than any agents we had at our command . . . I am sure I saved many lives by
their use."

Baer wrote of his own experiences treating wounded soldiers in Europe
during the First World War:

> At a certain battle during 1917, two soldiers with compound fractures
> of the femur and large flesh wounds of the abdomen and scrotum were
> brought into the hospital. These men had been wounded during an en-
> gagement and in such a part of the country, hidden by brush, that when
> the wounded of that battle were picked up they were overlooked. For
> seven days they lay on the battlefield without water, without food, and
> exposed to the weather and all the insects which were about that re-
> gion. On their arrival at the hospital I found that they had no fever and
> that there was no evidence of septicaemia or blood poisoning.
>
> . . . This unusual fact quickly attracted my attention. I could not un-
> derstand how a man who had lain on the ground for seven days with a
> compound fracture of the femur, without food and water, should be
> free of fever and of evidences of sepsis. On removing the clothing from
> the wounded part, much was my surprise to see the wound filled with
> thousands and thousands of maggots, apparently those of the blow fly
> . . . The sight was very disgusting and measures were taken hurriedly to
> wash out these abominable looking creatures. Then the wounds were
> irrigated with normal salt solution and the most remarkable picture was
> presented in the character of the wound which was exposed. Instead of
> having a wound filled with pus, as one would have expected, due to the
> degeneration of devitalized tissue and to the presence of the numerous
> types of bacteria, these wounds were filled with the most beautiful pink
> granulation tissue that one could imagine. [Granulation is a part of the
> process of healing.]

By the 1930s, maggot therapy had come into prominence. According to
Sherman and his colleagues, around 1,000 American surgeons used this
technique, and the Lederle Corporation sold sterile maggots for $5 per 1,000
(equivalent to 100 current dollars). But when sulfa drugs and antibiotics became available, the use of maggot therapy diminished. By the mid-1940s
it had all but disappeared, used only rarely as a last resort. For example,
the November 24, 1990, *Champaign News-Gazette* of Champaign, Illinois, re-

ported that after all else had failed, a local surgeon used sterile blow fly maggots to cure a chronic infection in a large, deep wound in the leg of a diabetic woman, thereby saving her from having her leg amputated.

During the last 10 years, however, maggot therapy has been making a strong comeback because of the ever-widening resistance of bacteria to antibiotics brought on by their overuse. It amazes me that the development of widespread resistance of pest insects to insecticides, which began in 1946 and is widely known, did not forewarn the medical profession. If an insect with only a few generations per year can become resistant through natural selection, just think how rapidly bacteria, which may have several generations per hour, can develop resistance.

At any rate, the use of medicinal maggots has been steadily increasing. In 2000, Sherman reported that "worldwide, the number of practitioners or centres employing this therapy has increased from less than a dozen, in 1995, to almost 1,000 today." In the United Kingdom, 10,000 batches of maggots were used by 700 centers during the previous 5 years. In the United States, progress has been slower. Currently, two laboratories produce sterile maggots and between them send over 300 batches of maggots per year to more than 70 American practitioners.

One "outlaw" species of blow fly is a parasite that feeds on healthy flesh in the wounds of mammals. This insect, the screwworm fly—named for the threadlike ridges on the body of the maggots—is one of the most horrifying of all the pest insects. Its scientific name, *Cochliomyia hominovorax,* means literally "the snail-like fly that devours people." Screwworms do sometimes attack people, but they more often attack cattle, dogs, other domestic animals, and wild animals ranging from rodents to deer. During her lifetime, a female may lay over 2,800 eggs in masses of up to 400 near a wound on an animal, perhaps nothing more than a tick bite or a nick from barbed wire. The maggots that hatch a day later enlarge the wound as they devour the healthy flesh at its edge. As the wound grows larger, it attracts more and more egg-laying females and grows ever larger. Without intervention, the result is often death. Screwworms once attacked tens of thousands of cattle and other animals in the United States each year. The only cure was the prohibitively expensive procedure of searching the range for each infested cow and smearing its wounds with an insecticidal ointment. The occasional human cases were gruesome. A text on medical entomology shows a photograph of a man whose face was almost entirely eaten away by screw-worm maggots as he lay unconscious in a field for several days.

When Edward Knipling and Raymond Bushland set out to control the screwworm, they went back to fundamentals rather than relying on insecti-cides. One of their first projects, begun in the 1940s, was a study of the sex life of the screwworm fly. This could sound a bit ridiculous to the uniniti-ated. But fortunately, Senator William Proxmire was not yet around to jeop-ardize their government support by presenting them with one of his infa-mous, self-proclaimed Golden Fleece Awards, meant to ridicule scientific research that Proxmire, trained in business administration, did not under-stand and considered to be frivolous and of no practical consequence.

Knipling and Bushland soon discovered that screwworm males mate many times, but that females mate only once. They then asked some ques-tions that no one had ever asked before: Is it possible to sterilize screwworm males, and will they afterward retain their interest in mating? Would mating with a sterile male turn off a female's interest in sex and thus condemn her to laying unfertilized eggs? Would it be possible to reduce or even elimi-nate screwworm populations by releasing into their environment sterile males that would compete for females with wild fertile males? The answer to all of these questions, most notably the last one, turned out to be a re-sounding yes.

Males, quickly and easily sterilized with radioactive cobalt, remained sex-ually active and females that mated with one did not mate again and laid only infertile eggs. The last question was definitively answered on the island of Curaçao in 1953. After setting up "screwworm factories" to produce ster-ile males, Knipling and Bushland released several succeeding waves of ster-ile males on the island. The result was spectacular. Many wild females mated with sterile males and laid infertile eggs. After a few months, screwworms were eradicated from Curaçao, and have not reappeared in almost 50 years.

Next the project was expanded and the screwworm was similarly eradi-cated from peninsular Florida. During the winter of 1958–59 hundreds of millions of sterile males were released from aircraft on 85,000 square miles of southern Florida. (No one was bothered by them because they were re-leased at the rate of only 200 per square mile per week.) The last screwworm to be seen in Florida was found on February 19, 1959. Every year, winter weather had wiped out all of the screwworms in the Southeast except for the population in southern Florida, but every summer they reinvaded the rest of the Southeast, usually transported on infested cattle, and often appeared as far north as Illinois. The cost of the control program was $10 million, but at that time the yearly loss to screwworms in Florida alone was $20 million. The subsequent eradication of screwworms from the Southwest freed the whole country of them and has saved our economy billions of dollars. Political problems have slowed the progress of further screwworm eradication, but these pests have just been eliminated all the way south to the 60-mile-wide Isthmus of Panama, and the South American population is prevented from invading Central and North America by a narrow, and therefore inexpensively maintained, barrier zone in Panama, like a cork in the neck of a bottle, a "no man's land" on which sterile males are frequently released.

Although I have digressed a bit to tell you about screwworms, which are essentially blow flies "gone bad," and the medical use of carrion-eating maggots, the ecological role of insects as disposers of dead animals is overwhelmingly more important than the ravages of screwworm maggots and the usefulness of blow fly maggots in medicine. Insects that feed on dung, which I come to next, are indispensable to the decomposition of the huge quantities of excreta that animals produce before they die.

11

RECYCLING DUNG

High in the leafy canopy of Central and South American tropical forests live the three-toed sloths, eating leaves as they hang upside down from a branch. Despite their high-fiber diet, sloths defecate only about once a week. Jeffrey Waage and G. Gene Montgomery described the unusually elaborate process. The sloth climbs down to the forest floor and, hanging from a vine by its forelegs, digs a small hole with its hind claws; deposits in it about a cupful of hard, egg-shaped fecal pellets about a third of an inch in diameter; and finally covers the hole loosely with a layer of leaf litter. This provides a golden opportunity for the dozen to more than a hundred small moths that have been patiently biding their time in the sloth's fur. Female moths now leave the sloth and fly to the fresh dung to lay eggs. The resulting caterpillars eat dung and pupate in the dung pile. Several weeks later, adults emerge. Both males and females fly up into the canopy to search for a sloth. If they find one, they board it and mate. Then both sexes remain in its fur until, when the sloth goes to the ground to defecate, the females get off briefly to lay their eggs. The adult moths have functional mouthparts and readily sip water in captivity, but no one knows if they feed while they are on a sloth. Waage and Montgomery suggest that they may drink secretions from the sloth's body, such as those from the eyes or nose.

Sloth moths have been a widely known entomological curiosity since they were discovered in the nineteenth century. But until 1976, when Waage and Montgomery published their study of these insects, their natural history was completely misunderstood. It was assumed, with no corroborating evidence, that they spent their whole lives, from egg to adult, on a sloth, and that the caterpillars feed on the green algae that grow on the fur of all sloths. As recently as 1971, Richard Askew wrote of the sloth moth, "Both larvae and adults live amongst the dense hairs of the host . . . The hairs of the sloth are unique in being covered in minute pits in which grow green algae, the possible diet of [the] caterpillars." Hugh Cott pointed out that the algae help to camouflage the sloth by coloring its hair greenish-gray. If the caterpillars really did eat the algae and if they were numerous, they would threaten their host's survival, and thereby their own, by diminishing its camouflage and making it more readily visible to predators.

We perceive dung only as filth and, except for a four-letter expletive, rarely mention it in conversation. But J. Henri Fabre wrote, "What are our ugliness or beauty, our cleanliness or dirt to [nature]? Out of filth, she creates the flower, from a little manure, she extracts the thrice-blessed grain of wheat." Most of this chapter will be about the recycling by beetles and flies of the dung of cattle and other large, herbivorous, grazing mammals, because most of the research on the decomposition of dung has involved these animals, especially cattle. There are reasons for the rather narrow focus of research on the recycling of dung. First, cattle are economically very important and, as you will see later, their dung can cause serious ecological and economic problems if it is not promptly recycled. Second, because these animals are large and generally numerous, as are North American caribou or bison and wildebeests, zebras, and other grazers of the African savanna, they produce a great deal of dung that is obvious to the eye and always much more abundant than the dung of their predators, which must always be much less numerous than their prey.

Eventually, the dung of all animals must be decomposed and returned to the soil, but except for the dung of the large grazers, relatively little is known about how this is accomplished. Small animals are often very numerous and their biomass per acre may be greater than that of the large grazers. Yet our knowledge of the processes that return their abundant dung to the soil is scant. But there are exceptions. In Colombia's Tinigua National Park, according to Maria Clara Castellanos and her coworkers, 19 species of dung beetles were attracted to the excrement of wooly monkeys, and as A. Estrada and two coworkers reported, in Mexican rainforests the dung of howler monkeys is a staple of at least 30 species of dung beetles. In Florida, wrote Samuel Crumb in a U.S. Department of Agriculture bulletin, the caterpillars of an owlet moth live deep within the burrows of gopher tortoises, feeding on the excrement of the tortoise that hosts them.

Nests of the Australian golden-shouldered parrot, which are in chambers the birds excavate in terrestrial termite mounds, are kept free of the nestlings' feces by an unidentified caterpillar. (Most parent birds prevent fouling of the nest by carrying off their nestlings' droppings and discarding them well away from the nest.) This very unusual three-way relationship between birds, termites, and caterpillars is graphically described in W. McLennan's unpublished field notes quoted by Donald Thomson:

> Reached the place where I found [golden-shouldered parrot] nest containing well feathered young on 1.5.22 . . . found it contained seven young, one of which was still in the downy state with scarcely a feather showing. Could not make out what was the matter with the bottom of the nest; it was heaving and undulating in constant movement; little heads were flickering out and in through trap door-like openings. At first I thought it was alive with maggots; a few seconds later a number of larvae came right out of the bottom of the nest and started to eat up the excreta which one of the young [parrots] had just voided. I then saw that the larvae were caterpillars. In a couple of seconds the excreta was eaten up and the caterpillars at once disappeared into the bottom of the nest through the opening from which they emerged. I sat and watched them for some time; every now and again one or two caterpillars would come out and go exploring round the bottom and sides of the nest. The young birds were frequently voiding excreta, which at times would get all over their feet and tail-feathers. Instantly the caterpillars would swarm out and devour it, eating up every scrap even off the feet and feathers of the young; thus the young birds were kept scrupulously clean. The young birds did not take any notice of the caterpillars . . . I now examined the bottom of the nest. The trap door-like openings proved to be the mouths of cocoons in which the caterpillars lived, and these cocoons were lightly bound and matted with web, the interstices being filled with the excreta of the caterpillars and the fine chipped dirt in the bottom of the nest.

McLennan's notes go on to make clear that the caterpillars are not in the nest by chance. He found them in other nests with nestling golden-shouldered parrots but never in nests with only eggs. But he did see a pair of moths, presumably parents of these caterpillars, copulating in a nest with unhatched eggs. In nests from which the young birds had fledged, McLennan found only pupae or empty cocoons. He discovered only one nest that had never contained caterpillars. It was the only one contaminated with feces, a large mass of them trampled underfoot by young parrots that had recently flown.

🐜 🐜 🐜 Vertebrates, particularly herbivorous mammals, excrete as dung much of what they eat. Sheep, as P. McDonald and his coauthors pointed out in *Animal Nutrition,* assimilate about 53 percent of the hay they eat but eliminate as dung about 47 percent of it. Ilkka Hanski concluded that, on average, about 40 percent of the food intake of mammals is excreted as dung. It is a nutritious food for insects, usually rich in the nutrients they require. For example, the dry matter of the excrement of a blesbok, an African antelope, consists of about 9 percent protein, comparable to the 10 percent protein content of rice and corn meal.

Although sloths are common, about three per acre according to Waage and Montgomery, they do not form groups. Thus the three cupfuls of dung per acre they produce each week, probably not much more than a pound, is widely scattered and certainly not easy to find. As ecologists say, it is "patchily distributed," occurring in small, discrete, and widely dispersed patches. Most insects locate dung from a distance by its odor, a strategy that is usually but not always successful. But the behavior of sloth moths virtually guarantees that a mother-to-be that has found and boarded a sloth will eventually have access to sloth dung, the only food on which her offspring can survive.

Other insects also locate dung by finding an animal and staying with it until it defecates. In the vicinity of Manaus, Brazil, Brett Ratcliffe found three species of scarab beetles, two of them new to science, that, like sloth moths, live in the fur of three-toed sloths. He watched a female of one of these species leave a sloth and move to its newly deposited dung, possibly to feed and certainly to lay eggs, because he later found larvae and pupae of this beetle in that dung pile. As you have read, the cattle-plaguing horn flies spend their adult lives on a cow, darting off only to deposit eggs when the cow drops a dung pat. According to E. G. Matthews, an Australian zoologist, certain Australian scarabs go directly to the source for access to dung. These beetles have prehensile front claws with which they grasp the fur of a wallaby, usually near the anus. When the wallaby produces a fresh pellet of dung, two or three female beetles grab onto the pellet and drop to the ground with it. Once on the ground, they engage in a "lively tussle" for it. The victor rolls the pellet as much as a foot away and buries it about an inch and a half deep by digging soil from under it. Then she lays one egg on the dung pellet, which is just large enough to feed one larva.

✻ ✸ ✹ Unlike sloths, many herbivorous mammals—cattle, sheep, bison, caribou, wildebeests, elephants, and others—live in herds, consolidated groups of individuals that stay in more or less close proximity to each other and graze together. Jim Smith of Homer, Illinois, a cattle grower and fellow birder, told me that if there are 50 cows in a 100-acre pasture, they will usually be so close to each other that all members of the herd will be within an area of only 3 to 5 acres. On a camping trip to Custer State Park in South Dakota, my daughter Susan and I saw that it is much the same with a free roaming herd of hundreds of bison. The animals are closely packed and move along slowly as they graze. According to Smith, on his farm it takes about 2 acres of pasture to support one cow. In Arizona, wrote S. Charles Kendeigh in *Animal Ecology*, it takes about 38 acres of grassland to support one bison, which weighs about the same as a cow. Obviously, grazing bison in Arizona must range over a larger area in a day than do grazing cows in Illinois. Nevertheless, dung is abundant where there is or has recently been a herd of either of these animals. Cows, D. F. Waterhouse reported, drop an average of 12 dung pats per day, which, Michael Hutjens of the University of Illinois Department of Animal Sciences told me, weigh between 1 and 2 pounds each. One cow will produce about 126 pounds of dung per week. Hanski wrote, "Generally, droppings in pastures are so numerous that between-patch movement is not a great problem for dung beetles."

✻ ✸ ✹ For many insects, mainly beetles and flies, dung is a precious resource, food for themselves and their offspring. But it is an ephemeral resource that may be washed away by a heavy rain, baked too hard by the sun, or—very frequently—coopted and monopolized by other dung-feeding insects. Consequently, there is considerable competition for dung, even where herds of herbivores provide a seemingly plentiful supply.

One way to beat the competition, used by many scarab dung beetles, is to get there first and bury the dung so that it is out of the reach of competitors. The first step is to locate the dung, perhaps from a considerable distance. Scarab dung beetles in an Ecuadorian rainforest, wrote Stewart Peck and Adrian Forsyth, have two rather distinct "foraging" strategies for locating dung by its odor. Some species perch on a leaf not far above the ground and wait for an air current carrying the scent of dung to come along. Others make cruising flights during which they may intercept an odor trail on the breeze. According to Peck and Forsyth, beetles began to arrive at human excrement within a minute of its being deposited. During the day or evening, they removed dung very rapidly. A mass of 7 ounces was invariably completely buried within 2 hours. J. M. Anderson and M. J. Coe described a similar and even more spectacular instance of dung removal at Tsavo National Park in Kenya. They reported that 16,000 beetles with a total biomass of over a pound were attracted to a pile of elephant dung that weighed over 3 pounds and removed it all within 2 hours. Scarabs of the genus *Heliocopris* are so large that one pair can monopolize a whole pat of cattle dung and bury all or most of it in tunnels. Large males of the scarab *Heliocopris gigas* of Africa and Asia may be, Gilbert Arrow wrote in his book on horned beetles, "almost the size of a cricket ball," which is just slightly smaller than a baseball.

Another way to beat the competition—at least from members of your own species—is to stake out a territory and to defend it against challengers. The behavior of a fly of the genus *Scatophaga* ("dung eater" in Greek) was studied by Gerald Borgia: If a pat of cattle dung is crowded with males, they are not territorial, but if there are no more than eight per pat, they try to defend territories against each other. "At most there is one territory per pat," Borgia observed, "and in the great majority of cases . . . , the male controlling the territory is larger than any other male in the vicinity of the pat." The dominant male patrols his pat and attacks other males that land on it. Subordinate males stand at the edge of the pat or in the surrounding grass. A small male cannot defend his territory against a larger one. A male 8.4 millimeters long abandoned his territory after one attack by a male 9.1 millimeters long. The new male occupied the territory for little more than 6 minutes before he was replaced by a still larger male, 9.85 millimeters long. After about 5 minutes, this last male mated with a gravid female that had probably been attracted to the dung pat to be inseminated and to deposit her eggs.

Some species that neither bury nor defend dung against competitors minimize the impact of competition by getting a head start, by beating the crowd—at least the thickest of it—to the dung and then gorging themselves and completing their growing stage as soon as possible. My vote for the champion practitioners of this strategy goes to the horn flies. They spend most of their adult life on a cow, from time to time crawling down into the hairs on the back, withers, or belly to suck blood. Female horn flies, Carl Mohr reported, are the first insects to visit newly dropped cow dung. As soon as the pat hits the ground females are crawling over it and laying eggs. Mohr said, "Females lay small numbers of eggs singly . . . on the shiny wet surface during the first two minutes and usually within a minute and a half." After the first 2 minutes a cow pat no longer attracts horn flies. The eggs hatch in about 20 hours, and under optimal conditions the maggots grow to full size in only 50 hours and then pupate in the soil under the dung pat, having avoided at least some of the potential competition.

A cow's dung pat, only about 9 inches in diameter, soon becomes the home of a community of insects, a tiny ecosystem within the greater ecosystem of a pasture or a grassland. Just as a dead animal is colonized by a succession of flies and beetles, a dung pat is occupied by a succession of insects that, broadly speaking, begins with flies and ends with beetles. But a small, amorphous, and semiliquid heap of dung is more susceptible to the vagaries of the weather than is the body of a dead animal. If it dries out, potential colonizers are turned away, but if its physical condition is favorable, the insect community burgeons and the dung pat is decomposed and returned to the soil in a matter of weeks. As Carl Mohr wrote, "Over a hundred and fifty insects more or less characteristic of dung make up a complete succession in the weeks during which droppings gradually lose their individuality and become restored to the soil, and inhabited by soil animals."

The community of insects and other organisms in a dung pat is more complex than you might expect. Fungi grow in the dung and derive their nourishment from it. In turn, certain springtails and mites feed on the fungi and some of the fly maggots in the community feed both on the fungi and on the dung. Many different kinds of beetles colonize dung pats. The majority, mostly scarabs or their close relatives, feed directly on the dung. Some dig burrows in the soil beneath the pat and pull dung down into them. Others form spherical dung balls, roll them away from the main mass, and bury them in shallow burrows. Yet other species simply feed in the main dung mass, or are kleptoparasites (*klepto* is Greek for thief) that steal dung from other beetles or lay their eggs in the dung mass of another beetle. After dung-feeding flies and beetles have become established, predaceous beetles—mainly rove beetles and hister beetles—arrive to prey on them as well as on the fungus-feeding springtails and mites. There are also predaceous maggots that feed on other maggots, and some insects eat a mixed diet. Certain maggots feed on both dung and maggots. Adult beetles of the water scavenger family eat dung, but the larvae prey on insects in the dung. Finally, several small wasps parasitize the maggots.

🐜 🦟 🪰 Although flies and beetles are by far the major consumers of dung under most conditions, termites are the primary dung eaters in arid areas and tropical savannas during dry seasons. Almost 50 species of native Australian termites eat dung, mainly in dry and warm areas, reported P. Ferrar and J. A. L. Watson. Only 3 of them include the dung of native marsupials in their diet; the rest feed on the droppings of nonnative domestic animals, almost all on the dung of cattle but a few on that of horses and sheep. The dung of these animals, which were brought to Australia little more than 200 years ago, is a new resource for Australian termites. On an evolutionary time scale, the termites have been very quick to adapt to it.

These termites are sometimes in fresh, moist dung, but more often utilize old, dry dung pats, usually during the dry season. Under these conditions, they have the dung virtually to themselves, because it is too dry for maggots and dung beetles. According to Ferrar and Watson, the termites eat undigested, cellulose-containing fibers in the dung. The same species of subterranean termite may handle dung in two different ways. Small pellets, such as those of rabbits, sheep, and some marsupials, may be covered with a dome of earth, and the dung removed through tunnels in the soil that open under the dome. The termites do not cover large dung pats, such as those of cattle, but burrow directly into them from the soil. The pat is soon riddled with many feeding tunnels, and the dung becomes thoroughly mixed with the soil because it is moved into tunnels in the ground, and because the termites often fill tunnels in the pat with soil. The dung pat is "gradually transformed into a heap of . . . tunneled soil, commonly encased in a thin, residual shell from the original [pat]. This is eventually eroded by wind and rain."

In the Chihuahuan Desert of New Mexico, flies and beetles have a relatively minor role in the decomposition of dung, because they are active only during the brief rainy season in August and September, and even then attack only fresh dung pats but not the many hard dry ones that accumulated during the dry season. Walter Whitford and two coauthors demonstrated experimentally that subterranean termites are the major decomposers of dung in the Chihuahuan Desert. They oven-dried 48 termite-free pats of cattle dung and placed them on the ground, half on insecticide-free patches of soil and half on patches that had been sprayed with an insecticide that prevented the soil-dwelling termites from entering the dung in the only way they can, by burrowing into it from below. Every 28 days the experimenters examined the pats. By the fifty-sixth day, termites had become established in dung pats on untreated soil but not in pats on soil that had been sprayed. After 116 days, pats on soil treated with the insecticide were still free of termites and had lost an average of only about 4 percent of their original weight. But pats on untreated soil were occupied by an average of 273 termites per pat and had lost an average of about 47 percent of their original weight. Ultimately, the dung pats utilized by these termites completely disintegrated and became mixed with the soil in the next heavy rainfall.

"When the first English colonists arrived in Australia in January, 1788," wrote the Australian entomologist D. F. Waterhouse, "they brought ashore with them five cows, two bulls, seven horses, and 44 sheep." The introduction of these animals eventually caused a serious ecological problem, because in Australia there were no dung beetles that had evolved with cattle or other bovines and were equipped to consume their wet, sloppy droppings. As long as there were just a few cows scattered over large areas, there was little concern about the "durability" of cow dung. But as the number of cattle increased, it became obvious that pastures were littered with far too many dung pats, some of which had been there for several years. Waterhouse calculated that the dung produced in one year by one cow would, if it did not decompose, completely cover and prevent the growth of plants on as much as a tenth of an acre of pasture land. To make things worse, each pat is closely surrounded by a broad ring of rank herbage that cattle rarely eat, and that persists for a year or more. Thus, in total, the droppings of each cow take out of production a fifth of an acre of pasture each year. Waterhouse calculated that the 30 million cattle in Australia produce at least 300 million dung pats a day, and thereby render useless about 6 million acres of pasture land each year. But the loss is even greater because dung pats often persist for several years. The effect is thus cumulative, and at any one time considerably more than 6 million acres will be out of production.

The native dung beetles were overwhelmed by this avalanche of cow droppings. In Australia, according to B. M. Doube and his coauthors, there are almost 500 indigenous species of beetles that feed on dung; about 65 percent are scarabs and the others belong to three closely related families.

These beetles are, however, anatomically, physiologically, and behaviorally adapted to utilize the dry, fibrous dung pellets—the largest about the size of a golf ball—defecated by kangaroos and other marsupials. With minor exceptions, the only mammals native to Australia are marsupials. By about 135 million years ago, there were no longer any land connections between Australia and Eurasia and Africa. With no competition from mammals that evolved on other continents, the marsupial inhabitants of Australia underwent an adaptive evolutionary radiation that produced marsupial equivalents of placental mammals such as moles, flying squirrels, cats, wolves and other carnivores, and, in the kangaroos and wallabies, the ecological counterparts of antelopes and other large grazers.

The only practical way to solve the problem of accumulating cattle dung was to use a biological control. In 1967 the Commonwealth Scientific and Industrial Research Organization (CSIRO) of the Australian government began to release scarab beetles from Europe and Africa that feed on the dung of cows and dispose of it swiftly. As of 1991, 41 species had been released in Australia and 22 of them had established breeding populations. A species of dung beetle may be active only at a specific time of year or may occupy only certain habitats. Some, for example, live in shaded woodlands while others prefer open grasslands. The beetles that were introduced had been carefully selected with respect to the climate they were accustomed to, the time of year during which they are active, and their preferred habitat. Those from tropical Africa were released in tropical northern Australia, and those from southern Europe and the Cape Province of South Africa were released in temperate southern Australia. CSIRO selected a suite of beetles that would complement one another, active in different seasons and used to the different habitats that cattle frequent.

Doube and his coauthors concluded that "the introduced dung beetles have made a major contribution toward solving the problem of dung accumulation." Many species built up large populations within a few years, and some have become hugely abundant. At least one species of introduced dung beetle occurs in almost every pasture in Australia, and as many as seven coexist in a few localities. But the beetles have not solved the whole problem. There will be new introductions of beetles that prefer habitats not now occupied by other dung beetles or that are active at times of the year when currently resident dung beetles are not. Waterhouse said that the only complaint about the introduced dung beetles was voiced by a cattle rancher who for years had used dry dung pats to level irrigation pipes in his pastures; after the introduction of the dung beetles he had to cart blocks of wood for that purpose. In addition to preventing the accumulation of dung that choked off the growth of forage plants, the dung beetles improved the fertility of the soil by incorporating dung in it, and as you will see next, they reduced populations of pestiferous flies that breed in cattle dung.

🐜 🐜 🐜 The introduction of cattle into Australia has had severe and very noticeable ecological repercussions other than the effect of their dung on the growth of vegetation. Cattle dung, plentiful virtually everywhere on the continent except in the central deserts, supports, as Ilkka Hanski and Yvoe Cambefort noted, huge populations of dung-breeding flies that attack people, cattle, and other animals. The most infamous of them, the appall-ingly abundant and excruciatingly annoying bush fly, is attracted to peo-ple, often in very large numbers, and can, as Waterhouse put it, "make out-door life in summer a misery for humans and domestic animals alike." These flies suck on human body exudations, get into the eyes, ears, nostrils, and mouth, and scrape at sores until the skin breaks open and lymph or pus exudes.

In southwestern Australia, according to T. J. Ridsdill-Smith and J. N. Matthiessen, bush fly adults are on the wing during the warm months of the year, from November to February, but are most numerous in December and January. Adults are active only in daylight, from about sunrise to sunset. Bush flies grow from egg to adult very quickly, and adults have an average lifespan of about 4 weeks, contributing to the build-up of huge populations.

In Australia bush flies are produced largely in cattle dung and if there were no cattle the bush fly population would plummet. Nevertheless, there were bush flies, presumably small populations, in Australia when the first English settlers arrived and there were almost no cattle. What did bush fly maggots eat before cattle appeared in Australia? The answer is probably that they fed mainly on the feces of humans. Before colonization by the Eng-lish, Australia had an estimated population of about 300,000 aborigines. They usually did not dispose of their feces, but moved their camp when the surroundings became too dirty for comfort. "As over 100 bushflies have been bred from one stool . . . a tribe could maintain" wrote K. N. Norris, an Australian entomologist, "a considerable local population of bushflies." But what did Australian bush fly maggots eat before the aborigines arrived some 50,000 years ago? It's a fair question, but no one knows the answer. Perhaps they fed on the dung of some native Australian bird or mammal, and it is even possible that the aborigines brought the bush fly to Australia with them.

After cattle dung became widely available, it was only a matter of time until natural selection converted bush flies to the exploitation of this new and superabundant resource. For about 180 years almost all of it was available to egg-laying bush flies, because there were no dung beetles to dispose of it. The introduction of dung beetles has had a welcome depressing effect on bush fly populations. But the control of cattle dung and its associated bush flies is not yet complete. In southwestern Australia, according to T. J. Ridsdill-Smith and J. N. Matthiessen, the introduction of two dung beetles, one from Africa and one from Europe, did not lessen the bush fly population early in the warm season but greatly diminished it late in the season. They proposed that the bush fly could be almost completely controlled by introducing a new dung beetle that is active early in the season.

The ball-rolling dung beetles have fascinated people ranging from unlettered peasants to trained and amateur entomologists. As a friend and I drove along a gravel road that passes through a rolling grassland in Custer State Park in South Dakota, we came upon a man and his wife parked in the middle of the road. He was on his knees with his nose almost touching the gravel at the edge of the road and was obviously watching something. We stopped and asked if he was looking at an insect. He said, "Yes, I'm watching a dung beetle." We got out to look and there was a large, black dung beetle pushing a spherical ball of dung about the size of a Concord grape. The beetle tried several times to dig a hole in the hard-packed soil of the road but always failed. It eventually abandoned its dung ball and crawled away. All four of us searched for the dung pat that had provided the makings of the beetle's ball. We found it in the grass just a few feet away. It could only have been bison dung because there are no cattle in Custer State Park.

The great Fabre, the most gifted writer among the early entomologists, spent countless hours in the pastures of southern France observing and recording the behavior of dung-rolling beetles. There are many species of dung rollers, but as Fabre wrote,

> The dung-manipulators have as head of their line the Sacred Beetle or Scarab, whose strange behavior had already attracted the attention of

the fellah in the valley of the Nile, some thousand years before the Christian era. As he watered his patch of onions in the spring, the Egyptian peasant would see from time to time a fat black insect pass close by, hurriedly trundling a ball of Camel-dung backwards. He would watch the queer rolling thing in amazement, even as the Provençal peasant watches it to this day.

Although we associate the sacred scarab, aptly known as *Scarabaeus sacer* to entomologists, with ancient Egypt and its religious practices, the range of this large, stocky black beetle rings the Mediterranean from northern Africa to southern Europe.

The size of the dung ball varies with the species of beetle. The average weight of balls formed by the large sacred scarab is about an ounce, while those of a tiny species weigh only about a thousandth of an ounce. Even dung balls made by members of the same species may vary greatly in size. Sacred scarabs, according to Fabre, made balls that ranged from the size of a walnut to that of a man's clenched fist. Furthermore, a cooperating pair can make larger balls than a lone beetle. A lone *Kheper platynotus* formed balls that weighed an average of about 0.8 ounce, 13 times its own body weight; but pairs working together made much larger balls with an average weight of about 5 ounces, 39 times their combined body weights.

Fabre described in detail the way in which a sacred scarab forms a dung ball:

> The clypeus, or shield, that is the edge of the broad, flat head, is notched with six angular teeth arranged in a semicircle. This constitutes the tool for digging and cutting up, the rake that lifts and casts aside the unnutritious vegetable fibres, goes for something better, scrapes and collects it. A choice is thus made, for these connoisseurs differentiate between one thing and another, making a rough selection when the Beetle is occupied with his own provender, but an extremely scrupulous one when it is a case of constructing the maternal ball, which has a central cavity in which the egg will hatch. Then every scrap of fibre is conscientiously rejected and only the stercoral [excremental] quintessence is gathered as the material for building the inner layer of the cell. The young larva, on issuing from the egg, thus finds in the very walls of its lodging a food of special delicacy which strengthens its digestion and enables it afterwards to attack the coarse outer layers.
>
> Where his own needs are concerned, the Beetle is less particular and contents himself with a very general sorting. The notched shield then

A sacred scarab pushing a ball of dung. Behind her is an
ancient Egyptian stone carving of this scarab.

does its scooping and digging, its casting aside and scraping together more or less at random. The fore-legs play a mighty part in the work. They are flat, bow-shaped, supplied with powerful nervures and armed on the outside with five strong teeth. If a vigorous effort be needed to remove an obstacle or to force a way through the thickest part of the heap, the Dung-beetle makes use of his elbows, that is to say, he flings his toothed legs to right and left and clears a semicircular space with an energetic sweep. Room once made, a different kind of work is found for these same limbs: they collect armfuls of the stuff raked together by the shield and push it under the insect's belly, between the four hinder legs. These are formed for the turner's trade. They are long and slender, especially the last pair, slightly bowed and finished with a very sharp claw. They are at once recognized as compasses, capable of embracing a globular body in their curved branches and of verifying and correcting its shape. Their function is, in fact, to fashion the ball.

Armful by armful, the material is heaped up under the belly, between the four legs, which, by a slight pressure, impart their own curve to it and give it a preliminary outline. Then, every now and again, the rough-hewn pill is set spinning between the four branches of the double pair of spherical compasses; it turns under the Dung-beetle's belly until it is rolled into a perfect ball.

As Fabre explained, the beetles prepare some balls of dung for their own consumption and others, usually larger ones, to feed to their larval offspring. In either case, they protect the ball from competing beetles and the heat of the sun by rolling it away from the dung pat by pushing it backward and then burying it in loose soil in a chamber about 4 inches below the surface. If it intends to eat the dung ball, the beetle, explained Glyn Evans, digs a chamber large enough to accommodate the ball and itself; seals the opening to the chamber; and then starts to eat, usually continuing until all the dung is gone. Balls for rearing larvae are usually rolled by a male and a female working together. The pair copulate after the ball is buried; the male then leaves; and the female molds the ball into a pear shape and lays a single egg in the "neck" of the pear. She then closes the nest and goes off to prepare more nests, as many as 40. The larvae not only eat the dung, but also squeeze out as much nourishment as possible by eating and reeating their own feces. Rabbits do much the same thing, eating their feces to run them through the digestive system a second time, a process eloquently described in Richard Adams's *Watership Down,* a lively and engaging novel about rabbits in England.

The larva eats almost all of the dung ball but remains within its intact hard outer rind. When fully grown it molts to the pupal stage and then metamorphoses to the adult stage. After about 28 days, reported Fabre, the adult molts its pupal skin and is ready to dig its way to the surface to feed and go about the business of producing another generation. But, as Fabre found, it cannot break through to the surface if the sun has baked the soil hard. Some beetles die in their underground cells, but others manage to break out after a rain has softened the soil. In Egypt, according to the entomologist Ronald Cherry, the scarabs emerge from the water-softened soil after the annual flood of the Nile has subsided. Throughout the valley, the sacred scarab is the first noteworthy creature to emerge from the layer of fertile mud that the flood leaves behind.

In the anatomy, life cycle, and behavior of the adult sacred scarab the ancient Egyptians saw symbols of their religion's explanations of the mysteries of nature. The head of the beetle, with its radiating points, resembled the rising sun, and the rolling of the dung ball symbolized the movement of the sun. As Cambefort and Hanski put it, "a dung ball is shaped early in the morning, and it is rolled across the plain like the sun travels across the sky." In a 1987 article in the *Revue de l'Histoire des Religions,* Cambefort, of the Musée National d'Histoire Naturelle in Paris, proposed that the Egyptians associated the scarab with Osiris, lord of the underworld, ruler of all the dead who live there, and the god of fertility who presides over the sprouting of grain and the emergence of other life from the earth, including the sacred scarab. To the Egyptians, the metamorphosis of the mummy-like pupa to a new beetle may have symbolized victory over death.

The ancient Egyptians carved scarab amulets from stone. They were of various sizes, but most were small. The earliest, but only a few, appeared in the time of the Old Kingdom ,over 4,000 years ago. They became common during the Middle Kingdom and were included in burials, often under the wrappings of the mummy. The small carved scarabs served both as religious amulets and seals. Their flat undersides are often incised with hieroglyphics that people could press into wet clay to make an impression, much as Europeans once pressed the face of their signet rings into hot sealing wax. During the New Kingdom, which persisted until about 3,000 years ago, new and much larger stone scarabs appeared. They were put in the eviscerated mummy to replace the heart.

12

RECYCLING DEAD PLANTS

When you look at a swamp, a woodland, a crop field, or almost any other ecosystem on land, it is perfectly obvious that the mass of green vegetation by far exceeds the mass of animals—including even such giants as moose, bison, or elephants and such small but teemingly abundant creatures as worms and insects and other arthropods. And so it must always be, for plants are the ultimate source of food in all terrestrial ecosystems, and life could not go on if animals ate all of them and did not spare an abundant breeding stock to produce food for future generations. Ecologists estimate that the biomass of the succeeding levels of a food chain usually decreases by a factor of roughly ten, going up the chain from the producers, the plants, to the top carnivore. In other words, the total weight of the plants is ten times that of the plant feeders, and the total weight of the plant feeders is ten times that of the insects and other insectivores that eat the plant feeders.

The amount of plant growth eaten by grazers and browsers is actually quite low. Eugene Odum and Lawrence Bevier noted that it varies from about 50 percent in a heavily grazed pasture to less than 5 percent in a forest. The remainder sooner or later dies and becomes detritus : dead grass and herbaceous plants, the trunks and branches of dead trees, and a great abundance of fallen leaves, all too familiar to people who spend hours raking them from lawns in autumn. An acre of fir forest in the Cascade Mountains of Washington, according to an article in *Biogeochemistry* by William Schlesinger, produces each year almost a ton of plant detritus.

All of this dead plant matter must ultimately be recycled, decomposed and returned to the soil in the form of minerals and to the atmosphere as carbon dioxide and other gasses. As J. van der Drift and Martin Witkamp explained in an article on oak litter, the leaves that fall every year in a deciduous forest are gradually attacked by insects and a variety of organisms from micro-

scopic bacteria to earthworms as much as 4 inches long. "This biological attack together with . . . physical and chemical influences forms the decomposition process," which frees minerals and produces carbon dioxide, the
former leached into the soil and the latter released to the atmosphere. The
rate of decomposition varies with factors such as climate, soil type, and the
composition of the litter. In some soils in temperate areas, C. A. Edwards and
G. W. Heath pointed out, oak litter disappears in 8 to 15 months, but in the
tropics a leaf may be completely decomposed only a few weeks after it falls.
Leaves of different species decompose at very different rates. From January
to June in northern England, birch leaves lost about 83 percent of their
weight to decomposition, lime (basswood) leaves about 56 percent, and oak
leaves only about 21 percent.

Bacteria are the ultimate decomposers, but they do not work efficiently
unless detritus is first prepared for them by larger organisms. Fungi assimilate dead plant material, but they eventually die and are decomposed by
bacteria. Insects, other arthropods, and earthworms eat plant litter, chewing
it or grinding it into tiny pieces, and passing it out as feces, which are efficiently decomposed by bacteria—largely because the dead plant material
has been chemically altered and fragmented into tiny particles, a process
that greatly increases its surface area. For example, according to van der
Drift and Witkamp, larvae of a terrestrial caddisfly, to be discussed below,
fragment an oak leaf of average size into about 3,000 fecal pellets containing
about 10 million tiny leaf particles, thereby vastly increasing the leaf's exposed surface area. Probably of equal benefit to the bacteria are chemical
changes plant detritus undergoes as it passes through the digestive system of
an insect or some other creature. Van der Drift and Witkamp demonstrated
the effect of these chemical changes by incorporating either fecal pellets or
leaf particles artificially ground to the same size as those in feces into nutri-
ent-free agar gels. Fifty times as many fungal colonies and almost 10,000
times as many bacterial colonies grew on agar containing feces of the caddisfly as on agar containing only ground leaves.

Van der Drift and Witkamp found that more litter was pro-
duced in an oak coppice in the Netherlands than in Schlesinger's fir forest in
Washington—an average of almost a ton and a half, mainly oak leaves, per
acre each year. Various insects, other arthropods, and earthworms consume
this litter. Among the most important of them are larvae of the terrestrial

caddisfly *Enoicyla pusilla*. The members of the European genus *Enoicyla* are oddities among the caddisflies. As John Henry Comstock noted in his entomology textbook, in the larval stage they live only on land. All other 5,000 known caddisfly larvae live in freshwater lakes and streams. Like many of the aquatic caddisflies, the *Enoicyla* larvae build and live in tubular cases made of silk and small particles of debris. The population density of *Enoicyla* varies from place to place, but in one area of the coppice van der Drift and Witkamp found an unusually large population of from 700 to 1,000 per square yard, about 6 million per acre. The annual litter production in this area was just over 2 tons per acre, and the *Enoicyla* larvae ate a total of 800 pounds of it per acre in a year. Thus just this one species decomposed about 20 percent of the annual leaf litter.

🐜 🐜 🐜 Martin Witkamp explained a clever experimental technique ecologists use to unravel the complex interrelationships among different decomposers, microorganisms and litter-eating animals such as insects and worms. Dead leaves are put into small bags made of nylon or fiberglass netting and then placed in natural environments. Bags made of netting with a coarse mesh can be entered by fragmenters, such as earthworms, insects, and other litter-dwelling animals, and by bacteria and all other microorganisms. But netting with an extremely fine weave, with openings only about a ten thousandth of an inch wide, excludes all animals but admits microorganisms.

C. A. Edwards and G. W. Heath used this technique in England in a woodland with a mixture of tree species and in a pasture that had recently been cultivated. In the woodland, the leaf samples were placed not in bags but on bare soil under frames of coarse netting covered with leaf litter. In the former pasture, mesh bags containing leaves were buried about an inch below the surface of the soil. In the woodland "earthworms fragmented and removed by far the largest proportion" of the leaves under the coarse netting. Nevertheless, some mites, springtails, and other arthropods were present— but only about 7,600 per square yard. There were only about 57 earthworms per square yard, but they were relatively gigantic compared to the tiny arthropods and ate much more than all the arthropods combined. The pasture soil had many more mites (over 20,000), springtails (over 15,000), and other arthropods, including fly maggots (about 1,300)—a total of over 37,000 arthropods per square yard. In the pasture Edwards and Heath used bags with mesh openings of 0.28, 0.02, or 0.0001 inch. From July to April they examined the leaves in the bags every 2 months. The pasture, they wrote, "had a more balanced fauna and the mesh bags [with openings of different widths] enabled the role of the different sections of the fauna in breakdown to be determined."

Bags with the coarsest mesh (0.28 inch) were entered by many soil animals, principally earthworms, tiny white worms of the genus *Enchytraeus*, springtails, fly maggots, and predatory mites that prey on springtails. After these animals had been in the bags for 9 months, about 90 percent of each sample of oak leaves had been decomposed and about 70 percent of each sample of beech leaves. Bags with a 0.02 inch mesh admitted springtails and other small arthropods, but prevented entry by most larger animals, including earthworms. There was a corresponding decrease in the rate of decomposition. Nevertheless, during the 9 months of the experiment, about 38 percent of oak leaves and 32 percent of beech leaves had been decomposed in these bags, presumably after being fragmented, mostly by insects and other arthropods. Bags with the finest mesh, 0.0001 inch, allowed the entry of bacteria but excluded all soil animals. Edwards and Heath reported that leaves in these bags, not first prepared by fragmenters, showed no visible signs of decomposition.

🐜 🐜 🐜 Just as some scarab beetles bury dung in the soil, others bury dead vegetation, among them horned scarabs known as elephant beetles. The males of this inch-long, shiny, black beetle have three long horns on the back just behind the head; females have only one horn. In 1908, Abram Manee reported his observations of this beetle in the sand hills of North Carolina:

> On the night of July 11, '06, I took my first females by electric light. That same month we investigated . . . a hole by a cart path and dug out a working male. July 26 I took my first pair from between two exposed roots of a large oak. They were pulverizing the surface soil, preparatory to shaft digging. After several such [captures] of pairs and singles, I came to know the peculiar mound of earth always pulverized to a depth of one to three inches . . . Beneath the mound of loosened soil an inch[-wide] shaft extends vertically for six or eight inches. At bottom of shaft a one-and-a-quarter inch chamber reaches horizontally from one to

five inches, and in this chamber, packed with finely broken bits of dec-adent [dead] oak leaves, a solitary egg is deposited. Sometimes two or rarely three such chambers diverge from the same shaft, but I believe with never more than one egg in each. A favorite haunt for nesting is by a pile of dead oak leaves wind blown in some hollow, from which I con-clude that the young larva feeds on leaf debris and later on decadent oak roots.

It appears that these larvae ultimately resort to feeding on dead roots be-cause the leaf fragments provided by the parents are not enough to sustain them throughout the entire larval stage. The larvae of a South American rhi-noceros beetle that lives in grasslands and stocks burrows in the soil with vegetation has a similar problem, according to Gilbert Arrow, an expert on horned beetles. When the larvae run out of food, they burrow into the sur-rounding soil to feed on living roots.

Although some termites eat plant matter in the soil, many feed mainly, but not quite exclusively, on dead plant matter on the surface. Their major nutrient is the cellulose that forms the walls of plant cells and is the main constituent of wood. Most of the over 2,000 known species of termites live in the tropics, and according to Donald Borror and his coauthors, there are only 41 species in America north of Mexico.

All termites are social and their colonies—like the colonies of honey bees and ants—persist from year to year, in some species for decades. But unlike the social ants, wasps, and bees, whose nonreproductive castes are com-posed only of females, all castes in a termite colony—basically reproductives, workers, and soldiers—consist of both males and females. A new colony is founded by a pair, a "king" and a "queen," who remain together in the col-ony, mate repeatedly, and beget sterile workers and soldiers. Usually they are the only members of the colony capable of reproducing. The white, soft-bodied workers are totally wingless and perform all the chores of the colony, from excavating tunnels to feeding the king and queen and immatures too small to feed themselves. Soldiers, the defenders of the colony, are also wingless but have hard and darkly pigmented heads that in most species bear a pair of large and powerful jaws that may be variously adapted for bit-ing, crushing, slashing, or piercing. In other species—mainly tropical ones—the soldier's head is modified as a "squirt gun" that can expel a sticky, toxic substance. Mature colonies produce swarms of winged reproductives that leave the nest through an exit prepared for them by the workers and then fly away. After landing on the ground, they break off their now useless wings at a basal fracture line by sharply pressing the wing tips to the ground. Then pairs form and look for a site that suits their nesting requirements.

In general, termites can be divided into two groups, according to their need for moisture: subterranean species that require moist conditions and nest in the soil, and dry-wood termites that can live in dry wood and require no access to moist soil. Several species of both kinds live in America north of Mexico. Subterranean termites occur throughout most of the United States and in southwestern British Columbia in Canada. Dry-wood termites are found along the Pacific coast of California, in the southwestern states, and from Texas to the Atlantic coast as far north as South Carolina.

To most Americans, termites are nothing more than destructive pests that insidiously tunnel in the wood of their homes and sometimes go undetected until after they have caused extensive damage. But most termites, especially those in the tropics, are indispensable recyclers of dead wood and even of fragments of dead grasses and herbaceous plants. In *Termites and Termite Control*, T. Charles Kofoid placed our conflict with termites in an ecological perspective: "The termite problem arises because of man's attempts to change the ordinary processes of nature by preserving for his own use, over considerable periods of time, wood and its products, which it has been the immemorial function of the termites and associated organisms to break down and return to the soil and the atmosphere." The eastern subterranean termite, which is very destructive if it invades the wood of buildings, is common in the United States east of the Mississippi except in the northernmost parts of the northern states. Even structures built on concrete, brick, or stone foundations are invaded by workers from colonies that were founded nearby where there is wood in or upon moist soil, perhaps a discarded board or the roots of a tree that had been felled.

Most termites seal their nests so that there are no openings to the outside, because they—certainly the subterranean species—abhor dry air, air currents, and light. Consequently, if they do not find an accessible crack in the foundation, they build closed passageways of excrement and soil about the diameter of a lead pencil and shaped like Quonset huts that extend up the surface of the concrete foundation to the wooden parts of the house. Although unseen because they do not penetrate its surface, termites can destroy so much of a wooden board that it is little more than a hollow shell. Some years ago, the wooden floor of a house in Champaign, Illinois, was so weakened by termites that the refrigerator fell through the floor into the basement.

My family and I made the acquaintance of dry-wood termites during a sabbatical leave in Cali, Colombia—fortunately long before the drug wars began. Hordes of winged kings and queens often flew around inside our rented house at night. There was a desk that I had hoped to use, but when I leaned on it with my hands, the top caved in. The entire desk had been hollowed out by dry-wood termites, and was held together by not much more than a thin shell of wood and several thick coats of paint.

Some years ago, an architect I know went to Liberia in west Africa to build a palace in Monrovia for William Tubman, then the nation's president. Several months after his return, he called me about an insect problem with some beautiful wood carvings he had brought from Africa. There were tiny holes in the wood and on the shelf beneath them were tiny piles of sawdust. I thought that the carvings had been infested in Africa with tiny tropical dry-wood termites known as powder-post termites, which are among the few termites that make openings from their nest to the outside. Eventually they would have destroyed the carvings. My friend had already consulted a commercial pest control company. Focused only on chemical control, the company recommended the expensive procedure of fumigating the carvings— under vacuum so that the fumigant would enter the termites' tunnels. Was there a simpler alternative? I suggested putting the carvings in a freezer for a week. It worked! Whatever insects were in the carvings died and sawdust was never again ejected from the holes in the carvings.

Cellulose is the major portion of almost all termites' diet, the only food of many species. Cellulose is a polymer of the sugar glucose, a long chain of as many as 3,000 glucose molecules strung together by a chemical bond that cannot be broken by the usual enzymes in the digestive systems of insects and other animals. According to Michael Martin, who has done extensive research on cellulose digestion by insects, except for a few species, termites do not themselves produce enzymes, cellulases, that can split apart the otherwise indigestible cellulose polymer to release its constituent glucose molecules, which are readily digested by insects, humans, and probably all other animals.

Not surprisingly, termites have evolved a way of getting around their inability to produce cellulases. They have established mutually beneficial relationships with certain microorganisms, bacteria and protozoans, that produce cellulases and live in the hindgut of their digestive system. The termites benefit because the microorganisms digest cellulose for them, and the microorganisms benefit because the termite gut is a cozy home through which passes a steady stream of their food. But every time a termite molts, it first empties its gut and sheds the lining of its hindgut, thereby losing the microorganisms upon which it depends. Termites get around this difficulty by eating anal secretions of other termites, which contain the requisite microorganisms. If experimenters deprive termites of this opportunity by removing them from their colony, they ingest wood but starve to death because they lack the microorganisms that digest cellulose.

The great majority of termites, as T. G. Wood and W. A. Sands reported, live in the tropics and subtropics, most of them in forests. Paul Eggleton and his coworkers found 114 species in just one tropical forest reserve in Cameroon, on the west coast of Africa. According to Eggleton and his coworkers and to David Bignell and his coworkers, in this forest the total biomass of all termites combined is far greater than that of any other insect including ants, other arthropods, or other invertebrates such as worms. In this Cameroon forest, the density of termites may be over 4 million per acre, a biomass of over 1,200 pounds, constituting 95 percent of the biomass of all soil insects. Data from the tropical forests of Asia, Africa, and South America suggest, as Martin Speight and his coauthors reported, that termites are ten times more numerous than ants, the next most abundant group of insects. Bignell and his coauthors said that in a tropical forest termites and ants together vastly outnumber and outweigh the combined total of all other insects. In some deserts, Gary Polis reported, the total biomass of termites is greater than that of any other species of animal.

The great abundance of termites is underscored by comparisons between their biomass and that of mammals in the same ecosystem or a similar one nearby. In a savanna in Nigeria, there were about 94 pounds of termites per acre, while the average biomass of aboveground mammals—including the large grazers—was only about 9 pounds per acre. Even in temperate zone ecosystems, termites can be impressively abundant. The total biomass of cattle on a semi-arid rangeland in Texas was about 50 pounds per acre. During a 3-year period, the biomass of termites was not far behind at an average of 46 pounds per acre, and in one of those years it was much more, almost 80 pounds per acre.

How important are termites as fragmenters and decomposers of plant detritus? The answer is that they often make huge contributions to the recy-

At the top are mounds of an Australian termite that harvests dead
plant matter. Below, a queen swollen with eggs is attended
by her mate and a small worker gathering eggs.

cling of dead plants and fallen leaves. The plants in the Nigerian savanna mentioned above produced a total of about 4,800 pounds of litter per acre per year. Termites alone ate over 1,700 pounds, about 36 percent of this detritus. The termites of the Texas rangeland consumed 55 percent of the plant detritus on the surface and 38 percent of the detritus that had been incorporated in the soil. Eggleton and his coauthors noted that in a Malaysian tropical forest from 14 to 33 percent of the annual production of plant litter is eaten by termites, and Walter Whitford and two coworkers found that termites are significant recyclers of plant litter and dung in the Chihuahuan desert of New Mexico.

As Speight and his coauthors pointed out, termites are responsible for about 20 percent of the carbon dioxide production of savannas and about 2 percent of the worldwide production. Evan DeLucia, a plant ecologist at the University of Illinois, told me that by burning fossil fuels, humans release about the same amount of carbon dioxide into the atmosphere. Don't think that we can alleviate global warming by exterminating the termites. Without termites, the recycling of plant detritus would be greatly diminished and there would be a concomitant and probably disastrous decrease in the growth of green plants. Green plants are "sponges" that soak up carbon dioxide from the atmosphere, and a decrease in their global biomass would accelerate the pace of global warming. Tropical forests, for example, take up large quantities of carbon dioxide. As was pointed out in the July/August 2001 Nature Conservancy magazine, the trees and other plants in a mature tropical forest store about 60 tons of carbon per acre, but if the forest is cleared for raising cattle, as is too often done, the pasture plants will store only about 2 tons per acre. Only 1 square mile of coastal forest in Brazil offsets the emissions of almost 29,000 automobiles in one year. Keep in mind that termites do not add "new" carbon dioxide to the atmosphere. They just recycle the current natural production of carbon dioxide. But burning fossil fuels releases what is essentially an "additional increment" of carbon dioxide, because the carbon in these fuels was "taken out of action" hundreds of millions of years ago.

🐜 🐜 🐜 Clive Jones and two coauthors referred to termites as "ecosystem engineers," organisms that physically change habitats and thereby increase or decrease the supply of resources available to plants or animals. Beavers are a well-known example. Their dams create ponds that are the home of fish, many kinds of aquatic insects, and aquatic plants eaten by

moose. The physical changes in habitats caused by termites—and ants, too—
are less obvious but on a global basis are much more important than those
caused by beavers or other larger animals. By tunneling in the soil and
building aboveground mounds and nests, termites and ants constantly mod-
ify the structure of the soil, maintaining it in a condition suitable for other
inhabitants of the ecosystem. Termites, explained Wood and Sands, affect
the soil in two ways. First, their underground tunnels, "commonly so numerous as to collapse under-foot," greatly increase the porosity of the soil,
facilitating the growth of roots, aeration, and the drainage and storage of
water. Second, mound-building termites and ants bring mineral soil from
below up to the surface, where it mixes with humus, forming a fertile medium for the growth of plants. Amazingly large quantities of subsoil are
brought to the surface by termites, as much as 1,000 tons per acre by an African termite that builds large mounds. Two Australian mound-building termites had brought to the surface well over 3,000 cubic yards of subsoil per
acre. According to my friend Mike Ducey, a building contractor, that is
enough to fill over 300 dump trucks.

Although termites are the predominant recyclers of dead
wood in many if not most ecosystems, other insects, including some cockroaches, beetles, caterpillars, flies, and wood wasps, also consume dead
wood. These insects, like termites, survive and grow on a diet rich in cellulose, but deficient in vitamins, fats, and substances containing nitrogen, the
distinctive and essential component of proteins. They have overcome the
difficulty of digesting cellulose in two ways. Like termites, a few other insects, among them cockroaches of the genus *Cryptocercus* and a few beetles,
harbor cellulose-digesting symbionts in the digestive system. Other insects,
such as many long-horned beetles and ambrosia beetles, inoculate dead
wood with cellulose-digesting fungi. According to Michael Martin, a very
few insects, including a few termites, a few long-horned beetles, and the
Australian cockroach *Panesthia cribata*, can themselves secrete the enzymes
that digest cellulose.

Another difficulty for wood-feeding insects is the paucity of nutrients
other than the glucose they derive from the cellulose. Many of these insects have in their guts bacteria or other microorganisms that synthesize the
required nutrients. Among them are certain termites that harbor bacteria
which have the rare ability to take nitrogen from the air and "fix" it in chemical compounds that can be utilized by the termites. The more familiar "ni-

trogen-fixing" bacteria that live in nodules on the roots of legumes such as beans and clovers perform the same service for these plants.

🐝 🐜 🦟 The life style of cockroaches of the primitive genus *Cryptocercus* is, according to Louis Roth, probably the foremost expert on cockroaches, similar to that of termites in that they are social, feed on dead wood, and depend upon protozoa in their guts to digest cellulose. Until recently, most biologists thought that these cockroaches are the direct ancestors of the termites. But P. J. Gullan and P. S. Cranston pointed out that *Cryptocercus* cannot be the ancestor of the termites, because recent anatomical studies and genetic studies at the molecular level show that it is evolutionarily distant from them. This is, therefore, an exceptionally interesting instance of convergent evolution.

Until Lemuel Cleveland's ground-breaking 1933 work on *Cryptocercus punctulatus,* the only representative of this small genus of cockroaches in the eastern United States, termites were the only insects known to have cellulose-digesting protozoa in their intestines. Cleveland searched for other wood-eating insects that harbor these protozoa at a research station in the Appalachian Mountains in Virginia:

> I must confess that I had [previously] examined so many genera and species of wood-feeding insects without finding these protozoa, or any forms remotely related to them, that I, like other workers on this uniquely interesting group of organisms, had become almost reconciled to the generally accepted view that they occurred only in termites, where their evolution, into many families and genera of highly complex flagellates [a group of protozoa], had taken place.
>
> The first dead log (which I found within a hundred yards of the laboratory) yielded twenty-five or thirty specimens of roaches, each of which harbored a fauna of wood-ingesting hypermastigotes [a group of protozoa] far greater than that of any termite. Specimens were sent to Mr. A. N. Caudell of the U.S. National Museum, who identified them as *Cryptocercus punctulatus* Scudder.

🐝 🐜 🦟 There are two basic ways in which organisms that depend in one way or another on symbiotic microorganisms transfer these symbionts from one individual to another, notably from parents to young. As Vincent

Wigglesworth put it, some insects have "hereditary microorganisms" that are transmitted from parents to offspring, through the egg. Among them are various plant feeders and blood feeders. But this is not the case with either *Cryptocercus* or termites.

Termites, as you have read, lose their symbionts when they molt, but replace them by "anal solicitation" from another member of their colony. According to Cleveland, *Cryptocercus* accomplishes the same end in a different way. When nymphs molt, they lose many of the symbionts but retain a few that will reproduce and repopulate their digestive tract. But this does not explain how nymphs just hatched from the egg obtain the cellulose-digesting protozoa without which they cannot live. No symbionts are transmitted within the egg or on its surface, and newly hatched nymphs do not solicit anal secretions from their parents or some other member of the colony. Instead, they "seed" themselves with symbionts by eating fecal pellets passed by individuals that are about to molt. As the molt approaches, many of the protozoa in the gut become dormant by forming cysts that are resistant to unfavorable environmental conditions and are passed out with the feces.

Cleveland made the interesting discovery that this system of transferring symbionts to newly hatched nymphs is facilitated by the simultaneous occurrence of the hatching of eggs and the molting of nymphs in the colony. He observed that "the height of the breeding season—that is, when most of the [egg cases] are formed and placed in grooves [in the wood] to await hatching—occurs three to six weeks before the height of the annual moulting season; or, judging from observations on [egg cases] of other roaches, just about long enough for hatching of the [eggs] to coincide with moulting and the passing of fecal pellets containing protozoa."

Some flower flies (family Syrphidae) are well known to entomologists and amateur naturalists because the adults of certain species are deceptively exact mimics of bumble bees or stinging wasps, and because the larvae of some species feed voraciously on aphids and are important in controlling their numbers. But it is less well known that the larvae of quite a few species are recyclers of dead wood. Among them are the 50 or more North American species of the genus *Xylota*. Larvae of this genus burrow in and feed on rotting wood. The larvae of the related genus *Temnostoma* also tunnel in and presumably feed on dead wood. In 1929, L. E. Hildebrand, a teacher at New Trier High School in Winnetka, Illinois, sent to Clell Metcalf, then head of the Department of Entomology at the University of Illinois, about 90 pounds of partially rotted wood from a fallen log of a sugar maple in which *Temnostoma* larvae were tunneling. Metcalf described the condition of this wood and the behavior of the hundred or more larvae tunneling in it:

> The wood in which the larvae were found was very moist, still firm, but so far decayed that it could be broken apart by the unaided hands with some difficulty. The larvae made clean galleries through the wood, those made by the mature larvae being about one-eighth inch in diameter. Within these the larval body fits very snugly. The galleries are extended or burrowing is done by . . . rakes [on the first segment of the thorax]. The anterior end of the body is forced against the blind end of the burrow, the rakes are brought toward the median line so that their teeth point forward. Then the anterior three or four larval segments are contracted, at the same time pulling the rakes [to the side] and crowding the torn wood along the sides of the body between them and the walls of the galleries. Such scrapings are repeated at intervals of 4 or 5 seconds, when the larva is actively burrowing, and the gallery may be extended twice the length of the larval body in about fifty minutes.

The adult flies that Metcalf reared from these larvae are high-fidelity mimics of the small yellow and black potter wasps, which have a painful sting. *Temnostoma*'s mimicry of these wasps tends to protect it from birds, which learn to avoid painful encounters with potter wasps. This fly not only has a convincingly wasplike color pattern, but also mimics several other prominent features of its potter wasp model. Its waist is somewhat narrowed, as is the waist of the wasp. Like most other flies, *Temnostoma* has short, three-segmented antennae that are barely visible to the naked eye. But it mimics the long, black, and highly mobile antennae of potter wasps by waving its black anterior legs in front of its head. When the wasps sit on flowers, they fold their lightly tinted wings and hold them out to the side. When folded lengthwise in several layers, the wings look like dark brown bands. The fly, too, holds its wings out to the sides, but it cannot fold them. It mimics the appearance of a wasp's folded wing with a band of dark brown pigment that runs the length of the leading edge of the otherwise transparent wing.

Larvae of all the long-horned beetles feed only on plants, a few on herbaceous ones such as milkweed, some on living trees and woody shrubs, but according to E. Gorton Linsley, an expert on these beetles, the majority—at least of the more than 1,200 North American species—burrow within and feed on the solid wood or, less commonly, on the decaying wood of dead or dying trees. These larvae are known as round-headed borers to distinguish them from the flat-headed borers, larvae of the metallic wood-boring beetles (family Buprestidae), many of which also burrow in the wood of dead trees. Linsley wrote that long-horned beetle larvae play an important part, at least in temperate forests, in decomposing dead and dying trees and broken branches. Most of the relatively few long-horned beetles that attack living trees are host-specific. But most of those that feed on dead wood will consume the wood of many different kinds of trees, including both coniferous trees such as firs and pines and broad-leaved trees such as oaks and maples.

Larvae of the long-horned beetle *Arhopalus ferus,* noted H. R. Wallace, are important decomposers of dead pine stumps in England. The females lay eggs in batches of up to 20 under the bark of dying and dead trees and on standing stumps. The newly hatched larvae follow the grain of the wood as they burrow downward in the outer sapwood until they eventually reach the roots. Wallace described observations he made in his laboratory:

> When a larva is placed in a superficial cavity cut in a piece of pine wood, and covered by glass, it is possible to observe the feeding and boring behaviour. The larva first breaks up the wood with its mandibles and packs the shreds into any space between the glass and wood. It then smooths the walls by biting off any projecting pieces of wood. The larva now starts to bore. At the "head" of the cavity, wood is broken up and ingested. Pellets of frass [feces] are produced and every two minutes or so the larva turns round and packs the pellets behind it . . . Eventually, a long tunnel crammed with frass at one end is formed. The frass has a smaller volume than the original wood, and so as the larva progresses through the wood the volume of the air space in the tunnel increases.

Wallace found that in nature some burrows in stumps are as much as 60 inches long, of which about one-third is air space and two-thirds is packed with frass. After 2 years, fully grown larvae bore a tunnel that opens to the outside in April. They seal its opening with shredded wood, move back from 1 to 4 inches, and molt to the pupal stage. About a month later the pupa molts to reveal the adult beetle. Wallace wrote, "when the adult female has assumed its final form it remains quiescent at the pupation site for a week or more, by which time it is sexually mature; the males become active as soon a sclerotization [hardening of the skin] is completed, and they are then sexually mature. The first adults to vacate the pupation sites are the males." (The emergence of males before females is common among the insects. Natural selection drives males to emerge early so as to gain an advantage over other males in the cut-throat competition for females.)

Wallace found larvae in stumps with sound, unrotted wood and in stumps in all stages of decay, but they clearly prefer wood that is only slightly decayed. *Arhopalus* larvae can be very numerous. Wallace found 600 in one semi-decayed stump only about 10 inches in diameter. "In highly decayed stumps the larval population concentrates in the least decayed wood." But, he goes on the say, "The presence of fungi appears to be essential for the existence of the larvae—they do not occur in healthy trees." In the laboratory he placed weighed larvae in holes bored in pieces of wood from sound, unrotted stumps, from slightly decayed stumps, from partially decayed stumps, and from highly decayed ones. After 35 days he removed these larvae, weighed them again, and calculated their weight loss or gain. Those from sound wood had lost a little weight; those feeding in slightly decayed wood gained the most weight; those in partially decayed wood gained less that half of that; and those in highly decayed wood gained almost nothing.

The wood-boring horntails (family Siricidae), which you have already met, are also known as wood wasps or sawflies. Sawfly larvae, unlike the legless, grublike larvae of the other Hymenoptera, are caterpillar-like and have short stubby legs. Most feed on leaves, but horntails burrow in wood. The pigeon horntail is common in much of North America. In the larval stage it burrows in dead and dying trees of many species. Females drill narrow tunnels in the wood with their ovipositors and lay from 2 to 5 eggs in each one, depositing with them spores of their symbiotic fungus. This fungus, noted an Australian expert on wood wasps, F. David Morgan, is the food of the larvae. They bite off tiny fragments of wood, but rather than ingesting them, they extract nutrients from the filamentous fungal growth that permeates them. As the larvae burrow, they pass the depleted "fragments of wood along the outside of their bodies to join the accumulation of frass behind them." The burrows of pigeon horntails may be more than 9 feet long. Although they do not ingest wood, they prepare it for bacterial decomposition by fragmenting it and mixing it with their feces.

MICROCOSM

A sunny day in August: I am strolling in one of my favorite central Illinois areas, along a path bordering a woodland on a bank of the Sangamon River. A flash of iridescent blue catches my eye, a large blue and black butterfly clinging to the trunk of a white oak. As it slowly raises and lowers its shining wings, it sips sap oozing from a wound on the trunk. It is a beautiful red-spotted purple. Conspicuous and eye-catching, it is a focal point in an inanimate mass of green foliage and gray bark. Focused only on it, I am barely aware of its surroundings. Momentarily, I see the butterfly as an isolated object—as if a painter had framed it so as to make me see it as a thing unto itself, something separate from the rest of life. So it is with van Gogh's sunflowers. Van Gogh makes me see and take pleasure in a sunflower as never before, but there is an aspect of the life of a flower or a butterfly that no painting or photograph can capture: its place in the web of life—its interactions with the community of multitudinous plants and animals that make its existence possible or that are inimical to it.

The life of a red-spotted purple is interwoven with the lives of all the members of its community, both plants and animals. Some are enemies: the bird that can snatch an adult red-spotted purple from the air, the ambush bug that can stab a caterpillar to suck its juices, or the Venus flytrap that can clamp an adult within its leaf. Other plants and animals benefit this butterfly. Some affect it directly and may be essential to its survival, as is the plant it feeds on when it is in the caterpillar stage. Other organisms act indirectly and may be less important, such as a parasite that kills an insect that competes with the caterpillar for food. Although there are dangers, one thing is certain, living within its community the red-spotted purple has a chance to survive and reproduce, but separated from its community it will surely perish.

The sugar in the sap oozing from the oak provides some of the energy the butterfly—a female, to judge by her ample 4-inch wing span, greater than a male's—uses as she flies from tree to tree distributing hundreds of eggs. Sap will ooze from a wound on any tree. But what made the wound on this particular white oak, thereby benefiting the red-spotted purple? It began when a brown and white long-horned beetle more than an inch long, a white oak borer, chewed a hole in the bark and used her ovipositor to force an egg between the bark and the wood. The larva that hatched from this egg tunneled into the wood but maintained an opening to the outside through which it discarded its frass. Alerted by the hole in the bark and the frass, a hairy woodpecker on the lookout for its next meal used its powerful bill to get at the larva by chiseling away chips of bark and wood. From this wound came the copious flow of sap that attracted the butterfly. As I watch, it occurs to me that the oak that benefits the butterfly owes its very existence to a host of organisms. Not the least of them are a robber fly that ate a weevil that would have destroyed the acorn from which it sprouted and a provident gray squirrel that buried the acorn in the soil.

But the butterfly will soon need more food. She will imbibe nectar from flowers, and will drink the fermenting juices of a rotting apple lying on the ground. The apple fell from the tree because it was seriously damaged by the burrowing of a codling moth caterpillar that would not have been there if its ancestors had not been brought to North America by European settlers. The butterfly will also, to the surprise of some of you, sip juices from the carcass of a road-killed opossum and from a pile of coyote excrement that is all that is left of a rabbit recently eaten by the coyote. By feeding on sap, rotting fruit, carrion, and feces, the butterfly not only benefits itself, but makes a small contribution to the all-important recycling of dead organic matter.

Its appearance and its wing-waving behavior make the harmless and edible red-spotted purple a convincing mimic of the toxic and decidedly inedible pipe vine swallowtail. Insect-eating birds are made ill by eating pipe vine swallowtails and many soon learn to shun them and other butterflies that resemble them, including the red-spotted purple. This toxic swallowtail is, therefore, an important "resource" for the red-spotted purple, because, as its model, it is the foundation of the red-spotted purple's mimicry, one of its most important defenses against insect-eating birds. The pipe vine swallowtail also shields several other edible, mimetic butterflies and a moth against avian predators: both sexes of the spicebush swallowtail, female black and Missouri woodland swallowtails, the dark female form of the tiger swallow-

tail, the female Diana fritillary, and even the day-flying male of the promethea moth.

From what is known about the interactions between insectivorous birds,
mimetic insects, and their toxic models, we know that an overabundance of
these other mimics will be detrimental to the red-spotted purple and, indeed, to any other mimic of the pipe vine swallowtail. Avoidance learning by
a population of birds proceeds most rapidly if every insect with a "warning"
color pattern is actually toxic. But if the population of toxic insects is diluted
with too many edible insects with the same color pattern, learning will proceed more slowly, and a population of untutored birds willing to sample a
mimic will persist longer than it would otherwise.

To paraphrase Sir Walter Scott and William Shakespeare. Oh, what a tan-
gled web nature weaves; the web of life is of a mingled yarn, good and ill together. As we broaden our perception of the intricate web of life in the red-
spotted purple's ecosystem, we see that the plants and animals that influ-
enced the course of this butterfly's life—either for better or for worse—often
acted indirectly and sometimes several steps removed. Thus the many factors that benefit or harm the pipe vine swallowtail indirectly affect the red-
spotted purple.

Preeminent among this swallowtail's needs is food. The caterpillars eat
only foliage of plants of the birthwort family, among them the widespread
native Virginia snakeroot and the Dutchman's pipe, which grows naturally
in the southern United States and is cultivated as an ornamental in the
north. Birthworts contain toxins that protect them against their insect enemies, but the pipe vine swallowtail has turned the tables on these plants. Not
only is the caterpillar immune to birthwort toxins, but it sequesters them in
its body and retains them to the adult stage as a protection against birds.

To take our analysis one step further, anything that benefits a birthwort
plant benefits pipe vine swallowtails, and that in turn benefits red-spotted
purples. Among the many organisms that aid birthworts are their pollinators, scavenging flies. The flowers of some species attract dung-eating flies
by emitting the odor of feces. Others attract carrion feeders to flowers that
smell like carrion and are the color of rotting flesh. Among these flies are the
blue bottles and flesh flies, whose maggots recycle dead animals. Scavenging
flies not only serve birthworts as pollinators but also recycle dead animals
and thereby return to the soil nutrients essential to the growth and survival
of birthworts and all other plants.

The red-spotted purple that I saw on the oak tree would survive to mate and produce progeny. To avoid crowding her offspring, she would distribute her eggs widely, laying them one at a time—each one carefully placed at the tip of its own leaf. Instinctively, she lays them on trees whose leaves her caterpillar offspring will be willing to eat: oaks, poplars, or aspens—but most often on the favorite host plants of her species, wild black cherries and chokecherries.

Because insects grow only during the nymphal or larval stage, the caterpillar of the red-spotted purple must eat a prodigious amount of food as it increases in weight a thousand or more times as it grows from newly hatched larva to sexually mature adult. It probably assimilates from 30 to 40 percent of what it eats, discarding up to 60 percent as feces, which fall to the ground and are food for scavenging insects and bacteria. As cold weather approaches, the partly grown caterpillar rolls a leaf into a tube in which it shelters as it diapauses during the winter. In the spring, it terminates diapause and resumes feeding, and when fully grown molts to the pupal stage, a naked chrysalis that hangs by a short silken strap from the underside of a leaf or branch. In late May it emerges as an adult and begins the cycle once again.

Red-spotted purple caterpillars, which, like their parents, are not toxic, are protected from insect-eating birds by a special form of camouflage. They are not green to blend in with the foliage they feed on, as are many edible, plant-feeding insects. Instead, they look remarkably like bird droppings, objects of no interest to a hungry bird. The behavior of the red-spotted purple caterpillar complements its resemblance to a bird dropping. Unlike conventionally camouflaged caterpillars, it does not hide on the underside of a leaf during the day; instead it remains exposed on the upper side, and if some disturbance alerts it to the possible approach of a bird, it does not flee, but trusting in its camouflage, freezes and remains motionless. But at night, when its camouflage is not visible, it will be on the underside of a leaf—perhaps to hide from white-footed mice and certain bats that glean insects from leaves. Both fruit- and insect-eating birds indirectly benefit these caterpillars—much as pipe vine swallowtails benefit the adults—by providing them with models, droppings that cling to the upper sides of leaves.

It goes without saying that whatever benefits the cherry tree indirectly benefits the red-spotted purple. The cherry on which the caterpillar feeds would not exist if one of its parent's blossoms had not been pollinated by an insect. On a warm, sunny day in May in central Illinois, wild cherry trees swarm with little, brown solitary bees that come to its blossoms for nectar and pollen. The hind legs of the bees are covered with bright yellow pollen that is conspicuously visible in the triangular notches of the constriction between the abdomen and the thorax. Among them are a few small brown hover flies that carry no pollen but mimic the pollen carried by the bees with pubescent yellow markings on their abdomens. The seeds of the wild black cherry and chokecherries are dispersed by robins and other fruit-eating birds that swallow the cherries whole, digest their outer, fleshy parts and later, perhaps miles away, eliminate the hard and indigestible seeds in their droppings.

Ants protect the cherry trees against some of the insects that eat their leaves. The ants are attracted by extrafloral nectar secreted by glands on the leaf stems in early April, when the leaves are unfurling. At that time, tent caterpillars, the major defoliators of these trees, are just hatching from egg masses laid the previous summer and are still small enough for the ants to subdue. But by early May the tent caterpillars are too large for the ants to cope with and, parsimoniously, the tree stops secreting the extrafloral nectar that attracts the ants. This seasonal schedule is ideally suited for red-spotted purple caterpillars. The ants kill at least some of the tent caterpillars that would later have competed with the caterpillars for food, but the ants do not kill red-spotted purple caterpillars; those that resume feeding in the early spring after spending the winter in diapause are too large for the ants, and newly hatched ones, which are small enough for the ants to eat, do not hatch until late May or early June, after the ants are long gone.

🐜 🐜 🐜 The red-spotted purple is just one thread in an intricate and finely woven tapestry of life. Many of the tapestry's threads are missing, but enough of the pattern is there to give us an appreciation of its breadth and complexity. If we knew all there is to know about this butterfly's life, we would see that it is affected at least in some small way by virtually every organism in the ecosystem's community, its web of life.

SELECTED READINGS

MACROCOSM

Forbes, S. A. 1925. The lake as a microcosm. *Bulletin of the Illinois Natural History Survey,* 15: 537–550. Reprint of a paper read to the Peoria Scientific Association in 1887.

Stork, N. E. 1988. Insect diversity: facts, fiction, and speculation. *Biological Journal of the Linnean Society,* 35: 321–337.

1. POLLINATING

Barnes, E. 1934. Some observations on the genus *Arisaema* on the Nilgiri Hills, South India. *Journal of the Bombay Natural History Society,* 37: 629–639.

Barth, F. G. 1985. *Insects and Flowers.* Trans.M. A. Biederman-Thorson. Princeton: Princeton University Press.

Bawa, K. S. 1990. Plant-pollinator interactions in tropical rain forests. *Annual Review of Ecology and Systematics,* 21: 399–422.

Bernhardt, P. 1999. *The Rose's Kiss.* Washington, D.C.: Shearwater Books

Bertin, R. L. 1989. Pollination biology. In W. G. Abramson, ed., *Plant-Animal Interactions.* New York: McGraw-Hill.

Borror, D. J., D. M. DeLong, and C. A. Triplehorn. 1981. *An Introduction to the Study of Insects.* 5th ed. Philadelphia: Saunders College Publishing.

Buchmann, S. L., and G. P. Nabhan. 1996. *The Forgotten Pollinators.* Washington, D.C.: Shearwater Books.

Coleman, E. 1932. Further observations on the fertilization of Australian orchids by the male ichneumonid, *Lissopimpla semipunctata,* Kirb. (communicated by Prof. E. B. Poulton). *Proceedings of the Royal Entomological Society of London,* 6: 22–24.

Condit, I. J. 1947. *The Fig.* Waltham, Mass.: Chronica Botanica.

Dafni, A. 1984. Mimicry and deception in pollination. *Annual Review of Ecology and Systematics,* 15: 259–278.

Darwin, C. 1878. *The Effects of Cross and Self Fertilization in the Vegetable Kingdom.* 2nd ed. London: John Murray.

————1884. *The Various Contrivances by Which Orchids Are Fertilized by Insects.* 2nd ed. New York: D. Appleton and Co.

Dixon, K. W. , J. S. Pate, and J. Kuo. 1990. The western Australian fully subterranean orchid *Rhizanthella gardneri.* In J. Arditti, ed., *Orchid Biology: Reviews and Perspectives.* Portland, Ore.: Timber Press.

Dressler, R. L. 1968. Observations on orchids and euglossine bees in Panama and Costa Rica. *Revista de Biologia Tropical,* 15: 143–183.

————1981. *The Orchids.* Cambridge, Mass.: Harvard University Press.

Eltz, T., W. M. Whitten, D. W. Roubik, and K. E. Linsenmair. 1999. Fragrance collection, storage, and accumulation by individual male orchid bees. *Journal of Chemical Ecology,* 25: 157–176.

Faegri, K., and L. van der Pijl. 1979. *The Principles of Pollination Ecology.* Oxford: Pergamon Press.

Frisch, K. von. 1967. *The Dance Language and Orientation of Bees.* Trans. L. E. Chadwick. Cambridge, Mass.: Harvard University Press.

————1971. *Bees: Their Vision, Chemical Senses, and Language.* Rev. ed. Ithaca: Cornell University Press.

Harley, R. 1991. The greasy pole syndrome. In C. R. Huxley and D. F. Cutler, ed. *Ant-Plant Interactions.* Oxford: Oxford University Press.

Hölldobler, B., and E. O. Wilson. 1990. *The Ants.* Cambridge, Mass.: Harvard University Press.

Johnson, S. D., and W. J. Bond. 1994. Red flowers and butterfly pollination in the fynbos of South Africa. In M. Arianoutsou and R. H. Groves, ed., *Plant-Animal Interactions in Mediterranean-Type Ecosystems.* Dordrecht, Netherlands: Kluwer Academic Publishers.

Knoll, F. 1926. Insekten und Blumen, die *Arum*—Blütenstände und ihre Besucher. (Insects and flowers, *Arum* inflorescences and their visitors.) *Anhandlungen der Zoologische und Botanische Gesellschaft in Wien,* 12: 379–481.

Koorders, S. H. 1913. *Exkursionsflora von Java.* (*A Handbook of the Flora of Java.*) Vol. 4: *Atlas.* Jena: Verlag von Gustav Fischer.

McGregor, S. E. 1976. *Insect Pollination of Cultivated Crop Plants.* Agricultural Handbook 496. Washington, D.C.: Agricultural Research Service of the U.S. Department of Agriculture.

Meeuse, B. J. D. 1961. *The Story of Pollination.* New York: Ronald Press.

Michener, C. D. 1974. *The Social Behavior of the Bees.* Cambridge, Mass.: Harvard University Press.

Müller, H. 1883. *The Fertilization of Flowers.* Trans. D'Arcy W. Thompson. London: MacMillan.

Nilsson, L. A. 1992. Orchid pollination biology. *Trends in Ecology and Evolution,* 7: 255–259.

Nilsson, L. A., L. Jonsson, L. Ralison, and E. Randrianjohany. 1987. Angraecoid orchids and hawkmoths in central Madagascar: specialized pollination systems and generalist foragers. *Biotropica,* 19: 310–318.

Peakall, R., A. J. Beattie, and S. H. James. 1987. Pseudocopulation of an orchid by male ants: a test of two hypotheses accounting for the rarity of ant pollination. *Oecologia,* 73: 522–524.

Pijl, L. van der ,and C. H. Dodson. 1966. *Orchid Flowers.* Coral Gables: University of Miami Press.

Proctor, M., and P. Yeo. 1972. *The Pollination of Flowers.* New York: Taplinger.

Ramirez, W. 1974. Host specificity of fig wasps (Agaonidae). *Evolution,* 24: 680–691.

Ribbands, C. R. 1953. *The Behaviour and Social Life of Honeybees.* London: Bee Research Association.

Roubik, D. W., ed. 1995. *Pollination of Cultivated Plants in the Tropics.* Bulletin 118. Rome: Food and Agricultural Organization of the United Nations.

Seeley, T. D. 1995. *The Wisdom of the Hive.* Cambridge, Mass.: Harvard University Press.

Silberglied, R. E. 1979. Communication in the ultraviolet. *Annual Review of Ecology and Systematics,* 10: 373–398.

Simon, H. 1975. *The Private Lives of Orchids.* Philadelphia: J. B. Lippincott.

Sprengel, C. K. 1793. *Das entdeckte Geheimnis der Natur im Bau und in der Befruchtung der Blumen. (Nature's Secret of the Structure and Fertilization of Flowers Revealed.)* Berlin: F. Vieweg.

Valdeyron, G. and D. G. Lloyd. 1979. Sex differences and flowering phenology in the common fig, *Ficus carica L. Evolution,* 38: 673–685.

Waldbauer, G. P. 1988. Aposematism and Batesian mimicry: measuring mimetic advantage in natural habitats. In M. K. Hecht, B. Wallace, and G. T. Prance, ed., *Evolutionary Biology,* vol. 22. New York: Plenum Press.

Wilson, E. O. 1992. *The Diversity of Life.* Cambridge, Mass.: Harvard University Press.

Wilson, M., and J. Ågren. 1989. Differential floral rewards and pollination by deceit in unisexual flowers. *Oikos,* 55: 23–29.

Wyatt, R. 1983. Pollinator-plant interactions and the evolution of breeding systems. In L. Real, ed., *Pollination Biology.* Orlando: Academic Press.

2. DISPERSING SEEDS

Andersen, A. N. 1987. Effects of seed predation by ants on seedling densities at a woodland site in SE Australia. *Oikos,* 48: 171–174.

————1991. Seed harvesting by ants in Australia. In C. R. Huxley and D. F. Cutler, ed., *Ant-Plant Interactions.* Oxford: Oxford University Press.

Beattie, A. J. 1985. *The Evolutionary Ecology of Ant-Plant Mutualisms.* New York: Cambridge University Press.

Berg, R. Y. 1975. Myrmecochorous plants in Australia and their dispersal by ants. *Australian Journal of Botany,* 23: 475–508.

Bond, W., and P. Slingsby. 1984. Collapse of an ant-plant mutualism: the Argentine ant (*Iridomyrmex humilis*) and myrmecochorous Proteaceae. *Ecology,* 65: 1031–1037.

Brown, J. H., D. W. Davidson, and O. J. Reichman. 1979. An experimental study of competition between seed-eating desert rodents and ants. *American Zoologist,* 19: 1129–1143.

Burger, B. V., and W. G. B. Petersen. 1991. Semiochemicals of the Scarabaeinae, III: identification of the attractant for the dung beetle *Pachylomerus femoralis* in the fruit of the spineless monkey orange tree, *Strychnos madagascariensis. Zeitschrift für Naturforschung,* 46c: 1073–1079.

Darwin, C. 1876. *The Origin of Species.* Ed. P. H. Barrett and R. B. Freeman. New York: New York University Press.

Davison, E. A. 1982. Seed utilization by harvester ants. In R. C. Buckley, ed., *Ant-Plant Interactions in Australia.* The Hague: W. Junk.

Handel, S. N., and A. J. Beattie. 1990. Seed dispersal by ants. *Scientific American,* 263: 76–83.

Hölldobler, B., and E. O. Wilson. 1990. *The Ants.* Cambridge, Mass.: Harvard University Press.

Hughes, L., and M. Westoby. 1992. Capitula on stick insect eggs and elaiosomes on seeds: convergent adaptations for burial by ants. *Ecology,* 6: 642–648.

Keeler, K. H. 1989. Ant-plant interactions. In W. G. Abrahamson, ed., *Plant-Animal Interactions.* New York: McGraw-Hill.

Laman, T. 1997. Borneo's strangler fig trees. *National Geographic,* 191(4): 38–55.

Lincecum, G. 1874. The agricultural ant. *American Naturalist,* 8: 513–517.

Milewski, A. G., and W. J. Bond. 1982. Convergence of myrmecochory in mediterranean Australia and South Africa. In R. C. Buckley, ed., *Ant-Plant Interactions in Australia.* The Hague: W. Junk.

Moggridge, J. T. 1873. *Harvesting Ants and Trap-Door Spiders.* London: L. Reeve.

Pijl, L. van der. 1982. *Principles of Dispersal in Higher Plants.* Berlin: Springer-Verlag.

Ridley, H. N. 1930. *The Dispersal of Plants throughout the World.* Ashford, Eng.: L. Reeve and Co.

Risch, S. J., and C. R. Carroll. 1986. Effects of seed predation by a tropical ant on competition among weeds. *Ecology,* 67: 1319–1327.

Rissing, S. W. 1986. Indirect effects of granivory by harvester ants: plant species composition and reproductive increase near ant nests. *Oecologia,* 68: 231–234.

Rissing, S. W., and J. Wheeler. 1976. Foraging responses of *Veromessor pergandei* to changes in seed production (Hymenoptera: Formicidae). *Pan-Pacific Entomologist,* 52: 63–72.

Taber, S. W. 1998. *The World of the Harvester Ants.* College Station: Texas A & M University Press.

Tschinkel, W. R. 1999. Sociometry and sociogenesis of colony-level attributes of the Florida harvester ant (Hymenoptera: Formicidae). *Annals of the Entomological Society of America,* 92: 80–89.

Vulinec, K. 2000. Dung beetles (Coleoptera: Scarabaeidae), monkeys, and conservation in Amazonia. *Florida Entomologist,* 83: 229–241.

Went, F. W., J. Wheeler, and G. C. Wheeler. 1972. Feeding and digestion in some ants (*Veromessor* and *Manica*). *BioScience,* 22: 82–88.

Westoby, M., L. Hughes, and B. L. Rice. 1991. Seed dispersal by ants: comparing infertile with fertile soils. In C. R. Huxley and D. F. Cutler, ed., *Ant-Plant Interactions.* Oxford: Oxford University Press.

Wheeler, J., and S. W. Rissing. 1975. Natural history of *Veromessor pergandei* I. The nest. *Pan-Pacific Entomologist,* 51: 205–216.

———1975. Natural history of *Veromessor pergandei* II. Behavior. *Pan-Pacific Entomologist,* 51: 303–314.

3. SUPPLYING FOOD

Albert, V. A., S. E. Williams, and M. W. Chase. 1992. Carnivorous plants: phylogeny and structural evolution. *Science,* 257: 1492–1495.

Beattie, A. J. 1985. *The Evolutionary Ecology of Ant-Plant Mutualisms.* Cambridge. Cambridge University Press.

Buckley, R. C. 1982. Ant-plant interactions: a world review. In R. C. Buckley, ed., *Ant-Plant Interactions.* Hingham, Mass.: Kluwer Boston.

Büsgen, M. 1883. Die Bedeutung des Insektenfanges für *Drosera rotundifolia* L. (The significance of insect trapping for *Drosera rotundifolia* L.). *Botanische Zeitung,* 41: 568–577.

Chapin, C. T., and J. Pastor. 1995. Nutrient limitations in the northern pitcher plant, *Sarracenia purpurea. Canadian Journal of Botany,* 73: 728–734.

Darwin, C. 1893. *Insectivorous Plants.* 2nd ed., revised by Francis Darwin. London: John Murray.

Davidson, D. W. 1988. Ecological studies of Neotropical ant gardens. *Ecology,* 69: 1138–1152.

Forel, A. 1898. La parabiose chez les fourmis. (Symbiosis in the ants.) *Bulletin de la Société Vaudoise des Sciences Naturelles,* 34: 380–384.

Givnish, T. J. 1989. Ecology and evolution of carnivorous plants. *In* W. G. Abrahamson, ed., *Plant-Animal Interactions.* New York: McGraw-Hill

Heslop-Harrison, Y. 1978. Carnivorous plants. *Scientific American,* 238: 104–115.

Huxley, C. R. 1978. The ant-plants *Myrmecodia* and *Hydnophytum* (Rubiaceae) and the relationships between their morphology, ant occupants, physiology, and ecology. *New Phytologist,* 80: 231–268.

Huxley, C. R., and D. F. Cutler, ed. 1991. *Ant-Plant Interactions.* Oxford: Oxford University Press.

Janzen, D. H. 1974. Epiphytic myrmecophytes in Sarawak: mutualism through the feeding of plants by ants. *Biotropica,* 6: 237–257.

Joel, D. M., B. E. Juniper, and N. Dafni. 1985. Ultraviolet patterns in the traps of carnivorous plants. *New Phytologist,* 101: 585–593.

Jolivet, P. 1998. *Interrelationship between Insects and Plants.* Boca Raton, Fla.: CRC Press.

Juniper, B. E., and J. K Buras. 1962. How pitcher plants trap insects. *New Scientist,* 13: 75–77.

Kleinfeldt, S. E. 1978. Ant-gardens: the interaction of *Codonanthe crassifolia* (Gesneriaceae) and *Crematogaster longispina* (Formicidae). *Ecology,* 59: 449–456.

———1986. Ant-gardens: mutual exploitation. In B. Juniper and R. Southwood, ed., *Insects and the Plant Surface.* London: Edward Arnold.

Lloyd, F. E. 1942. *The Carnivorous Plants.* Waltham, Mass.: Chronica Botanica Co.

Macedo, M., and G. T. Prance. 1978. Notes on the vegetation of Amazonia II. The dispersal of plants in Amazonian white sand campinas: the campinas as functional islands. *Brittonia,* 30: 203–215.

Prance, G. T. 1973. Gesneriads in the ant gardens of the Amazon. *Gloxinian,* 23: 27–28.

Schnell, D. E. 1976. *Carnivorous Plants of the United States and Canada.* Winston-Salem, N.C.: John F. Blair.

Slack, A. 1980. *Carnivorous Plants.* Cambridge, Mass.: MIT Press.

Ule, E. 1902. Ameisengärten im Amazonasgebiet. (Ant gardens in the Amazon area.) *Botanische Jahrbücher,* 30 (supplement): 45–52.

Wheeler, W. M. 1921. A new case of parabiosis and the "ant gardens" of British Guiana. *Ecology,* 2: 89–103.

4. PROVIDING DEFENSE

Agrawal, A. A., and M. T. Rutter. 1998. Dynamic anti-herbivore defense in ant-plants: the role of induced responses. *Oikos,* 83: 227–236.

Alborn, H. T., T. C. J. Turlings, T. H. Jones, G. Stenhagen, J. H. Loughrin, and J. H. Tumlinson. 1997. An elicitor of plant volatiles from beet armyworm oral secretion. *Science,* 276: 945–949.

Belt, T. 1874. *The Naturalist in Nicaragua.* London: Edward Bumpus.

Bentley, B. L. 1977. Extrafloral nectaries and protection by pugnacious bodyguards. *Annual Review of Ecology and Systematics,* 8: 407–427.

———1977. The protective function of ants visiting the extrafloral nectaries of *Bixa orelana* (Bixaceae). *Journal of Ecology,* 65: 27–38.

Cook, O. F. 1904. An enemy of the cotton boll weevil. *Science,* 19: 862–864.

Godfray, H. C. J. 1994. *Parasitoids: Behavioral and Evolutionary Ecology.* Princeton: Princeton University Press.

Hölldobler, B., and E. O. Wilson. 1990. *The Ants.* Cambridge, Mass.: Harvard University Press.

Huxley, C. R., and D. F. Cutler, ed. 1991. *Ant-Plant Interactions.* Oxford: Oxford University Press.

Janzen, D. H. 1966. Coevolution of mutualism between ants and acacias in Central America. *Evolution,* 20: 249–275.

———1967. Interaction of the bull's horn acacia (*Acacia cornigera* L.) with an ant inhabitant (*Pseudomyrmex ferruginea* F. Smith) in eastern Mexico. *University of Kansas Science Bulletin,* 47: 315–558.

Keeler, K. H. 1977. The extrafloral nectaries of *Ipomea carnea* (Convolvulaceae). *American Journal of Botany,* 64: 1182–1188.

————1980. The extrafloral nectaries of *Ipomea leptophylla* (Convolvulaceae). *American Journal of Botany,* 67: 216–222.

Kessler, A., and I. T. Baldwin. 2001. Defensive function of herbivore-induced plant volatile emissions in nature. *Science,* 291: 2141–2144.

Koptur, S. 1984. Experimental evidence for defense of *Inga* (Mimosidae) saplings by ants. *Ecology,* 65: 1787–1793.

————1985. Alternative defenses against herbivores in *Inga* (Fabaceae: Mimosidae) over an elevational gradient. *Ecology,* 66: 1639–1650.

Letourneau, D. K. 1998. Ants, stem borers, and fungal pathogens: experimental tests of a fitness advantage in *Piper* ant-plants. *Ecology,* 79: 593–603.

Messina, F. J. 1981. Plant protection as a consequence of an ant-membracid mutualism: interactions on goldenrod (*Soldiago* sp.). *Ecology,* 62: 1433–1440.

Nealis, V. G. 1986. Responses to host kairomones and foraging behavior of the insect parasite *Cotesia rubecula. Canadian Journal of Zoology.* 64. 2393–2398.

O'Dowd, D. J. 1979. Foliar nectar production and ant activity on a neotropical tree, *Ochroma pyramidale. Oecologia,* 43: 223–248.

Pickett, C. H., and W. D. Clark. 1979. The function of extrafloral nectaries in *Opuntia acanthocarpa* (Cactaceae). *American Journal of Botany,* 66: 618–625.

Ross, G. N. 1966. Life-history studies on Mexican butterflies. IV. The ecology and ethology of *Anatole rossi,* a myrmecophilous metalmark (Lepidoptera: Riodinidae). *Annals of the Entomological Society of America,* 49: 985–1004.

Stephenson, A. G. 1982. The role of extrafloral nectaries of *Catalpa speciosa* in limiting herbivory and increasing fruit production. *Ecology,* 63: 663–669.

Tilman, D. 1978. Cherries, ants, and tent caterpillars: timing of nectar production in relation to susceptibility of caterpillars to ant predation. *Ecology,* 59: 686–692.

Turlings, T. C. J., and J. H. Tumlinson. 1992. Systemic release of chemical signals by herbivore-injured corn. *Proceedings of the National Academy of Sciences,* 89: 8399–8402.

Turlings, T. C. J., J. H. Tumlinson, and W. J. Lewis. 1990. Exploitation of herbivore-induced plant odors by host-seeking parasitic wasps. *Science,* 250: 1251–1253.

Vanderplank, F. L. 1960. The bionomics and ecology of the red tree ant, *Oecophylla* sp., and its relationship to the coconut bug *Pseudotheraptus wayi* Brown (Coreidae). *Journal of Animal Ecology,* 29: 15–33.

Way, M. H. 1963. Mutualism between ants and honeydew-producing Homoptera. *Annual Review of Entomology,* 8: 307–344.

Wheeler, W. M. 1910. *Ants: Their Structure, Development and Behavior.* New York: Columbia University Press.

5. GIVING SUSTENANCE

Akre, R. D., A. Greene, J. F. MacDonald, P. J. Landolt, and H. G. Davis. 1980. *The Yellowjackets of America North of Mexico.* Agriculture Handbook no. 552. Washington, D.C.: U.S. Department of Agriculture.

Askew, R. R. 1971. *Parasitic Insects.* New York: American Elsevier.

Autumn, K., Y. A. Liang, S. Tonia Hsieh, Wolfgang Zesch, W. P. Chan, T. W. Kenny, R. Fearing, and R. J. Full. 2000. Adhesive force of a single gecko foot hair. *Nature,* 405: 681–684.

Babbitt, L. H. 1937. *Amphibia of Connecticut.* Bulletin of the Connecticut State Geological and Natural History Survey, no. 57. Hartford.

Balduf, W. V. 1939. *The Bionomics of Entomophagous Insects.* St. Louis: John S. Swift.

Bellairs, A. 1970. *The Life of Reptiles.* 2 vols. New York: Universe Books.

Bodenheimer, F. S. 1951. *Insects as Human Food.* The Hague: W. Junk.

Brodeur J., and L. E. M. Vet. 1994. Usurpation of host behaviour by a parasitic wasp. *Animal Behaviour,* 48: 187–192.

Cahalane, V. H. 1947. *Mammals of North America.* New York: MacMillan.

Cloudsley-Thompson, J. L. 1958. *Spiders, Scorpions, Centipedes, and Mites.* New York: Pergamon Press.

Comstock, J. H. 1950. *An Introduction to Entomology.* 9th ed. Ithaca: Comstock Publishing Co.

Conover, A. 2000. Sloth bears. *Smithsonian,* 30: 88–95.

Corbet, P. S. 1963. *The Biology of Dragonflies.* Chicago: Quadrangle Books.

DeFoliart, G. R. 1975. Insects as a source of protein. *Bulletin of the Entomological Society of America,* 21: 161–163.

————1989. The human use of insects as food and as animal feed. *Bulletin of the Entomological Society of America,* 35: 22–35.

Dybas, H. S., and D. D. Davis. 1962. A population census of seventeen-year periodical cicadas (Homoptera: Cicadidae: *Magicicada*). *Ecology,* 43: 432–444.

Eberhard, W. G. 2000. Spider manipulation by a wasp larva. *Nature,* 406: 255–256.

Ewert, J. P. 1974. The neural basis of visually guided behavior. *Scientific American,* 230: 34–42.

————1980. *Neuroethology.* Trans.Transemantics. New York: Springer-Verlag.

Fenton, M. B. 1995. Natural history and biosonar signals. In A. N. Popper and R. R. Fay, ed., *Hearing by Bats.* New York: Springer-Verlag.

Foelix, R. F. 1982. *Biology of Spiders.* Cambridge, Mass.: Harvard University Press.

Gauld, I., and B. Bolton. 1988. *The Hymenoptera.* New York: Oxford University Press.

Gill, F. B. 1994. *Ornithology.* 2nd ed. New York: W. H. Freeman.

Golley, F. B., and J. B. Gentry. 1964. Bioenergetics of the southern harvester ant, *Pogonomyrmex badius. Ecology,* 45: 217–225.

Goodall, J. 1963. Feeding behaviour of wild chimpanzees. *Symposia of the Zoological Society of London,* 10: 39–47.

Hickson, S. J. 1889. *A Naturalist in North Celebes.* London: John Murray.

Hocking, B. 1971. *Six-Legged Science.* Cambridge, Mass.: Schenkman Publishing.

Hölldobler, B., and E. O. Wilson. *The Ants.* Cambridge, Mass.: Harvard University Press.

Kick, S. A. 1982. Target-detection by the echolocating bat, *Eptesicus fuscus. Journal of Comparative Physiology,* A, 145: 431–435.

Kyle, H. M. 1926. *The Biology of Fishes.* New York: MacMillan.

Lack, D. 1947. *Darwin's Finches.* New York: Harper and Brothers.

Lankester, E. R. 1889. The Muybridge photographs. *Nature,* 40: 78–80.

Lawson, F. R. , R. L. Rabb, F. E. Guthrie, and T. G. Bowery. 1961. Studies of an integrated control system for hornworms on tobacco. *Journal of Economic Entomology,* 54: 93–97.

Linsenmaier, W. 1972. *Insects of the World.* Trans. L. E. Chadwick. New York: McGraw-Hill.

Marsh, F. L. 1937. Ecological observations upon the enemies of *cecropia,* with particular reference to its hymenopterous parasites. *Ecology,* 18: 106–112.

Marshall, A. J., and R. Hook. 1960. The breeding biology of equatorial vertebrates: reproduction of the lizard *Agama agama lionotus* Boulenger at lat. 0º01′N. *Proceedings of the Royal Zoological Society of London,* 134: 197–205.

Martin, I. G. 1981. Venom of the short-tailed shrew (*Blarina brevicauda*) as an insect immobilizing agent. *Journal of Mammalogy,* 62: 189–192.

Metcalf, R. L., and R. A. Metcalf, 1993. *Destructive and Useful Insects.* 5th ed. New York: McGraw Hill.

Moyle, P. B., and J. J. Cech, Jr. 1982. *Fishes: An Introduction to Ichthyology.* Englewood Cliffs, N.J.: Prentice-Hall.

Normile, D. 2000. New feathered dino firms up bird links. *Science,* 288: 1721.

Noyes, H. 1937. *Man and the Termite.* London: Peter Davies.

Odum, E. P., C. E. Connell, and L. B. Davenport. 1962. Population energy flow of three primary consumer components of old-field ecosystems. *Ecology,* 43: 88–96.

Oliver, J. 1955. *The Natural History of North American Amphibians and Reptiles.* Princeton, N. J.: Van Nostrand.

Peterson, R. T. 1963. *The Birds.* New York: Time Inc.

Polis, G. A. 1979. Prey and feeding phenology of the desert sand scorpion *Paruroctonus mesaensis* (Scorpionidae: Vaejovidae). *Journal of the Zoological Society of London,* 188: 333–346.

Preston-Mafham, R, and K. Preston-Mafham. 1984. *Spiders of the World.* New York: Facts on File.

———1993. *The Encyclopedia of Land Invertebrate Behaviour.* Cambridge, Mass.: MIT Press.

Rabb, R. L. 1960. Biological studies of *Polistes* in North Carolina (Hymenoptera: Vespidae). *Annals of the Entomological Society of America,* 53: 111–121.

Reeve, H. K. 1991. *Polistes.* In K. G. Ross and R. W. Matthews, ed., *The Social Biology of Wasps.* Ithaca: Comstock Publishing Associates.

Rieppel, O. C. 1992. Reptiles and amphibians through the ages. In H. C. Cogger and R. G. Zweifel, ed., *Encyclopedia of Reptiles and Amphibians,* 2nd ed. San Diego: Academic Press.

Schmidt, K. P., and D. D. Davis. 1941. *Field Book of Snakes of the United States and Canada.* New York: Putnam.

Schneirla, T. C. 1956. The army ants. In *Annual Report of the Board of Regents of the Smithsonian Institution for 1955.* Washington, D.C.: Smithsonian Institution.

Shine, R. 1992. Snakes. In H. G. Cogger and R. G. Zweifel, ed., *Encyclopedia of Reptiles and Amphibians.* San Diego: Academic Press.

Speight, M. R., M. D. Hunter, and A. D. Watt. 1999. *Ecology of Insects.* London: Blackwell Science.

Stowe, M. K., J. H. Tumlinson, and R. R. Heath. 1987. Chemical mimicry: bolas spiders emit components of moth prey species sex pheromones. *Science,* 236: 964–967.

Turillazzi, S., and M. J. West-Eberhard, ed. 1996. *Natural History and Evolution of Paper-Wasps.* New York: Oxford University Press.

U.S. Commissioner of Agriculture. 1870. Food products of North American Indians. In *The Annual Report of the Commissioner of Agriculture.* Washington, D.C.: U.S. Department of Agriculture.

Waldbauer, G. P. 1968. The consumption and utilization of food by insects. *Advances in Insect Physiology,* 5: 229–288.

Waldbauer, G. P., and J. K. Sheldon. 1971. Phenological relationships of some aculeate Hymenoptera, their dipteran mimics, and insectivorous birds. *Evolution,* 25: 371–382.

Waldbauer, G. P., and J. G. Sternburg. 1982. Cocoons of *Callosamia promethea* (Saturniidae): adaptive significance of differences in mode of attachment to the host tree. *Journal of the Lepidopterist's Society,* 36: 192–199.

Waldbauer, G. P., J. G. Sternburg, W. G. George, and A. G. Scarbrough. 1970. Hairy and downy woodpecker attacks on cocoons of urban *Hyalophora cecropia* (L.) and other saturniids. *Annals of the Entomological Society of America,* 63: 1366–1369.

Wheeler, A. 1985. *The World Encyclopedia of Fishes.* London: Macdonald and Co.

Weis, A. E., and M. R. Berenbaum. 1989. Herbivorous insects and green plants. In W. G. Abrahamson, ed., *Plant-Animal Interactions.* New York: McGraw-Hill.

Wiegert, R. G., and F. C. Evans. 1967. Investigations of secondary productivity in grasslands. In K. Petrusewicz, ed., *Secondary Productivity of Terrestrial Ecosystems.* Warsaw, Poland: Panstwowe Naukowe.

Winston, M. L. 1997. *Nature Wars.* Cambridge, Mass.: Harvard University Press.

6. GIVING PROTECTION

Auclair, J. L. 1963. Aphid feeding and nutrition. *Annual Review of Entomology,* 8: 439–490.

Baylis, M., and N. E. Pierce. 1992. Lack of compensation by final instar larvae of the myrmecophilous lycaenid butterfly, *Jalmenus evagoras,* for the loss of nutrients to ants. *Physiological Entomology,* 17: 107–114.

———1993. The effects of ant mutualism on the foraging and diet of lycaenid caterpillars. In H. E. Stamp and T. M. Casey, ed., *Caterpillars.* New York: Chapman and Hall.

Beebe, M. B., and C. W. Beebe. 1910. *Our Search for a Wilderness.* New York: Henry Holt.

Belt, T. 1888. *The Naturalist in Nicaragua.* London: Edward Bumpus.

Bristow, C. M. 1984. Differential benefits from ant attendance to two species of Homoptera on New York ironweed. *Journal of Animal Ecology,* 3: 715–726.

Chisholm, A. H. 1944. The problem of "anting". *Ibis,* 86: 389–405.

Clark, J. T. 1976. The eggs of stick insects (Phasmida): a review with descriptions of the eggs of eleven species. *Systematic Entomology,* 1: 95–105.

Clayton, D. H. 1991. Coevolution of avian grooming and ectoparasite avoidance. In J. E. Loye and M. Zuk ed., *Bird-Parasite Interactions.* Oxford: Oxford University Press.

Clayton, D. H., and J. G. Vernon. 1993. Common grackle anting with lime fruit and its effect on ectoparasites. *The Auk,* 110: 951–952.

Condry, W. 1947. Behaviour of young carrion crow with ants. *British Birds,* 40: 114.

Edwards, W. H. 1878. On the larvae of *Lyc. pseudoargiolus* and attendant ants. *The Canadian Entomologist,* 10: 131–136.

Fernald, H. T., and H. H. Shepard. 1942. *Applied Entomology.* 4th ed. New York: McGraw-Hill.

Fiedler, K., and U. Maschwitz. 1989. The symbiosis between the weaver ant, *Oecophylla smaragdina,* and *Athene emolus,* an obligate myrmecophilous lycaenid butterfly. *Journal of Natural History,* 23: 833–846.

Flint, W. P. 1919. *The Corn Root-Aphis.* Entomological Series Circular 4. Urbana: Illinois State Natural History Survey.

Forbes, S. A. 1887. The lake as a microcosm. *Bulletin of the Illinois Natural History Survey,* 15: 537–550.

———1906. The corn-root aphis and its attendant ant (*Aphis maidiradicus* Forbes and *Lasius niger* L. variety *americanus* Emery). *Bulletin of the United States Department of Agriculture, Division of Entomology,* 60: 29–39.

Hindwood, K. A. 1959. The nesting of birds in the nests of social insects. *The Emu,* 59: 1–36.

Hölldobler, B. 1984. The wonderfully diverse ways of the ant. *National Geographic,* 165: 778–813.

Hölldobler, B., and E. O. Wilson. 1990. *The Ants.* Cambridge, Mass.: Harvard University Press.

Hughes, L., and M. Westoby. 1992. Capitula on stick insect eggs and elaiosomes on seeds: convergent adaptations for burial by ants. *Functional Ecology,* 6: 642–648.

Janzen, D. H. 1969. Birds and the ants × acacia interaction in Central America, with notes on birds and other myrmecophytes. *The Condor,* 71: 240–256.

Joyce, F. J. 1993. Nesting success of rufous-naped wrens (*Campylorhynchus rufinucha*) is greater near wasp nests. *Behavioral Ecology and Sociobiology,* 32: 71–77.

Kelso, L., and M. M. Nice. 1963. A Russian contribution to anting and feather mites. *The Wilson Bulletin,* 75: 23–26.

Kennedy, J. S., and T. E. Mittler. 1953. A method of obtaining phloem sap via the mouth-parts of aphids. *Nature,* 171: 528.

MacLaren, P. I. R. 1950. Bird-ant nesting associations. *The Ibis,* 92: 564–566.

Maschwitz, U., K. Dumpert, and G. Schmidt. 1985. Silk pavilions of two *Camponotus (Karavaievia)* species from Malaysia: description of a new nesting type in ants (Formicidae: Formicinae). *Zeitschrift für Tierpsychologie,* 69: 237–249.

Maschwitz, U., M. Wüst, and K. Schurian. 1975. Blaulingsraupen als

Zuckerlieferanten für Ameisen. (Lycaenid caterpillars as sugar sources for caterpillars.) *Oecologia*, 18: 17–21.

Metcalf, R. L., and R. A. Metcalf. 1993. *Destructive and Useful Insects*. 5th ed. New
York: McGraw-Hill.

Myers, J. G. 1935. Nesting associations of birds with social insects. *Transactions of the
Royal Entomological Society of London*, 83: 11–22.

Newcomer, E. J. 1912. Some observations on the relations of ants and lycaenid caterpillars and a description of the relational organs of the latter. *Journal of the New
York Entomological Society*, 20: 31–36.

Pierce, N. E., R. L. Kitching, R. C. Buckley, M. F. J. Taylor, and K. F. Benbow. 1987.
The costs and benefits of cooperation between the Australian lycaenid butterfly,
Jalmenus evagoras, and its attendant ants. *Behavioral Ecology and Sociobiology*, 21:
237–248.

Robinson, S. K. 1985. Coloniality in the yellow-rumped cacique as a defense against
nest predators. *The Auk*, 102: 506–519.

Ross, G. N. 1966. Life-history studies on Mexican butterflies. IV. The ecology and
ethology of *Anatole rossi*, a myrmecophilous metalmark (Lepidoptera:
Riodinidae). *Annals of the Entomological Society of America*, 59: 985–1004.

Simmons, K. E. L. 1966. Anting and the problem of self-stimulation. *Journal of Zoology*, 149: 145–162.

Thomas, J., and R. Lewington. 1991. *The Butterflies of Britain and Ireland*. London:
Dorling Kindersley.

Von Hagen, W. 1938. Contribution to the biology of *Nasutitermes* (sensu strictu). *Proceedings of the Zoological Society of London*, 108: 39–48.

Whitaker, L. M. 1957. A résumé of anting, with particular reference to a captive orchard oriole. *The Wilson Bulletin*, 69: 195–262.

Wilson, E. O. 1971. *The Insect Societies*. Cambridge, Mass.: Harvard University Press.

Wunderle, J. M., Jr., and K. H. Pollock. 1985. The bananaquit-wasp nesting association and a random choice model. *Ornithological Monographs*, 36: 595–603.

Zoebelein, G. 1956. Der Honigtau als Nahrung der Insketen. (Honeydew as food for
insects.) *Zeitschrift für angewandte Entomologie*, 38: 129–167.

7. CONTROLLING PLANT POPULATIONS

Abrahamson, W. G., and A. W. Weis. 1997. *Evolutionary Ecology across Three Trophic
Levels: Goldenrods, Gallmakers, and Natural Enemies*. Princeton: Princeton University
Press.

Andersen, A. N. 1987. Effects of seed predation by ants on seedling densities at a
woodland site in SE Australia. *Oikos*, 48: 171–174.

———1991. Seed harvesting by ants in Australia. In C. R. Huxley and D. F. Cutler,
ed., *Ant-Plant Interactions*. Oxford: Oxford University Press.

Anderson, T. E., and G. P. Waldbauer. 1977. Development and field testing of a
quantitative technique for extracting bean leaf beetle larvae and pupae from soil.
Environmental Entomology, 6: 633–636.

Bailey, L. H. 1949. *Manual of Cultivated Plants*. Rev. ed. New York: MacMillan.

Benson, W. W., K. S. Brown, Jr., and L. E. Gilbert. 1975. Coevolution of plants and herbivores: passion flower butterflies. *Evolution,* 29: 659–680.

Berenbaum, M. R. 1995. *Bugs in the System*. Reading, Mass.: Addison-Wesley.

Bogdandy, St. v. 1927. Ausrottung von Bettwanzen mit Bohnenblättern. (Extermination of bed bugs with bean leaves.) *Die Naturwissenschaften,* 15: 474.

Bongers, J. and W. Eggerman. 1971. Der Einfluss des Subsocialverhaltens der spezialisierten Samensauger *Oncopeltus fasciatus* Dall. und *Dysdercus fasciatus* Sign. auf ihre Ernährung. (Influence on their nutrition of the subsocial behavior of the specialized seed suckers *Oncopeltus fasciatus* Dall. And *Dysdercus fasciatus* Sign.) *Oecologia,* 93: 293–302.

Broersma, D. B., R. L. Bernard, and W. H. Luckmann. 1972. Some effects of soybean pubescence on populations of potato leafhopper. *Journal of Economic Entomology,* 65: 78–82.

Brown, J. H., O. J. Reichman, and D. W. Davidson. 1979. Granivory in desert ecosystems. *Annual Review of Ecology and Systematics,* 10: 201–227.

Brues, C. T. 1946. *Insect Dietary*. Cambridge, Mass.: Harvard University Press.

Carson, W. P., and R. B. Root. 1999. Top-down effects of insect herbivores during early succession: influence on biomass and plant dominance. *Oecologia,* 121: 260–272.

Carter, W. 1973. *Insects in Relation to Plant Disease*. 2nd ed. New York: John Wiley and Sons.

Clausen, C. P. ed. 1977. *Introduced Parasites and Predators of Arthropod Pests and Weeds: A World Review*. Agricultural Handbook no. 480. Washington, D.C.: U.S. Department of Agriculture.

Comstock, J. H. 1950. *An Introduction to Entomology*. 9th ed. Ithaca: Comstock Publishing Co.

Cotton, R. T. 1956. *Pests of Stored Grain and Grain Products*. Minneapolis: Burgess Publishing.

Craighead, F. C. 1950. *Insect Enemies of Eastern Forests*. Washington, D.C.: U.S. Department of Agriculture.

DeBach, P., ed. 1964. *Biological Control of Insect Pests and Weeds*. New York: Reinhold.

Dussourd, D. E. 1993. Foraging with finesse: caterpillar adaptations for circumventing plant defenses. In N. E. Stamp and T. M. Casey, ed., *Caterpillars*. New York: Chapman and Hall.

Dussourd, D. E., and T. Eisner. 1987. Vein-cutting behavior: insect counterploy to the latex defense of plants. *Science,* 237: 898–901.

Ehrlich, P. R., and P. H. Raven. 1965. Butterflies and plants: a study in coevolution. *Evolution,* 18: 586–608.

Feeny, P. 1970. Seasonal changes in oak leaf tannins and nutrients as a cause of spring feeding by winter moth caterpillars. *Ecology,* 51: 565–581.

Felt, E. P. 1905. *Insects Affecting Park and Woodland Trees*, vol. 1. Albany: New York State Education Department.

———1940. *Plant Galls and Gall Makers*. Ithaca: Comstock Publishing Co.

Forcella, F. 1981. Twig nitrogen content and larval survival of twig-girdling beetles, *Oncideres singulata* (Say) (Coleoptera: Cerambycidae). *The Coleopterists Bulletin,* 35: 211–212.

———1982. Why twig-girdling beetles girdle twigs. *Naturwissenschaften,* 69: 398–400.

Fraenkel, G. S. 1959. The raison d'être of secondary plant substances. *Science,* 129: 1466–1470.

Fraenkel, G. S., and F. Fallil. 1981. The spinning (stitching) behaviour of the rice leaf folder, *Cnaphalocrosis medinalis. Entomologia Experimentalis et Applicata,* 29: 138–146.

Friend, R. B. 1931. The squash vine borer. Connecticut Agricultural Experiment Station Bulletin 328. New Haven: Connecticut Agricultural Experiment Station.

Frost, S. W. 1942. *Insect Life and Insect Natural History.* New York: Dover.

Gilbert, L. E. 1971. Butterfly-plant coevolution: has *Passiflora adenopoda* won the selectional race with heliconiine butterflies? *Science,* 172: 585–586.

Haack, R. A., K. R. Law, V. C. Mastro, H. S. Ossenbruggen, and B. J. Raimo. 1997. New York's battle with the Asian long-horned beetle. *Journal of Forestry,* 95(12): 11–15.

Hartnett, D. C., and W. G. Abrahamson. 1979. The effects of stem gall insects on life history patterns in *Solidago canadensis. Ecology,* 60: 910–917.

Heinrich, B. 1971. The effect of leaf geometry on the feeding behaviour of the caterpillar of *Manduca sexta* (Sphingidae). *Animal Behaviour,* 19: 119–124.

———How avian predators constrain caterpillar foraging. In N. E. Stamp and T. M. Casey, ed., *Caterpillars.* New York: Chapman and Hall.

Heinrich, B., and S. L. Collins. 1983. Caterpillar leaf damage and the game of hide-and-seek with birds. *Ecology,* 64: 592–602.

Hölldobler, B., and E. O. Wilson. 1990. *The Ants.* Cambridge, Mass.: Harvard University Press.

Janzen, D. H. 1969. Seed-eaters versus seed size, number, toxicity, and dispersal. *Evolution,* 23: 1–27.

———1971. Seed predation by animals. *Annual Review of Ecology and Systematics,* 2: 465–492.

———1976. Effect of defoliation on fruit-bearing branches of the Kentucky coffee tree, *Gymnocladus dioicus* (Leguminosae). *American Midland Naturalist,* 95: 474–478.

Jolivet, P. 1998. *Interrelationship between Insects and Plants.* Boca Raton: Chemical Rubber Company Press.

Kennedy, J. S., and T. E. Mittler. 1953. A method for obtaining phloem sap via the mouthparts of aphids. *Nature,* 171: 258.

Kogan, M., G. P. Waldbauer, G. Boiteau, and C. E. Eastman. 1980. Sampling bean leaf beetles on soybean. In M. Kogan and D. C. Herzog, ed., *Sampling Methods in Soybean Entomology.* New York: Springer-Verlag.

Levin, D. A. 1973. The role of trichomes in plant defense. *The Quarterly Review of Biology,* 48: 3–15.

Linsenmaier, W. 1972. *Insects of the World.* Trans.L. E. Chadwick. New York: McGraw-Hill.

MacFarlane, J. H., and A. J. Thorsteinson. 1980. Development and survival of the twostriped grasshopper, *Melanoplus bivittatus* (Say) (Orthoptera: Acrididae), on various single and multiple plant diets. *Acrida,* 9: 63–76.

Marquis, R. J., and C. J. Whelan. 1994. Insectivorous birds increase growth of white oak through consumption of leaf-chewing insects. *Ecology,* 75: 2007–2014.

Metcalf, R. L., and R. A. Metcalf. 1993. *Destructive and Useful Insects.* 5th ed. New York: McGraw-Hill.

Mitchell, R. 1977. Bruchid beetles and seed packaging by palo verde. *Ecology,* 58: 644–651.

Needham, J. G., S. W. Frost, and B. H. Tothill. 1928. *Leaf-mining Insects.* Baltimore: Williams and Wilkins

Nilsson, L. A. 1992. Orchid pollination biology. *Trends in Ecology and Evolution,* 7: 255–259.

Price, P. W. 1997. *Insect Ecology.* 3rd ed. New York: John Wiley and Sons.

Price, P. W., and T. P. Craig. 1984. Life history, phenology, and survivorship of a stem-galling sawfly, *Burra lasiolepis* (Hymenoptera: Tenthredinidae) on arroyo willow, *Salix lasiolepis,* in northern Arizona. *Annals of the Entomological Society of America,* 77: 712–719.

Ralph, C. P. 1976. Natural food requirements of the large milkweed bug, *Oncopeltus fasciatus* (Hemiptera: Lygaeidae), and their relation to gregariousness and host plant morphology. *Oecologia,* 26: 157–175.

Rathcke, B. J., and R. W. Poole. 1975. Coevolutionary race continues: butterfly larval adaptation to plant trichomes. *Science,* 187: 175–176.

Richardson, H. H. 1943. The action of bean leaves against the bedbug. *Journal of Economic Entomology,* 36: 543–545.

Risch, S. J., and C. R. Carroll. 1986. Effects of seed predation by a tropical ant on competition among seeds. *Ecology,* 67: 1319–1327.

Risebrow, A., and A. F. G. Dixon. 1987. Nutritional ecology of phloem-feeding insects. In F. Slansky, Jr., and J. G. Rodriguez, ed., *Nutritional Ecology of Insects, Mites, Spiders, and Related Invertebrates.* New York: John Wiley and Sons.

Roland, J., S. J. Hannon, and M. A. Smith. 1986. Foraging pattern of pine siskins and its influence on winter moth survival in an apple orchard. *Oecologia,* 69: 47–52.

Rosenthal, G. A., and M. R. Berenbaum, ed. 1991. *Herbivores: Their Interactions with Secondary Plant Metabolites.* 2nd ed. San Diego: Academic Press.

Schillinger, J. H., Jr., and R. L. Gallun. 1968. Leaf pubescence of wheat as a deterrent to the cereal leaf beetle, *Oulema melanopus. Annals of the Entomological Society of America,* 61: 900–903.

Schoener, T. W., and C. A. Toft. 1983. Spider populations: extraordinarily high densities on islands without top predators. *Science,* 219: 1353–1355.

Seigler, D. S. 1998. *Plant Secondary Metabolism.* Boston: Kluwer Academic Publishers.

Sláma, K., and C. M. Williams. 1966. The juvenile hormone. V. The sensitivity of the bug, *Pyrrhocoris apterus,* to a hormonally active factor in American paper pulp. *Biological Bulletin,* 130: 235–246.

Slansky, F., Jr., and A. R. Panizzi. 1987. Nutritional ecology of seed-sucking insects. In F. Slansky, Jr., and J. G. Rodriguez, ed., *Nutritional Ecology of Insects, Mites, Spiders, and Related Invertebrates.* New York: John Wiley and Sons.

Sork, V., and J. Bramble. 1993. Ecology of mast-fruiting in three species of North American deciduous oaks. *Ecology,* 74: 528–541.

Southgate, B. J. 1979. Biology of the Bruchidae. *Annual Review of Entomology,* 24: 449–473.

Spiller, D. A., and T. W. Schoener. 1990. A terrestrial field experiment showing the impact of eliminating top predators on foliage damage. *Nature,* 347: 469–471.

Toong, Y. C., D. A. Schooley, and F. C. Baker. 1988. Isolation of insect juvenile hormone III from a plant. *Nature,* 333: 170–171.

Turchin, P., P. L. Lorio, Jr., A. D. Taylor, and R. F. Billings. 1991. Why do populations of southern pine beetles (Coleoptera: Scolytidae) fluctuate? *Environmental Entomology,* 40: 401–409.

Waldbauer, G. P. 1975. Position of bean leaf beetle eggs in soil near soybeans determined by a refined sampling procedure. *Environmental Entomology,* 4: 375–380.

Weis, A. E., and M. R. Berenbaum. 1989. Herbivorous insects and green plants. In W. G. Abrahamson, ed., *Plant-Animal Interactions.* New York: McGraw-Hill.

Wigglesworth, V. B. 1972. *The Principles of Insect Physiology.* 7th ed. London: Chapman and Hall.

Wilf, P., C. C. Labandeira, W. J. Kress, C. L. Staines, D. M. Windsor, A. L. Allen, and K. R. Johnson. 2000. Timing the radiations of leaf beetles: hispines on gingers from latest cretaceous to recent. *Science,* 289: 291–294.

Williams, K. S., and L. E. Gilbert. 1981. Insects as selective agents on plant vegetative morphology: egg mimicry reduces egg laying by butterflies. *Science,* 212: 467–469.

Zeng, Y. 1995. Longest life cycle. In T. J. Walker, ed., *University of Florida Book of Insect Records.* Gainesville: University of Florida Press.

8. CONTROLLING INSECT POPULATIONS

Askew, R. R. 1971. *Parasitic Insects.* New York: American Elsevier.

Barker, R. J. 1958. Notes on some ecological effects of DDT sprayed on elms. *Journal of Wildlife Management,* 22: 269–274.

Bent, A. C. 1948. *Life Histories of North American Nuthatches, Wrens, Thrashers, and Their Allies.* Smithsonian Institution, United States National Museum Bulletin 195. Washington, D.C.: Smithsonian Institution.

Berenbaum, M., and E. Miliczky. 1984. Mantids and milkweed bugs: efficacy of aposematic coloration against invertebrate predators. *American Midland Naturalist,* 111: 64–68.

Borror, D. J., D. M. DeLong, and C. A. Triplehorn. 1981. *An Introduction to the Study of Insects.* 5th ed. Philadelphia: Saunders College Publishing.

Bowdish, T. I., and T. L. Bultman. 1993. Visual cues used by mantids in learning aversion to aposematically colored prey. *American Midland Naturalist,* 129: 215–222.

Brower, L. P. 1969. Ecological chemistry. *Scientific American,* 220: 22–29.

Campbell, R. W., and T. R. Torgersen. 1982. Some effects of predaceous ants on western spruce budworm pupae in north central Washington. *Environmental Entomology,* 11: 111–114.

Campbell, R. W., T. R. Torgersen, and N. Srivastava. 1983. A suggested role for predaceous birds and ants in the population dynamics of the western spruce budworm. *Forest Science,* 29: 779–790.

Carson, R. 1962. *Silent Spring.* Boston: Houghton Mifflin.

DeBach, P. 1974. *Biological Control by Natural Enemies.* Cambridge: Cambridge University Press.

Denlinger, D. L., R. K. Jann, and M. F. B. Chaudhury, 1983. Parturition in the tsetse fly *Glossina morsitans:* pattern of activity, sound production, and evidence for control by the mother's brain. *Journal of Insect Physiology,* 29: 715–721.

Denlinger, D. L., and J. Zdárek. 1997. A hormone from the uterus of the tsetse fly, *Glossina morsitans,* stimulates parturition and abortion. *Journal of Insect Physiology,* 43: 135–142.

Dunlap, T. R. 1981. *DDT.* Princeton: Princeton University Press.

Ehrlich, P. R., and A. H. Ehrlich. 1970. *Population, Resources, Environment.* San Francisco: W. H. Freeman.

Eibl-Eibesfeldt, I., and E. Eibl-Eibesfeldt. 1968. The workers' bodyguard. *Animals,* 11: 16–17.

Eisner, T. 1965. Defensive spray of a phasmid insect. *Science,* 148: 966–968.

Eisner, T., and J. Dean. 1976. Ploy and counterploy in predator-prey interactions: orb-weaving spiders versus bombardier beetles. *Proceedings of the National Academy of Sciences, U.S.A.,* 73: 1365–1367.

Eisner, T., I. Kriston, and D. J. Aneshansley. 1976. Defensive behavior of a termite (*Natsutitermes exitiosus. Behavioral Ecology and Sociobiology,* 1: 83–125.

Eisner, T., E. van Tassell, and J. E. Carrel. 1967. Defensive use of a "fecal shield" by a beetle larva. *Science,* 158: 1471–1473.

Evans, H. E. 1966. The accessory burrows of digger wasps. *Science,* 152: 465–471.

Feir, D., and J-S Suen. 1971. Cardenolides in the milkweed plant and feeding by the milkweed bug. *Annals of the Entomological Society of America,* 64: 1173–1174.

Godfray, H. C. J. 1994. *Parasitoids.* Princeton: Princeton University Press.

Gross, P. 1993. Insect behavioral and morphological defenses against parasitoids. *Annual Review of Entomology,* 38: 251–273.

Huang, H. T., and P. Yang. 1987. The ancient cultured citrus ant. *BioScience,* 37: 665–670.

Hogue, C. L. 1972. Protective function of sound perception and gregariousness in *Hylesia* larvae (Saturniidae: Hemileucinae). *Journal of the Lepidopterist's Society,* 26: 33–34.

James, M. T., and R. F. Harwood. 1969. *Medical Entomology.* New York: Macmillan.

Jeanne, R. L. 1970. Chemical defense of brood by a social wasp. *Science,* 168: 1465–1466.

Kaiser, J. 2000. How an old tree outwits its foes. *Science,* 289: 2032.

Lavine, M. D., and N. E. Beckage. 1995. Polydnaviruses: potent mediators of host insect immune dysfunction. *Parasitology Today,* 11: 368–378.

Luckmann, W. H., and R. L. Metcalf. 1994. The pest management concept. In R. L. Metcalf and W. H. Luckmann, ed., *Introduction to Insect Pest Management,* 3rd ed. New York: John Wiley and Sons.

Malthus, T. R. 1817. *An Essay on the Principle of Population.* 5th ed. London: John Murray.

McCook, H. C. 1882. Ants as beneficial insecticides. *Proceedings of the Academy of Natural Sciences of Philadelphia,* 34: 263–271.

Metcalf, R. L., and R. A. Metcalf. 1993. *Destructive and Useful Insects.* 5th ed. New York: McGraw-Hill.

Milius, S. 2000. When ants squeak. *Science News,* 157: 92–94.

Minnich, D. E. 1925. The reactions of the larvae of *Vanessa antiopa* Linn. to sounds. *Journal of Experimental Zoology,* 42: 443–469.

Moran, M. D., and L. E. Hurd. 1998. A trophic cascade in a diverse arthropod community caused by a generalist arthropod predator. *Oecologia,* 113: 126–132.

Moran, M. D., T. P. Rooney, and L. E. Hurd. 1996. Top-down cascade from a biotrophic predator in an old-field community. *Ecology,* 77: 2219–2227.

Nafus, D. 1991. Biological control of *Penicillaria jocosatrix* (Lepidoptera: Noctuidae) on mango on Guam with notes on the biology of its parasitoids. *Environmental Entomology,* 20: 1725–1731.

Nutting, W. L. 1970. Free diurnal foraging by the North America nasutiform termite, *Tenuirostritermes tenuirostris. Pan-Pacific Entomologist,* 46: 39–42.

Nutting, W. L., M. S. Blum, and H. M. Fales. 1974. Behavior of the North American termite *Tenuirostritermes tenuirostris* with special reference to the soldier frontal gland secretion, its chemical composition, and use in defense. *Psyche,* 81: 167–177.

Oldroyd, H. 1964. *The Natural History of Flies.* New York: W. W. Norton.

Powell, D., S. C. Chandler, and V. W. Kelley. 1947. *Pest Control in Commercial Fruit Planting.* Circular 610 of the University of Illinois College of Agriculture Extension Service in Agriculture and Home Economics. Urbana: University of Illinois College of Agriculture Extension Service in Agriculture and Home Economics.

Roces, F., and B. Hölldobler. 1995. Vibrational communication between hitchhikers and foragers in leaf-cutting ants (*Atta cephalotes*). *Behavioral Ecology and Sociobiology,* 37: 297–302.

Schreiner, I. H., and D. M. Nafus. 1992. Changes in a moth community mediated by biological control of the dominant species. *Environmental Entomology,* 21: 664–668.

Strand, M. R., and L. L. Pech. 1995. Immunological basis for compatibility in parasitoid-host relationships. *Annual Review of Entomology,* 40: 31–56.

Tothill, J. D., T. H. C. Taylor, and R. W. Paine. 1930. *The Coconut Moth in Fiji.* London: The Imperial Bureau of Entomology.

Turchin, P., P. L. Lorio, Jr., A. D. Taylor, and R. F. Billings. 1991. Why do populations of southern pine beetles (Coleoptera: Scolytidae) fluctuate? *Environmental Entomology,* 40: 401–409.

Turchin, P., A. D. Taylor, and J. D. Reeve. 1999. Dynamical role of predators in population cycles of a forest insect: an experimental test. *Science,* 285: 1068–1071.

Volker, K. C., and R. G. Simpson. 1975. Behavior of alfalfa weevil larvae affecting the establishment of *Tetrastichus incertus* in Colorado. *Environmental Entomology,* 4: 742–744.

Winston, M. L. 1997. *Nature Wars.* Cambridge, Mass.: Harvard University Press.

9. CONTROLLING VERTEBRATE POPULATIONS

Askew, R. R. 1971. *Parasitic Insects.* New York: American Elsevier.

Bates, H. W. 1892. *The Naturalist on the River Amazons.* London: John Murray.

Bennett, G. F. 1961. On three species of Hippoboscidae (Diptera) on birds in Ontario. *Canadian Journal of Zoology,* 39: 379–406.

Bent, A. C. 1948. *Life Histories of North American Nuthatches, Wrens, Thrashers, and Their Allies.* Smithsonian Institution, U.S. National Museum Bulletin 195. Washington, D.C.: Smithsonian Institution.

Bishopp, F. C., E. W. Laake, and R. W. Wells. 1949. *Cattle Grubs or Heel Flies with Suggestions for their Control.* Farmer's Bulletin 1596. Washington, D.C.: U.S. Department of Agriculture.

Brown, C. R., and M. B. Brown. 1986. Ectoparasitism as a cost of coloniality in cliff swallows *(Hirundo pyrrhonota). Ecology,* 67: 1206–1218.

Christian, K. A. 1980. Cleaning/feeding symbiosis between birds and reptiles of the Galapagos Islands: new observations of inter-island variability. *The Auk,* 97: 887–889.

Clark, L. 1991. The nest protection hypothesis: the adaptive use of plant secondary compounds by European starlings. In J. E. Loye and M. Zuk, ed., *Bird-Parasite Interactions.* Oxford: Oxford University Press.

Clark, L., and J. R. Mason. 1985. Use of nest material as insecticidal and anti-pathogenic agents by the European starling. *Oecologia,* 67: 169–176.

———1988. Effect of biologically active plants used as nesting material and the derived benefit to starling nestlings. *Oecologia,* 77: 174–180.

Clay, T. 1949. Piercing mouth-parts in the biting lice (Mallophaga). *Nature,* 164: 617–619.

———1949. Some problems in the evolution of a group of ectoparasites. *Evolution,* 3: 279–299.

Clayton, D. H. 1991. Coevolution of avian grooming and ectoparasite avoidance. In J. E. Loye and M. Zuk, ed., *Bird-Parasite Interactions.* Oxford: Oxford University Press.

Cott, H. B. 1975. *Looking at Animals.* New York: Charles Scribner's Sons.

Division of Entomology, Commonwealth Scientific and Industrial Research Organization of Australia. 1991. *The Insects of Australia.* Ithaca: Cornell University Press.

Enserink, M. 2000. The enigma of West Nile. *Science,* 290: 1482–1484.

Fenner, F., and F. N. Ratcliffe. 1965. *Myxomatosis.* Cambridge: Cambridge University Press.

Finkel, E. 1999. Australian biocontrol beats rabbits, but not rules. *Science,* 285: 1842.

Gold, C. S., and D. L. Dahlsten. 1983. Effects of parasitic flies (*Protocalliphora* spp.) on nestlings of mountain and chestnut-backed chickadees. *The Wilson Bulletin,* 95: 560–572.

Halford, F. J. 1954. *Nine Doctors and God.* Honolulu: University of Hawaii Press.

Hamilton, W. D., and M. Zuk. 1982. Heritable true fitness and bright birds: a role for parasites? *Science,* 218: 384–387.

Harrison, H. H. 1975. *A Field Guide to the Birds' Nests.* Boston: Houghton Mifflin.

Harwood, R. F., and M. T. James. 1979. *Entomology in Human and Animal Health.* New York: Macmillan.

Henshaw, H. W. 1902. *Birds of the Hawaiian Islands, Being a Complete List of the Birds of the Hawaiian Possessions with Notes on Their Habits.* Honolulu: Thomas G. Thrum.

Horsfall, W. R. 1962. *Medical Entomology.* New York: Ronald Press.

Humphries, D. A. 1969. Behavioural aspects of the ecology of the sand-martin flea *Ceratophyllus styx jordani* Smit (Siphonaptera). *Parasitology,* 59: 311–334.

Jellison, W. L. 1940. The burrowing owl as a host to the argasid tick, *Ornithodorus parkeri.* U.S. Public Health Service, Public Health Reports, 55, part 1. Washington, D.C.: U. S. Public Health Service.

Johnson, L. L., and M. Boyce. 1991. Female choice of males with parasite loads in sage grouse. In J. E. Loye and M. Zuk, ed., *Bird-Parasite Interactions.* Oxford: Oxford University Press.

Johnson, L. S., and D. J. Albrecht. 1993. Effects of haematophagous ectoparasites on nestling house wrens, *Troglodytes aedon:* who pays the cost of parasitism? *Oikos,* 66: 255–262.

Karstad, L. 1971. Pox. In J. W. Davis, R. C. Anderson, L. Karstad, and O. Trainer, ed., *Infectious and Parasitic Diseases of Wild Birds.* Ames: Iowa State University Press.

Kemper, H. E., and H. O. Peterson. 1953. *Cattle Lice and How to Eradicate Them.* Farmer's Bulletin no. 909. Washington, D.C.: U.S. Department of Agriculture.

Linduska, J. P., and A. W. Lindquist. 1952. Some insect pests of wildlife. In *Insects: The Yearbook of Agriculture.* Washington, D.C.: U.S.Department of Agriculture.

Marshall, A. G. 1981. *The Ecology of Ectoparasitic Insects.* London: Academic Press.

Metcalf, R. L., and R. A. Metcalf. 1993. *Destructive and Useful Insects.* 5th ed. New York: McGraw-Hill.

Morris, C. D. 1988. Eastern equine encephalomyelitis. In T. P. Monath, ed., *The Arboviruses: Epidemiology and Ecology,* vol. 3. Boca Raton: Chemical Rubber Company Press.

Mowat, F. 1952. *People of the Deer.* Boston: Little Brown.

Rogers, C. A., R. J. Robertson, and B. J. Stutchbury. 1991. Patterns and effects of parasitism by *Protocalliphora sialia* on tree swallow nestlings. In J. E. Loye and M. Zuk, ed., *Bird-Parasite Interactions.* Oxford: Oxford University Press.

Ross, J. H. 2000. People of the reindeer. *Smithsonian,* 31: 54–64.

Rothschild, M., and T. Clay. 1952. *Fleas, Flukes, and Cuckoos.* London: Collins.

Royte, E. 1995. On the brink: Hawaii's vanishing species. *National Geographic,* 188: 1–37.

Savage, T. S. 1847. On the habits of the "drivers" or visiting ants of West Africa. *Transactions of the Entomological Society of London,* 5: 1–15.

Schmidt, G. D., and L. S. Roberts. 1981. *Foundations of Parasitology.* St. Louis: C. V. Mosby.

Schmidt, K. P., and D. D. Davis. 1942. *Field Book of Snakes of the United States and Canada.* New York: G. P. Putnam's Sons.

Schneirla, T. C. 1956. The army ants. In *Annual Report of the Board of Regents of the Smithsonian Institution for 1955.* Washington, D.C.: Smithsonian Institution.

Schneirla, T. C., ed., 1971. *Army Ants: A Study in Social Organization.* San Francisco: W. H. Freeman.

Sengupta, S. 1981. Adaptive significance of the use of margosa leaves in nests of house sparrows, *Passer domesticus. Emu,* 81: 114–115.

Spurrier, M. F., M. S. Boyce, and B. F. J. Manly. 1991. Effects of parasites on mate choice by captive sage grouse. In J. E. Loye and M. Zuk, ed., *Bird-Parasite Interactions.* Oxford: Oxford University Press.

Steelman, C. D. 1976. Effects of external and internal arthropod parasites on domestic livestock production. *Annual Review of Entomology,* 21: 155–178.

Travis, B. V. 1938. The fire ant (*Solenopsis* spp.) as a pest of quail. *Journal of Economic Entomology,* 31: 649–652.

Warner, R. E. 1968. The role of introduced diseases in the extinction of the endemic Hawaiian avifauna. *Condor,* 70: 101–120.

Wehr, E. E. 1971. Nematodes. In J. W. Davis, R. C. Anderson, L. Karstad, and D. O. Trainer, ed., *Infectious and Parasitic Diseases of Wild Birds.* Ames: Iowa State University Press.

Wickler, W. 1968. *Mimicry in Plants and Animals.* Trans. R. D. Martin. New York: McGraw-Hill.

Wigglesworth, V. B. 1972. *The Principles of Insect Physiology.* 7th ed. London: Chapman and Hall.

10. RECYCLING DEAD ANIMALS

Baer, W. S. 1931. The treatment of chronic osteomyelitis with the maggot (larva of the blow fly). *Journal of Bone and Joint Surgery,* 13: 438–475.

Catts, E. P., and M. L. Goff. 1992. Forensic entomology in criminal investigations. *Annual Review of Entomology,* 37: 253–272.

Eggert, A. K., and J. K. Müller. 1997. Biparental care and social evolution in burying beetles: lessons from the larder. In T. C. Choe and B. J. Crespi, ed., *The Evolution of Social Behavior in Insects and Arachnids.* Cambridge: Cambridge University Press.

Erzinçlioglu, Z. 2001. *Maggots, Murder, and Men.* Colchester, Eng.: Harley.

Fabre, J. H. 1919. *The Glow-Worm and Other Beetles.* Trans. A. T. de Matos. New York: Dodd, Mead, and Co.

————1920. *The Life of the Fly.* Trans. A. T. de Matos. New York: Dodd, Mead, and Co.

Goff, M. L. 2000. *A Fly for the Prosecution.* Cambridge, Mass.: Harvard University Press.

Hanski, I. 1987. Nutritional ecology of dung- and carrion-feeding insects. In F. Slansky, Jr., and H. G. Rodriguez, ed., *Nutritional Ecology of Insects, Mites, Spiders, and Related Invertebates.* New York: John Wiley and Sons.

Hölldobler, B., and E. O. Wilson. 1990. *The Ants.* Cambridge, Mass.: Harvard University Press.

Knipling, E. F. 1967. Sterile technique: principles involved, current application, limitations, and future application. In J. W. Wright and R. Pal, ed., *Genetics of Insect Vectors of Disease.* Amsterdam: Elsevier.

Lacordaire, J. T. 1834–1838. *Introduction á l'entomologie.* Paris: Librairie Encyclopédique de Roret.

Lord, W. D. 1990. Case histories of the use of insects in investigations. In E. P. Catts and N. H. Haskell, ed., *Entomology and Death: A Procedural Guide.* Clemson, S.C.: Joyce's Print Shop.

Metcalf, R. L., and W. H. Luckmann. 1994. *Introduction to Insect Pest Management.* 3rd ed. New York: John Wiley and Sons.

Metcalf, R. L., and R. A. Metcalf. 1993. *Destructive and Useful Insects.* 5th ed. New York: McGraw-Hill.

Norris, K. R. 1965. The bionomics of blow flies. *Annual Review of Entomology,* 10: 47–68.

Payne, J. A. 1965. A summer carrion study of the baby pig *Sus scrofa* Linnaeus. *Ecology,* 46: 592–602.

Pickering, R. B. 1997. Maggots, graves, and scholars. *Archaeology,* 50: 46–47.

Reed, H. B., Jr. 1958. A study of dog carcass communities in Tennessee, with special reference to the insects. *American Midland Naturalist,* 59: 213–245.

Robinson, W. 1935. Allantoin, a constituent of maggot excretions, stimulates healing of chronic discharging wounds. *Journal of Parasitology,* 21: 354–358.

Scott, M. P. 1996. Communal breeding in burying beetles. *American Scientist,* 84: 376–382.

————1998. The ecology and behavior of burying beetles. *Annual Review of Entomology,* 43: 595–618.

Sherman, R. A. 2000. Maggot therapy—the last five years. *European Tissue Repair Society,* 7: 97–98.

Sherman, R. A., M. J. R. Hall, and S. Thomas. 2000. Medicinal maggots: an ancient remedy for some contemporary afflictions. *Annual Review of Entomology,* 45: 55–81.

Smith, K. G. V. 1986. *A Manual of Forensic Entomology.* Ithaca: Cornell University Press.

Wigglesworth, V. B. 1972. *The Principles of Insect Physiology.* 7th ed. London: Chapman and Hall.

Wilson, D. S., and J. Fudge. 1984. Burying beetles: intraspecific interactions and reproductive success in the field. *Ecological Entomology,* 9: 195–203.

Wilson, D. S., and W. G. Knollenberg. 1987. Adaptive indirect effects: the fitness of burying beetles with and without their phoretic mites. *Evolutionary Ecology,* 1: 139–159.

Wilson, E. O. 1975. *Sociobiology.* Cambridge, Mass.: Harvard University Press.

11. RECYCLING DUNG

Adams, R. 1972. *Watership Down.* New York: Avon.

Anderson, J. M., and M. J. Coe. 1974. Decomposition of elephant dung in an arid, tropical environment. *Oecologia,* 14: 111–125.

Arrow, G. J. 1951. *Horned Beetles.* The Hague: W. Junk.

Askew, R. R. 1971. *Parasitic Insects.* New York: American Elsevier.

Blueweiss, L., H. Fox, V. Kudzma, D. Nakashima, R. Peters, and S. Sams. 1978. Relationships between body size and some life history parameters. *Oecologia,* 37: 257 272.

Borgia, G. 1980. Sexual competition in *Scatophaga stercoraria:* size- and density-related changes in male ability to capture females. *Behaviour,* 75: 185–206.

Cambefort, Y. 1987. Le scarabée dans l'Égypte ancienne. (The scarab in ancient Egypt). *Revue de l'Histoire des Religions,* 204: 3–46.

Cambefort, Y., and I. Hanski. 1991. Dung beetle population biology. In I. Hanski and Y. Cambefort, ed., *Dung Beetle Ecology.* Princeton: Princeton University Press.

Castellanos, C. R., F. Escobars, and P. R. Stevenson. 1999. Dung beetles (Scarabaeidae: Scarabaeinae) attracted to wooly monkey (*Lagothrix lagotricha* Humboldt) dung at Tinigua National Park, Colombia. *Coleopterists Bulletin,* 53: 155–159.

Cherry, R. H. 1985. Insects as sacred symbols in ancient Egypt. *Bulletin of the Entomological Society of America,* 31: 15–16.

Cott, H. B. 1957. *Adaptive Coloration in Animals.* London: Methuen.

Crumb, S. E. 1956. *The Larvae of the Phalaenidae.* Technical Bulletin no. 1135. Washington, D.C.: U.S. Department of Agriculture.

Doube, B. M., A. Macqueen, T. J. Ridsdill-Smith, and T. M. Weir. 1991. Native and introduced dung beetles in Australia. In I. Hanski and Y. Cambefort, ed., *Dung Beetle Ecology.* Princeton: Princeton University Press.

Estrada, A., A. Anzures D., and R. Coates-Estrada. 1999. Tropical rain forest fragmentation, howler monkeys (*Alouatta palliata*), and dung beetles at Los Tuxtlas, Mexico. *American Journal of Primatology,* 48: 253–262.

Evans, G. 1975. *The Life of Beetles.* New York: Hafner Press.

Fabre, J. H. 1918. *The Sacred Beetle and Others.* Trans. A. T. de Matos. New York: Dodd, Mead, and Co.

Ferrar, P., and J. A. L. Watson. 1970. Termites (Isoptera) associated with dung in Australia. *Journal of the Australian Entomological Society,* 9: 100–102.

Hanski, I. 1987. Nutritional ecology of dung- and carrion-feeding insects. In F. Slansky, Jr., and J. G. Rodriguez, ed., *Nutritional Ecology of Insects, Mites, Spiders, and Related Invertebrates.* New York: John Wiley and Sons.

Hanski, I. 1991. The dung insect community. In I. Hanski and Y. Cambefort, ed., *Dung Beetle Ecology.* Princeton: Princeton University Press.

Hanski, I., and Y. Cambefort. 1991. Species richness. In I. Hanski and Y. Cambefort, ed. *Dung Beetle Ecology.* Princeton: Princeton University Press.

Kendeigh, S. C. 1961. *Animal Ecology.* Englewood Cliffs, N.J.: Prentice-Hall.

Matthews, E. G. 1972. *A Revision of the Scarabaeinae Dung Beetles of Australia. I. Tribe Onthophagini. Australian Journal of Zoology,* supplement no. 9.

McDonald, P., R. A. Edwards, J. F. D. Greenhalgh, and C. A. Morgan. 1995. *Animal Nutrition.* 5th ed. New York: John Wiley and Sons.

Mohr, C. O. 1943. Cattle droppings as ecological units. *Ecological Monographs,* 13: 275–298.

Norris, K. R. 1966. Notes on the ecology of the bushfly, *Musca vertustissima* (Diptera: Muscidae) in the Canberra district. *Australian Journal of Zoology,* 14: 1139–1156.

Peck, S. B., and A. Forsyth. 1982. Composition, structure, and competitive behaviour in a guild of Ecuadorian rain forest dung beetles (Coleoptera; Scarabaeidae). *Canadian Journal of Zoology,* 60: 1624–1634.

Ratcliffe, B. C. 1980. New species of coprini (Coleoptera: Scarabaeidae: Scarabaeinae) taken from the pelage of three-toed sloths (*Bradypus tridactylus* L.) (Edentata: Bradypodidae) in central Amazonia with a brief commentary on scarab-sloth relationships. *Coleopterists Bulletin,* 34: 337–350.

Ridsdill-Smith, T. J., and J. N. Matthiessen. 1988. Bush fly, *Musca vetustissima* Walker (Diptera: Muscidae), control in relation to seasonal abundance of scarabaeine dung beetles (Coleoptera: Scarabaeidae) in south-western Australia. *Bulletin of Entomological Research,* 78: 633–639.

Thomson, D. F. 1934. Some adaptations in the disposal of feces: the hygiene of the nest in Australian birds. *Proceedings of the Zoological Society of London,* 1934: 701–706.

Waage, J. K., and G. G. Montgomery. 1976. *Cryptoses choloepi:* a coprophagous moth that lives on a sloth. *Science,* 193: 157–158.

Waterhouse, D. F. 1977. The biological control of dung. In T. Eisner and E. O. Wilson, ed., *The Insects.* San Francisco: W. H. Freeman.

Whitford, W. G., Y. Steinberger, and G. Ettershank. 1982. Contributions of subterranean termites to the "economy" of Chihuahuan desert ecosystems. *Oecologia,* 55: 298–302.

12. RECYCLING DEAD PLANTS

Arrow, G. J. 1951. *Horned Beetles.* The Hague: W. Junk.

Bignell, D. E., P. Eggleton, L. Nunes, and K. L. Thomas. 1997. Termites as mediators of carbon fluxes in tropical forest. In A. D. Watt, N. E. Stork, and M. D. Hunter, ed., *Forests and Insects.* London: Chapman and Hall.

Borror, D. J., D. M. DeLong, and C. A. Triplehorn. 1981. *An Introduction to the Study of Insects.* 5th ed. Philadelphia: Saunders.

Cleveland, L. R. 1934. The wood-feeding roach *Cryptocercus,* its protozoa, and the symbiosis between protozoa and roach. *Memoirs of the American Academy of Arts and Sciences,* 17: 185–342.

Comstock, J. H. 1950. *An Introduction to Entomology.* 9th ed. Ithaca: Comstock Publishing Co.

Crossley, D. A., Jr., and M. P. Hoglund. 1962. A litter-bag method for the study of microarthropods inhabiting leaf litter. *Ecology,* 43: 571–573.

Edwards, C. A., and G. W. Heath. 1963. The role of soil animals in breakdown of leaf material. In J. Doeksen and J. van der Drift, ed., *Soil Organisms.* Amsterdam: North-Holland Publishing Co.

Eggleton, P., D. E. Bignell, W. A. Sands, N. A. Mawdsley, J. H. Laughton, T. G. Wood, and N. C. Bignell. 1996. The diversity, abundance, and biomass of termites under differing levels of disturbance in the Mbalmayo Forest Reserve, southern Cameroon. *Philosophical Transactions of the Royal Society of London,* series B, 351: 51 68

Gullan, P. J., and P. S. Cranston. 1994. *The Insects: An Outline of Entomology.* London: Chapman and Hall.

Jones, C. J., J. H. Lawton, and M. Shachak. 1994. Organisms as ecosystem engineers. *Oikos,* 69: 373–386.

Kofoid, C. A., ed. 1934. *Termites and Termite Control.* Berkeley: University of California Press.

Linsley, E. G. 1959. Ecology of Cerambycidae. *Annual Review of Entomology,* 4: 99–138.

———1961. *The Cerambycidae of North America.* Berkeley: University of California Press.

Manee, A. H. 1908. Some observations at Southern Pines, N. Carolina. *Entomological News,* 19: 286–289.

Martin, M. M. 1991. The evolution of cellulose digestion in insects. *Philosophical Transactions of the Royal Society of London,* series B, 333: 281–288.

Metcalf, C. L. 1933. An obscure *Temnostoma* differentiated by its larval characters. *Annals of the Entomological Society of America,* 26: 1–7.

Morgan, F. D. 1968. Bionomics of Siricidae. *Annual Review of Entomology,* 13: 239–256.

Odum, E. P., and L. J. Bevier. 1984. Resource quality, mutualism, and energy partitioning in food chains. *American Naturalist,* 124: 360–376.

Polis, G. A. 1991. Food webs in desert communities: complexity via diversity and omnivory. In G. A. Polis, ed., *The Ecology of Desert Communities.* Tucson: University of Arizona Press.

Romoser, W. S., and J. G. Stoffolano, Jr. 1998. *The Science of Entomology.* Boston: WCB/McGraw-Hill.

Roth, L. M. 1982. Blattaria. In S. B. Parker, ed., *Synopsis and Classification of Living Organisms,* vol. 2. New York: McGraw-Hill.

Schlesinger, W. H. 1991. *Biogeochemistry.* San Diego: Academic Press.

Speight, M. R., M. D. Hunter, and A. D. Watt. 1999. *Ecology of Insects.* Oxford: Blackwell Science Ltd.

Van der Drift, J., and M. Witkamp. 1960. The significance of the breakdown of oak litter by *Enoicyla pusilla* Burm. *Netherlands Journal of Ecology,* 13: 486–492.

Wallace, H. R. 1954. Notes on the biology of *Arhopalus ferus* Mulsant (Coleoptera: Cerambycidae). *Proceedings of the Royal Entomological Society of London,* 29(A): 99–113.

Whitford, W. G., Y. Steinberger, and G. Ettershank. 1982. Contributions of subterranean termites to the "economy" of Chihuahuan desert ecosystems. *Oecologia,* 55: 298–302.

Wigglesworth, V. B. 1972. *The Principles of Insect Physiology.* 7th ed. London: Chapman and Hall.

Witkamp, M. 1971. Soils as components of ecosystems. *Annual Review of Ecology and Systematics,* 2: 85–110.

Wood, T. G. 1978. Food and feeding habits of termites. In M. V. Brian, ed., *Production Ecology of Ants and Termites.* Cambridge: Cambridge University Press.

Wood, T. G., and W. A. Sands. 1978. The role of termites in ecosystems. In M. V. Brian, ed., *Production Ecology of Ants and Termites.* Cambridge: Cambridge University Press.

ACKNOWLEDGMENTS

I owe a great debt to the many friends and colleagues who helped me by giving generously of their advice, encouragement, and knowledge: Gary Alpert, Carol Augspurger, George Batzli, May Berenbaum, Julia Berger, Stewart Berlocher, Sam Beshers, John Bouseman, Stephen Buchman, Thomas Burke, David Cowan, Evan DeLucia, David Denlinger, Michael Ducey, Susan Fahrbach, Susan Gabay-Laughnan, the late Arthur Ghent, Deborah Gordon, Jason Hamilton, Robert Harwood, Michael Hutjens, Michael Irwin, James Kaler, Kathleen Keeler, Emerson Nafziger, James Nardi, Robert Novak, Jimmy Olson, Philip Parrillo, Peter Price, Susan Ratcliffe, Gene Robinson, David Roemer, William Rose, David Seigler, James Smith, James Sternburg, and Stephen Trumbo.

Meredith Waterstraat wielded her pencil with skill and flare to draw the illustrations that so beautifully complement the text. My heartfelt thanks go to Dorothy Nadarski, who patiently and meticulously typed and retyped the manuscript. This book benefited greatly from the untailing encouragement of Michael Fisher and Ann Downer-Hazell, the painstaking editing of Nancy Clemente, the boundless design talents of Marianne Perlak, and the skilled and creative typesetting of the people at Technologies 'N Typography.

INDEX

Note: Page numbers in *italics* refer to illustrations.